SPACE PHYSICS WITH ARTIFICIAL SATELLITES

ISKUSSTVENNYE SPUTNIKI V RAZREZHENNOI PLAZME

ИСКУССТВЕННЫЕ СПУТНИКИ В РАЗРЕЖЕННОЙ ПЛАЗМЕ

SPACE PHYSICS WITH ARTIFICIAL SATELLITES

by

Ya. L. Al'pert, A. V. Gurevich,

and

L. P. Pitaevskii

Authorized translation from the Russian by

H. H. Nickle

Springer Science+Business Media, LLC

1965

The Russian text was published by Nauka in Moscow in 1964.

Яков Львович Альперт,
Александр Викторович Гуревич,
Лев Петрович Питаевский.

Искусственные спутники
в разреженной плазме.

Library of Congress Catalog Card Number 64-23253

ISBN 978-1-4899-4919-6 ISBN 978-1-4899-4917-2 (eBook)
DOI 10.1007/978-1-4899-4917-2

Originally published by Plenum Press in 1965.
Softcover reprint of the hardcover 1st edition 1965

PREFACE

The launching of artificial satellites and cosmic rockets and the investigations of the ionosphere and of interplanetary space carried out with their help have intensified in contemporary physics interest in the investigation of phenomena arising in a highly rarefield plasma in the vicinity of bodies moving in it. The present monograph is devoted to a theoretical investigation of a number of problems appearing in connection with the investigation of this problem.

For a description of the motion of bodies in a highly rarefied medium, where the mean free paths of the particles are larger than the dimensions of the body, the usual methods of hydrodynamics or aerodynamics, as is well known, are already inapplicable. Hence for the solution of these problems it is necessary to use kinetic theory, which takes account of the fact that a gas is not a continuous medium, but an aggregate of individual particles. A systematic kinetic approach to the questions considered below is the basic feature of this book.

The case, realized in the upper ionosphere, of a rapidly moving body whose velocity is much larger than the thermal velocity of neutral particles and ions, that is, larger than the velocity of sound, and whose dimension is larger than the Debye radius, is examined in most detail. However, certain sections of the book are also devoted to an investigation of effects which occur when the size of the body becomes comparable to or smaller than the Debye radius, and the body's velocity is comparable to the velocity of the particles. Such conditions apparently occur at the limits of the earth's ionosphere and in interplanetary space.

The results of calculations of the disturbances in the concentrations of neutral particles, ions, and electrons, and also the disturbances, appearing as a consequence of this, of the electric and magnetic fields in the vicinity of the body are presented in Chapters II through VI. Chapter VII is devoted to an investigation of the scattering of radiowaves on the thus-formed trail of a body. The disturbances of the plasma in the vicinity of a body at rest are calculated in Chapters VIII and IX; this enables us, in particular, to develop a more rigorous theory of a probe in a rarefied plasma; the results of this theory differ substantially from the results of the approximate theories which are usually used (Chapter X). Also the question of disturbance of the plasma by bodies whose dimensions are smaller than the Debye radius is investigated (§17 and 40), and the question of the influence of an alternating electric field (field of the transmitting antennas – §18) on the plasma is considered. However, these questions, as well as the problem of disturbance of the magnetic field (Chapter VI) are considered in less detail. This is related to the account of the interesting question of bounded motion of charged particles in the vicinity of a small body (§35).

In order to solve the problems indicated above, it is important to know the characteristic properties of the interaction of the body's surface with the plasma and with the incident electromagnetic radiation and corpuscular fluxes. Unfortunately, at the present time, these questions have almost not been investigated at all, the more so since effects arising in this connection strongly depend on the material and method of processing of the surface. For this reason, in calculations we shall everywhere assume that the particles interact with the surface in the simplest way, namely, they can only be reflected or absorbed. In particular, we shall neglect such processes as evaporation and destruction of the body's surface upon collision with particles of the medium or under the action of radiation fluxes, the photoeffect, etc. The great interest presently developing with regard to investigations by artificial satellites and cosmic rockets enables one to hope that the study of these important questions will advance substantially in the near future, especially on the experimental side.

The monograph is only a first attempt at an investigation and generalization of an extremely complex and manifold problem, which is of great importance in contemporary physics and technology. It is natural, therefore, that there are a number of deficiencies in it.

It should be noted that in many cases only those questions whose solution can be obtained in analytic form or require a minimum amount of numerical calculations are considered. Of course, a broader application of numerical methods enables one to considerably generalize later on and make our results more precise.

The monograph was basically written from the results of articles [1-10, 63-66, 70-72] by the authors. In those cases when the results of other investigators are used, the corresponding references are given in the text. One can find extensive lists of references to the literature devoted to the interaction of a body with a plasma and adjoining problems in the review articles [10, 11, 12] (also see [7]).

In conclusion the authors express their gratitude to the reviewer of the book, V. P. Silin, for a number of remarks, to V. I. Krayushkina and V. L. Morozova for substantial help in the preparation of the book, and to L. V. Pariiska for performing a number of numerical calculations.

BASIC NOTATION

n – concentration of neutral particles.
n_0 – undisturbed concentration of neutral particles.
N – concentration of electrons.
N_0 – undisturbed concentration of electrons and of ions.
N_i – concentration of ions.
δn, δN, δN_i – disturbances of the particle concentrations.
l_n – mean free path of the neutral particles.
$l_{i,e}$ – mean free path of ions, electrons.
ν_{nn} – number of collisions between neutral particles.
$\nu_{e,in}$ – number of collisions between electrons or ions and neutral particles.
ν_{ii} – number of collisions between ions.
$v_{n,i,e}$ – thermal velocity of neutral particles, ions, electrons $[v_e = (2\varkappa T/m)^{1/2}]$.
V_0 – velocity of the body.
$\varkappa$ – Boltzmann's constant.
$k_0 = \omega/c$ – wave vector in free space.
$\mathbf{k}$ – wave vector of the incident wave.
$\mathbf{k}'$ – wave vector of the scattered wave.
$\mathbf{q} = \mathbf{k} - \mathbf{k}'$ – difference between the wave vectors of the incident and scattered waves.
T – temperature.
$-e$ – electron's charge.
m – mass of an electron.
M – mass of a heavy particle (neutral particle).
M_i – mass of ion.
$\mathcal{M}$ – angular momentum.
ω_0 – frequency of plasma oscillations.
D – Debye radius.
$\mathbf{H}_0$ – intensity of the external constant magnetic field of the earth.
Ω – angular Larmor frequency of an ion.
ρ_H – average Larmor radius of an ion (v_i/Ω), ρ_{He} – the same for an electron (v_e/ω_H).
ω_H – angular Larmor frequency of electrons.
$f_{n,i,e}$ – distribution functions of neutral particles, ions, and electrons.
f_0 – Maxwellian distribution function.
α – the angle between the direction of motion of the body ($\mathbf{V}_0$) and the magnetic field ($\mathbf{H}_0$).
ψ – the angle between the wave vectors $\mathbf{k}$ and $\mathbf{k}'$.
ψ_1 – the angle between the electric field E_0 of the incident wave and the wave vector $\mathbf{k}'$ of the scattered wave.
χ – the angle between $\mathbf{q}$ and $\mathbf{V}_0$.
z – the coordinate along the direction of motion of the body.
x, y – coordinates along directions which are orthogonal to the motion of the body; $\rho = (x^2 + y^2)^{1/2}$
$\mathbf{r}$ – radius vector from the body's center in an arbitrary direction.
R_0 – linear dimension of the body, radius of a sphere.
S_0 – transverse cross section of the body.
dS – differential surface area element.

$\mathbf{r}_S$ – radius vector to a point on the surface of the body.
$\mathbf{n}$ – normal to the body's surface.
$d\sigma$ – effective differential scattering cross section.
ε – dielectric permittivity of the plasma.
$\mathcal{E}_{n,i}$ – energy of the particle (molecule, ion).
U – potential energy of the particle.
$\mathbf{u}$ – velocity of the particle in a fixed coordinate system.
$\mathbf{v}$ – velocity of the particle relative to the body.

$\Phi(x) = \frac{2}{\sqrt{\pi}} \int_0^x e^{-u^2} du$ – probability function.

$$Q(a) = \left(\sqrt{\pi} + 2i \int_0^a e^{x^2} dx \right) e^{-a^2}.$$

Y – collision integral.
I – collision integral referring to collisions with the body.
$j_{n,i,e}$ – flux of neutral particles, ions, electrons on a unit element of area.
$J = \int j dS$ – total particle flux.
$\hat{\mathbf{j}}$ – electric current density.
φ – potential of the field.
φ_0 – potential of the body.
$\varphi^* = |\varphi| e/\varkappa T$ – value of the potential in dimensionless units.

CONTENTS

CHAPTER I

INTRODUCTION

This book, as already indicated, is devoted to a theoretical investigation of a number of phenomena which arise in connection with the motion of bodies (artificial satellites and rockets) in the upper ionosphere or an interplanetary medium. Therefore, before turning to an exposition of the essential nature of the problems studied, it is appropriate to present here certain reference data about the properties of these media. We shall use this data, in particular, for various estimates and numerical calculations.

§ 1. Data on the Properties of the Upper Ionosphere

The upper atmosphere is, as is well known, an extremely rarefied, partially ionized gas (plasma). The concentration of neutral particles, molecules, and atoms in it decreases sharply with increase of altitude: from 10^{13} particles/cm^3 at an altitude of 100 km to $\sim 10^5$ particles/cm^3 at an altitude of 1000 km (see Table I). The concentration of ions and electrons at not too great altitudes varies strongly depending on the time of year and even during a twenty-four hour period. With increasing height the concentration of ions and electrons decreases considerably more slowly than the molecular concentration. Accordingly, the degree of ionization of the ionosphere increases abruptly with increasing height. At an altitude of 100 km the plasma is still very weakly ionized – the ratio N_0/n_0 amounts altogether to only 10^{-8} to 10^{-10}, at an altitude of 300 km the ratio $N_0/n_0 \sim 10$, but at an altitude of 3000 km the ratio N_0/n_0 is already $\sim 10^4$. Here the atmosphere is already almost completely ionized.

The molecular composition of the atmosphere changes very little up to an altitude of approximately 100 km. With increase of height, first the oxygen molecules, and then the nitrogen molecules begin to dissociate. Ions of nitric oxide also appear in this region of the ionosphere. According to the latest data, helium apparently dominates in the range of altitudes from 700 to 1000 km, above 1200 to 1400 km molecular hydrogen predominates. The relative concentration of atomic hydrogen increases substantially at large distances from the earth.

The average temperature in the ionosphere increases with height, from approximately 230° at an altitude of 100 km to 3-4,000° at greater altitudes. The earth's magnetic field turns out to have a substantial effect on physical processes in the ionosphere. Its magnitude decreases slowly with increasing height (see Table I).

At the present time it is still far from clear where the ionosphere really ends, i.e., where the gaseous envelope which moves together with the earth ends, and where interplanetary space begins. It is frequently assumed that interplanetary space begins where protons (ionized hydrogen) are the principal constituent of the heavy particles in the plasma. However, the altitude of this boundary is also unknown up to now.

According to various data, one can assume that at distances of three to four times the earth's radius we already have interplanetary gas. Therefore, we also give in the tables the expected properties of the medium at these distances from the earth and in cosmic (interstellar) space.

Physical processes in the ionospheric plasma are characterized by a complete set of parameters. To begin with, there are the usual parameters of the kinetic theory of gases: the mean free path, the average thermal velocity, the collision frequency. At least three kinds of particles are always present in a plasma: neutral particles, ions, and electrons. Consequently, each of these is characterized by its own mean free path, average velocity, collision frequency (see Table II). Another group of parameters, relating to only charged particles, characterize the electromagnetic processes. These are the plasma (Langmuir) frequency ω_0, the Debye radius D, the Larmor frequency of revolution, and the average Larmor radii of electrons and ions. Numerical values for all of these quantities are given in Table III at different altitudes in the ionosphere.

TABLE I *

Height, km	n_0, cm^{-3}	N_0, cm^{-3}	$T°$	N_0/n_0	M_0	H_0, oersteds
Ionosphere						
100	$(2,5{\div}4)\cdot10^{13}$	$(2–100)\cdot10^{3}$	~230	$10^{-10}–10^{-8}$	28	0.49
200	$3\cdot10^{10}$	$(3–50)\cdot10^{4}$	450–800	$10^{-5}–10^{-6}$	24 (200–300)	0.45
300	$3\cdot10^{9}$	$(1–20)\cdot10^{5}$	1000	$10^{-3}–10^{-4}$		0.44
400	$5\cdot10^{8}$	$(5–15)\cdot10^{5}$	1500	$10^{-2}–10^{-3}$	~20 (400–500)	0.40
500	$5\cdot10^{7}$	$(4–10)\cdot10^{5}$	~1800	$10^{-2}–10^{-1}$		0.37
700	$6\cdot10^{6}$	$(2–5)\cdot10^{5}$	~2000	$\sim10^{-1}$	~16	0.35
1000	10^{5}	10^{5}	~3000	~1	16 [4]	0.33
3000	1 (?)	$(5–7)\cdot10^{3}$	~4000	$\sim10^{4}$	4 [2]	0.16
Interplanetary Gas						
(3–4) R^0	?	(3–5) 10^{2}	$5\cdot10^{3}$ (?)	?	1	$(5–15)\cdot10^{-3}$
$100\cdot R^0$	?	10^{2}	—	—	1	$(5–10)\cdot10^{-4}$
Cosmic Space						
—	—	~1	10^{4} (?)	?	1	10^{-5}

*R^0) Radius of the earth; n_0) concentration of neutral particles; N_0) concentration of electrons, ions; T) temperature in degrees Kelvin; H_0) intensity of the earth's magnetic field in oersteds for the case when $H_0 \simeq 0.5$ oersted near the earth's surface; M_0) average molecular weight of neutral particles, ions.

TABLE II *

Height, km	l_n, cm	$l_{e,i}$, cm	ν_n, sec^{-1}	$\nu_{e(n+i)}$ sec^{-1}	ν_{ii}, sec^{-1}	$v_{n,i}$, $\frac{cm}{sec}$	v_e, $\frac{cm}{sec}$
			Ionosphere				
100	10^1	50	$4 \cdot 10^3$	$2 \cdot 10^5$	10^3	$4 \cdot 10^4$	$9.4 \cdot 10^6$
200	$8 \cdot 10^3$	$9 \cdot 10^3$	9	$2 \cdot 10^3$	10^2	$7 \cdot 10^4$	$1.5 \cdot 10^7$
300	10^5	$7 \cdot 10^3$	0,9	$3 \cdot 10^3$	15	$9 \cdot 10^4$	$1.9 \cdot 10^7$
400	$7 \cdot 10^5$	$2 \cdot 10^4$	$\sim 10^{-1}$	10^3	7	10^5	$2.2 \cdot 10^7$
500	$6 \cdot 10^6$	$4 \cdot 10^4$	$\sim 10^{-2}$	$7 \cdot 10^2$	3	$1.4 \cdot 10^5$	$2.5 \cdot 10^7$
700	$5 \cdot 10^7$	10^5	$\sim 10^{-3}$	$2 \cdot 10^2$	1	$1.6 \cdot 10^5$	$2.9 \cdot 10^7$
1000	$8 \cdot 10^8$	$8 \cdot 10^5$	$\sim 10^{-4}$	40	~ 0.2	$2 \cdot 10^5$ $(4 \cdot 10^5)$	$3.4 \cdot 10^7$
3000	$2 \cdot 10^{14}$	$3 \cdot 10^6$	$\sim 10^{-9}$	14	~ 0.1	$6 \cdot 10^5$ (10^6)	$3.9 \cdot 10^7$
			Interplanetary Gas				
(3—4) R°	—	$3 \cdot 10^9$	~ 0	$\sim 10^{-2}$	$\sim 5 \cdot 10^{-4}$	$\sim 2 \cdot 10^6$	$4.4 \cdot 10^7$
			Cosmic Space				
—	—	$7 \cdot 10^{11}$	~ 0	$\sim 10^{-4}$	$\sim 3 \cdot 10^{-6}$	$\sim 3 \cdot 10^6$	$5.4 \cdot 10^7$

* ν_n) Number of collisions between neutral particles; $\nu_{e(n+i)}$) number of collisions of electrons with ions and neutral particles; ν_{ii}) number of collisions between ions; $v_{n,i}$) thermal velocity of neutral particles, ions; v_e) thermal velocity of electrons; l_n) mean free path of neutral particles; $l_{e,i}$) mean free path of electrons, ions.

TABLE III *

Height, km	ω_0, sec^{-1}	D, cm	ω_H, sec^{-1}	ω_H/ω_0	ϱ_{He}, cm	Ω, sec^{-1}	ϱ_H, cm
Ionosphere							
100	$(3-20)\cdot 10^6$	1	$8\cdot 10^6$	~ 1	~ 1		$2\cdot 10^2$
200	$(1-4)\cdot 10^7$	0.2—1	$7.9\cdot 10^6$	~ 0.3	~ 2		$4.0\cdot 10^2$
300	$(2-8)\cdot 10^7$	0.1—0.7	$7.7\cdot 10^6$				$5.6\cdot 10^2$
400	$(4-7)\cdot 10^7$	0.2—0.4	$7.3\cdot 10^6$	~ 0.1 (300–400)	~ 3 (300–500)	$\sim 2\cdot 10^2$ (100–700)	$5.5\cdot 10^2$
500	$6\cdot 10^7$	0.3—0.6	$7\cdot 10^6$				
700	$(2-4)\cdot 10^7$	0.4—0.7	$6.4\cdot 10^6$	~ 0.2 (500–700)	~ 4		$8.2\cdot 10^2$
1000	$2\cdot 10^7$	1	$5.7\cdot 10^6$	~ 0.3	~ 5	$2\cdot 10^2 (8\cdot 10^2)$	$\sim 10^3 (5\cdot 10^2)$
3000	$5\cdot 10^6$	4	$2.8\cdot 10^6$	~ 0.5	~ 14	$4\cdot 10^2 (8\cdot 10^2)$	$1.2\cdot 10^3$
Interplanetary Gas							
$(3-4)\,R^\circ$	$\sim 10^6$	30	$(1-3)\cdot 10^5$	~ 0.2	$2\cdot 10^2$	10^2	$2\cdot 10^4$
$100\,R^\circ$	$\sim 5\cdot 10^5$	50	$\sim 2\cdot 10^4$	$\sim 3\cdot 10^{-2}$	$2\cdot 10^3$	8	$2.5\cdot 10^5$
Cosmic Space							
—	$\sim 5\cdot 10^4$	700	$\sim 2\cdot 10^2$	$\sim 3\cdot 10^{-3}$	$\sim 3\cdot 10^5$	$\sim 5\cdot 10^{-2}$	$6\cdot 10^7$

* ω_0) Frequency of plasma oscillations; D) Debye radius; ω_H) Larmor frequency of electrons; ρ_{He}) Larmor radius of electrons; Ω) Larmor frequency of ions; ρ_H) Larmor radius of ions.

It is clear from Table II that the mean free path of neutral particles in the upper ionosphere is extremely large. Even at the comparatively low altitude of 200 km, the mean free path $l_n \sim 80$ m. With increase of height, l_n increases very rapidly and $l_n \sim 1$ km at an altitude of 300 km, $l_n \sim 8000$ km at an altitude of 1000 km. Thus, at distances from the earth's surface of the order of the earth's radius or less, l_n significantly depends on the direction. The mean free path of ions and electrons is also rather large; thus, $l_e \sim 0.1$ km at an altitude of 200 km, $l_e \sim 8$ km at an altitude of 1000 km. However, l_e varies considerably more slowly with altitude. The reason for this is that the mean free path of electrons and ions at altitudes exceeding 200 km is by this time determined by collisions between charged particles, but their concentration and thermal velocity, as is evident from Tables I and II, changes less sharply with altitude than the concentration of neutral particles does.

The thermal velocities of neutral particles and of ions are of the same order of magnitude. They increase with altitude, owing to the increase of temperature and the decrease in the average mass of the particles, from $M_0 \sim 24$ at an altitude of the order of 200 km to $M_0 \sim 16$ at an altitude ~ 700 km. According to the latest data, one can assume that the value of M_0 rapidly falls to $M_0 = 4$ (helium) above 700 km. Furthermore, at altitudes greater than 1200 to 1400 km, $M_0 = 2$, i.e., molecular hydrogen now predominates. The thermal velocity of electrons is, on the whole, extremely large – the order of 100 to 500 km/sec.

The Debye radius (Debye shielding distance), i.e., the distance over which, as is well known, a static electric charge is screened by polarization of the plasma, is extremely small in the ionosphere. It varies from 0.1 to 4 cm with increase of altitude from 100 to 3000 km. The Debye radius in interplanetary or interstellar gas attains a length of several meters.

The Larmor radius of revolution of an electron or an ion in the earth's magnetic field characterizes the influence of a magnetic field on different processes in a plasma. The average Larmor radius of electrons in the ionosphere is not large, of the order of a few centimeters; it increases with increasing altitude and is equal to several meters in interplanetary gas. The average radius of the Larmor orbits of ions is always a rather large quantity; in the ionosphere it varies from 2 to 20 meters.

A large number of fast particles, forming radiation belts, exist in regions located at considerable distances from the earth's surface. Data on the number of fast particles in these radiation belts and their energy is still incomplete at the present time. The inner radiation belt begins at altitudes of 400 to 1000 km and extends to an altitude ~10,000 km. In the region of greatest intensity of the inner belt, the radiation flux of protons with energies greater than 1 Mev amounts to $3 \cdot 10^4$ protons/cm$^2 \cdot$ sec. The flux of electrons with energies over 20 kev in the inner belt is of order 10^{10} electrons/cm$^2 \cdot$ sec. The outer radiation belt begins at altitudes of the order of 1.5 to 2 times the earth's radius and extends to an altitude of 10 to 14 times the earth's radius. The maximum intensity in this belt occurs at an altitude of three to four times the earth's radius. Here the flux of electrons with energies above 20 kev is of the order of 10^8 to 10^9 electrons/cm$^2 \cdot$ sec.

The intensity of the sun's electromagnetic radiation incident on the earth reaches approximately 0.5 to 1 erg/cm$^2 \cdot$ sec in the interval of frequencies capable of ionizing the upper ionosphere (800 to 1100 A). The flux of gamma rays from the sun is the same order of magnitude.

The reader can find detailed information about the structure and properties of the upper ionosphere, methods of investigating it, etc., in monographs [13, 14] which are especially devoted to the ionosphere, and in the periodic literature.

§ 2. Properties of the Interaction of Moving Bodies with the Ionosphere

The characteristic dimensions of rockets or artificial satellites moving through the upper ionosphere are of the order of one or a few meters. The mean free path of the particles here, as is evident from Table II, is considerably larger. This means, as already indicated above, that for an investigation of processes in the vicinity of a moving body, it is impossible to regard the plasma as a continuous medium, i.e., in order to describe such effects one cannot use the methods of hydrodynamics or aerodynamics usually applicable under terrestrial conditions or in the lower part of the ionosphere. It is necessary by this time to take the fact that the plasma consists of an aggregate of individual particles into consideration. Therefore, it is necessary to use kinetic theory in order to describe processes in the vicinity of moving bodies.

TABLE IV

Height, km	$V_0, \frac{cm}{sec}$	$V_0/v_{n,i}$	V_0/v_e	D/R_0	ϱ_H/R_0	ϱ_{He}/R_0
Ionosphere						
100		20	0,1	10^{-2}	2	10^{-2}
200		~ 10	$6\cdot 10^{-2}$		4	$2\cdot 10^{-2}$
300			$4\cdot 10^{-2}$		5.5	$3\cdot 10^{-2}$
400	$\sim 8\cdot 10^5$	8				
500			$3\cdot 10^{-2}$	$(1-10)\cdot 10^{-3}$	8	
700		5				$4\cdot 10^{-2}$
1000		4	$2\cdot 10^{-2}$	$4\cdot 10^{-2}$	10	$5\cdot 10^{-2}$
3000		3.5			20	$\sim 10^{-1}$
Interplanetary Gas						
$(3-4)R^0$	$\sim 6\cdot 10^5$	~ 0.3	$\sim 10^{-2}$	0.3	$4\cdot 10^2$	~ 2
$100R^0$	$\sim 2\cdot 10^5$	~ 0.1	$\sim 5\cdot 10^{-3}$	0.5	$4\cdot 10^3$	~ 20
Cosmic Space						
—		$\sim 3\cdot 10^2$	$\sim 2\cdot 10^{-3}$	7	$6\cdot 10^5$	$\sim 3\cdot 10^3$

Collisions between particles generally do not turn out to have an important influence on the distrubance of the concentration of particles in the vicinity of a body whose dimensions are much smaller than the mean free path. They begin to manifest themselves only at large distances from the body, and just at distances of the order of the mean free path.* During collisions, the particles reflected from a body can in principle ionize and heat the gas in front of a body. However, the energy of the reflected neutral particles and ions is only of the order of 10 to 20 eV, and their velocity is much smaller than the velocity of electrons in atoms. Under such conditions the ionization probability is small.† Heating of the gas by reflected particles is small, as one can easily see, upon fulfillment of the condition

$$l_n \gg \frac{V_0}{v_n} R_0, \tag{1.1}$$

where $v_n = (2\varkappa T/M)^{1/2}$ is the thermal velocity of the gas molecules, R_0 is the size of the body. In fact, the total energy of particles reflected from the surface of a body during the time interval Δt is, in order of magnitude, obviously equal to

$$\mathscr{E}_{\text{refl}} \sim MV_0^2 \cdot n_0 R_0^2 \; (V_0 \Delta t).$$

These particles travel without collisions over a distance of the order of the mean free path l_n. Therefore, they are slowed down within a volume $\sim l_n^2 V_0 \Delta t$ if it is assumed that $V_0 \Delta t \gg l_n$. The total number of particles in this volume is the order of $n_0 l_n^2 (V_0 \Delta t)$. Therefore, on the average the energy of the particles of the unperturbed gas will increase by the amount $\Delta\mathscr{E} \sim MV_0^2 (R_0/l_n)^2$ due to collisions with the reflected particles. The

*Collisions play a determining role for particles moving in finite orbits around the body. Charged particles (electrons or ions), attracted by the electric field of the body, may undergo such motion. It is possible, however, only in the vicinity of a body whose dimensions do not exceed the Debye radius (see Table III). These questions will be investigated in more detail in Chapter VIII.

†The large values of the additional ionization obtained in article [15] stem from the fact that the gas-kinetic cross section, 10^{-16} cm^2, was erroneously taken as the ionization cross section in this article. As a matter of fact, it appears that the ionization cross section under the indicated conditions does not exceed 10^{-20} cm^2 (precise data is not available).

conditions $\Delta\mathcal{E} \ll \varkappa T$ also leads to the inequality (1.1). For bodies whose dimensions are the order of 100 cm, condition (1.1) is well satisfied in the ionosphere at altitudes exceeding 200 km.

The characteristic size of the objects of interest to us in the ionosphere is large in comparison with the Debye radius, as is clear from Table IV. This, in particular, means that the Debye screening turns out to have great influence on the distribution of the electric field and of the charged particles near a body. Furthermore, comparing the characteristic velocity of bodies with the thermal velocities $v_{n,\,i}$ of ions or molecules (see Table IV), we convince ourselves that in the ionosphere the body's velocity V_0 is much larger than the thermal velocities of neutral particles and ions. This condition may already not be satisfied in interplanetary gas. Therefore, the conditions which apply to a moving body during passage into interplanetary space vary strongly.

We shall see that in the ionosphere the case of most interest is when the dimension of the body is smaller than the mean free path and larger than the Debye radius, but the velocity of the body is larger than the thermal velocity of neutral particles or ions, and smaller than the thermal velocity of electrons. It is precisely this case which we shall consider in more detail below.

We shall also make some remarks about the radiation of waves by material objects moving through the ionosphere. Since the velocity of these objects is supersonic ($V_0 \gg v_{n,\,i}$), they can radiate sound waves [16]. However, these waves are weakly damped if their wavelength is greater than the mean free path of the particles. Since here the dimension of the body is much smaller than the mean free path, so the amplitude of the weakly damped sound waves, radiated during motion of the body, turns out to be small in general. Ionic plasma waves may be excited in the region of wavelengths smaller than the mean free path. However, their attenuation in an isothermal plasma is very large (see [17]). Electronic plasma waves, generally speaking, cannot be generated by a body, because the body's velocity is much smaller than the thermal velocity of the electrons, and therefore also much smaller than the velocity of the corresponding waves. It is true that scattering of the electric field of the body by inhomogeneities, which in principle can also lead to a certain weak excitation of plasma waves or even electromagnetic waves [18], may play a role here. Side by side with this, it is also necessary to keep in mind that, due to the presence of the earth's magnetic field, the question of both the spectrum of the excited waves as well as their attenuation is much more complicated; in particular, the case of longitudinal motion of the body may be singled out. Therefore, it is hardly possible to reach any kind of final, even qualitative, conclusions with regard to the cause of radiation of waves by a moving body until a consistent theory is developed.

The stability of the disturbed region near a moving body is a problem of considerable interest. This question, as in ordinary hydrodynamics, reduces to an investigation of character of development of small deviations from the stationary (steady-state) distribution: if these deviations increase with time, then the region under consideration is unstable; if, however, arbitrary disturbances damp out, then the situation is stable. At the present time, such an investigation has still not been carried out; therefore, here we shall only present certain general considerations.

Instabilities of the various nonequilibrium states of a plasma are of two basic types. In the first place, instabilities exist which are associated with the deviation of the plasma particles' velocity distribution from a Maxwellian distribution. If the distribution of electrons is non-Maxwellian, then the plasma will be unstable in the case when this distribution is nonmonotonic, i.e., when the velocity distribution function has other maxima in addition to the maximum for the velocity equal to zero. Instabilities of this type are called two-stream instabilities in English, while the Russian term is "beam" instabilities, since the instability of an electron beam in a plasma [45] is a typical example of them. In this case, the distribution function has an additional maximum at the point where the velocity of the beam corresponds to the ordered motion. If the velocity distribution of the ions is non-Maxwellian, for example, if an ionic beam exists in the plasma, then an instability appears only when the velocity of ions (which corresponds to the ordered motion) with respect to the electrons becomes of the order of the thermal velocity of the electrons [46]. In our case, an instability of the "beam" type is probably not present. This is associated with the fact that the velocity of motion of the body is much less than the thermal velocity of the electrons; therefore the velocity distribution of the electrons is almost undisturbed. The ion distribution is perturbed rather strongly; however, the maximum velocities of the nonequilibrium ions (for example, the ions reflected from the body) prove to be the same order as the velocity of the body, i.e., all of these velocities are considerably smaller than the thermal velocity of the electrons. The case of a metallic body or probe, on which a positive potential is present, is a special case. In this case, the body attracts electrons, and an electron beam directed towards the body appears in the plasma; under certain conditions this leads to an instability [81].

The type of instability described above is characteristic of a spatially homogeneous plasma. Recently instabilities of another type, associated with an inhomogeneity in the spatial distribution of the particle densities and of the temperature [47-49], have been studied. It was found that a nonhomogeneous plasma in a magnetic field is unstable. The initial deviations from a stationary state lead to the growth of a different type of wave in the plasma. In our case the plasma is essentially inhomogeneous. However, it is impossible to direcently apply the results of articles [47-49] to the disturbed region near the body. This is associated, in the first place, with the fact that the body is moving rapidly, so that the distribution investigated for instability is nonstationary in the coordinate system where the plasma is at rest. Such nonstationarity will probably turn out to have a stabilizing effect, since not all of the disturbances are able to increase during the time of flight of the body. Here the motion of a body along a magnetic field is a special case. In this case, as we shall see below, the disturbed region turns out to be very long; therefore, the distribution changes very slowly with time, i.e., it is almost stationary. It is possible to think that in this case an instability actually exists. It is also necessary to keep in mind that the equations in articles [47-49] are derived under the assumption that the characteristic dimension of the inhomogeneity in a plasma is large in comparison with the Larmor radius of the ions. In our case, this condition is usually not satisfied.

If, as the result of further investigations, it turns out that the disturbed region is unstable and an instability is able to develop, then this will mean that a region of irregular nonstationary oscillations of the plasma – a "turbulent trail" – is generated behind the body. Such a turbulent trail will radiate electromagnetic "noise" with frequencies of the order of the characteristic frequencies of those oscillations which turn out to be unstable. In addition, the turbulence leads to an anomalous increase of the coefficient of diffusion of the plasma across a magnetic field [50]. As a result, reabsorption of the body's trail will take place more rapidly, i.e., the length of the disturbed region is decreased.

In concluding this section, we remark that the question of the interaction of the plasma particles, the electromagnetic radiation of the sun, and corpuscular protons with the surface of a body is of major importance. These processes are extremely diverse. They significantly depend on the properties of the surface: material, temperature, and method of manufacture. These phenomena have still been quite inadequately investigated, especially theoretically. In the majority of cases these processes are too complicated for a reliable theoretical calculation. On the other hand, it is extremely difficult and sometimes impossible in general to realize the same conditions in the laboratory which occur in the ionosphere. The results of investigations of these effects directly on artificial satellites are unknown to us up to now. Because of the reasons mentioned, we shall not dwell on this sphere of questions here, referring the reader to the review article [12], where the known (from the world literature) experimental data dealing with this problem are summarized. Many of the problems mentioned, specifically with application to the motion of objects in the ionosphere and in interplanetary space, are explained in detail in this article. We also refer to the monographs [19, 20] in which a sequence of data on the interaction of particles with the surface of bodies is presented.

§ 3. Basic Equations

In order to investigate the disturbances produced in a medium in the vicinity of a moving object, we must use the methods of the kinetic theory of gases. This stems from the fact that the free-path length, in the part of the ionosphere of interest to us, is considerably greater than the dimensions of the body, i.e.,

$$l_n,\ l_i,\ l_e \gg R_0, \tag{1.2}$$

where the subscripts n, i, and e refer to a neutral particle, an ion, and an electron. Let us investigate the corresponding equations in turn.

a. Equation for the Distribution Function of the Neutral Particles. Neutral particles (molecules, atoms) will be described by the distribution function*

$$f_n(r,\ v,\ t),$$

whose meaning is as follows. The product of f_n times the volume element in configuration space – in the space of velocities and coordinates – i.e.,

$$f_n(r,\ v,\ t)\, d^3v\, d^3r$$

(where $d^3v = dv_x dv_y dv_z$, $d^3r = dxdydz$), is the number of particles whose coordinates lie between x and x + dx, y and y + dy, z and z + dz, and whose velocity components are between v_x and $v_x + dv_x$, v_y and $v_y + dv_y$, v_z and $v_z + dv_z$. The particle density n and other average quantities can be expressed in terms of the function f_n,

$$n(r,\ t) = \int f_n(r,\ v,\ t)\, d^3v.$$

We shall carry out the investigation in a coordinate system moving together with the body. Here **v** is the velocity of the particle relative to the body. It is related to the velocity **u** in a fixed coordinate system by the formula

$$v = u - V_0,$$

where $\mathbf{V}_0$ is the velocity of the body. Since it is possible to regard the body's motion as rectilinear and uniform, the picture is stationary in our coordinate system and the function f_n does not depend on the time

$$f_n = f_n(r,\ v). \tag{1.3}$$

As is well known, the distribution function obeys the Boltzmann equation

$$\frac{\partial f_n}{\partial t} + v\frac{\partial f_n}{\partial r} + \frac{1}{M} F \frac{\partial f_n}{\partial v} = Y, \tag{1.4}$$

where M is the molecular mass, **F** is the force acting on a molecule, Y is the collision integral. However, as long as distances smaller than l_n are important in the problem, by virtue of the conditions (1.2) one can neglect the effect of collisions on the distribution function, i.e., it is possible to assume Y = 0. In addition, outside the body no forces act on the neutral particles. Further, it is obvious from Eq. (1.3) that $(\partial f/\partial t) = 0$. Therefore, Eq. (1.4) for all space outside the body simply reduces to the equation

$$v\frac{\partial f_n}{\partial r} = 0. \tag{1.5}$$

The gas is undisturbed at large distances from the body, so that as $r \to \infty$ the function f_n must go over into the Maxwell distribution function f_0 (**u**).

$$f_n \to f_0(v + V_0) = n_0\left(\frac{M}{2\pi\varkappa T}\right)^{3/2} \exp\left\{-\frac{Mu^2}{2\varkappa T}\right\} = n_0\left(\frac{M}{2\pi\varkappa T}\right)^{3/2} \exp\left\{-\frac{M(v+V_0)^2}{2\varkappa T}\right\}, \tag{1.6}$$

where n_0 is the unperturbed density of molecules in the ionosphere, T is the temperature of the gas, $\varkappa$ is the Boltzmann constant. In the case when no collisions between particles occur, the boundary condition (1.6) may cause some confusion, since usually the establishment of a Maxwellian distribution in a gas is associated with collisions. In the absence of collisions, however, the particles rapidly disperse after having collided with the

*Of course, if neutral particles of different types exist in the medium (for example, N_2, O_2, O, N, H, etc., in the ionosphere), then each of these is described by its own distribution function f_n.

body. Therefore, at a sufficiently large distance from the body there is a very small number of those particles whose distribution was perturbed by the body, so that the total distribution is very close to Maxwellian. The electric field also falls off rather rapidly with distance and cannot change the distribution of particles at infinity. Of course, by infinitely large distances in (1.6), one should understand distances that are large in comparison with the disturbed region near the body, but much smaller than the mean free path.

b. Boundary Conditions on the Surface of the Body. The boundary conditions on the surface of the body are determined by the nature of the reflection or scattering of particles by this surface and are represented by an integral relation, connecting (on the surface of the body) the distribution function of the particles incident on the surface with the distribution function of the scattered particles. In this connection, if the gas is sufficiently rarefied, this relationship will be linear. The particles incident on the surface are characterized by the fact that $\mathbf{nv} < 0$ for them, where $\mathbf{n}$ is the outward normal to the surface of the body, since, of course, only particles traveling in a direction toward the surface are incident on the surface. Conversely, $\mathbf{nv} > 0$ for the scattered particles.

Let $w(\mathbf{v}, \mathbf{v}_1, \mathbf{r}_S)$ be the probability that, having hit with velocity $\mathbf{v}_1$ at the point $\mathbf{r}_S$ of the surface, the particle acquires the velocity $\mathbf{v}$ after scattering. Then it is obvious that the flux of scattered particles per unit surface area will be expressed in terms of the number of incident particles by the formula

$$-\int\limits_{(nv_1)<0} w(v, v_1, r_S)(nv_1) f(v_1)\, d^3v_1, \tag{1.7}$$

where $\mathbf{r}_S$ is a point on the surface of the body. On the other hand, this flux is equal to

$$f(v)(nv). \tag{1.8}$$

Therefore, the desired boundary condition has the form

$$(nv) f(r_S, v) = -\int\limits_{(nv_1)<0} w(v, v_1, r_S)(nv_1) f(r_S, v)\, d^3v_1, \tag{1.9}$$

where $(\mathbf{nv}) > 0$.

It is often convenient to solve the kinetic equation without boundary conditions on the body, i.e., in unlimited space, but the presence of the surface is taken into account by the addition to the right side of the equation of a special term, different from zero only on this surface. If the surface of the body is given by the equation

$$F(r_S) = 0, \tag{1.10}$$

then the general form of such a term is

$$A(r_S, v)\,\delta(F), \tag{1.11}$$

where $\delta(F)$ is a delta function and $A(\mathbf{r}_S, \mathbf{v})$ is the function describing the interaction of particles with the surface of the body.

Let us uncover the meaning of the function $A(\mathbf{r}_S, \mathbf{v})$. For this, we assume that the left side of the kinetic equation is the total time derivative of the distribution function. Therefore, the integral of expression (1.11) over all space is the change in the number of molecules, having velocity $\mathbf{v}$, per unit time. One can reduce the integration of (1.11) over volume to an integration over dS dn, where dS is the (differential) element of surface area on the body, and dn is the (differential) element of length in the direction normal to the surface. Thus

$$\int A(r_S, v)\,\delta(F)\, d^3r = \int A(r_S, v)\,\delta(F)\, dS\, dn = \int A(r_S, v)\,\delta(F) \frac{dF}{|\nabla F|}\, dS = \oint \frac{A(r_S, v)}{|\nabla F|} dS. \tag{1.12}$$

It is obvious from (1.12) that $A/|\nabla F|$ is the change per unit time in the number of particles having velocity **v** due to oscillations with unit surface (area) of the body at the point $\mathbf{r}_S$. This number of particles is obviously equal to the difference between the number, determined by expression (1.7), of particles which have acquired the velocity **v** as a result of scattering, and the number, determined by formula (1.8), of particles incident with this velocity. Hence

$$\frac{A}{|\nabla F|}=\begin{cases} -\int\limits_{(nv_1)<0} w(v, v_1, r_S)(nv_1)f(r_S, v_1)\,d^3v_1 & \text{for } nv>0; \\ (nv)f(r_S, v) & \text{for } nv<0. \end{cases} \tag{1.13}$$

Therefore

$$A(r_S, v)=-\nabla F\begin{cases} \int\limits_{(v_1\nabla F)_S<0} w(v, v_1, r_S)v_1 f(r_S, v_1)\,d^3v_1 & \text{for } (v\nabla F)>0; \\ -vf(r_S, v) & \text{for } (v\nabla F)<0. \end{cases} \tag{1.14}$$

Here the sign of the function F is chosen so that the direction of ∇F coincides with the direction of the outward normal **n**; $(\mathbf{v}\nabla F)_S$ means that the quantity $\mathbf{v}\nabla F$ is evaluated on the surface S.

Now let us consider a particular form of the function w in various cases. If specular reflection of particles from the surface occurs, i.e., if the angle of incidence of the particle on the surface (the angle between $\mathbf{v}_1$ and **n**) is equal to its angle of reflection (the angle between **v** and **n**), then upon scattering the component of the particle's velocity normal to the surface simply changes sign, so that

$$w(v, v_1, r_S)=\delta\,[v_1-v+2(vn)n], \tag{1.15}$$

where **n** is the normal to the surface at the point $\mathbf{r}_S$. In this case the function $A(\mathbf{r}_S, \mathbf{v})$ has the form

$$A(r_S, v)=\begin{cases} (\nabla F v)f(r_S, v-2(vn)n) & \text{for } (nv)_S>0; \\ -(\nabla F v)f(r_S, v) & \text{for } (nv)_S<0. \end{cases} \tag{1.16}$$

In particular, for a spherical surface $F = r - R_0$, and for specular reflection the term on the right side of the kinetic equation (1.5), describing the interaction of particles with the body's surface, assumes the form

$$A(r_S, v)\delta(F)=\begin{cases} \frac{rv}{r}\delta(r-R_0)f\left(r, v-\frac{2r(rv)}{r^2}\right) & \text{for } rv>0; \\ \frac{rv}{r}\delta(r-R_0)f(r, v) & \text{for } rv<0. \end{cases} \tag{1.17}$$

If particles are diffusely scattered upon collision with the body's surface, i.e., if the surface of the body is very rough so that the particle may, with equal probability, be reflected at any angle to the surface, then

$$w(v, v_1, r_S)=\frac{1}{2\pi v^2}\delta(v-v_1) \quad \text{for } nv_1<0. \tag{1.18}$$

Here the scattering is assumed to be elastic, i.e., the magnitude of a particle's velocity does not change upon reflection. In this case the function $A(\mathbf{r}_S, \mathbf{v})$ has the form

$$A(r_S, v)=\begin{cases} -\frac{\nabla F}{2\pi}\int\limits_{(\nabla F v_1)<0} do_1 v_1 f(r_S, v_1) & \text{for} \quad (v\nabla F)>0; \\ (v\nabla F) f(r_S, v) & \text{for} \quad (v\nabla F)<0, \end{cases} \tag{1.19}$$

where do_1 is the element of solid angle in the direction $\mathbf{v}_1$. If the particle scattering affiliated with diffuse reflection is not elastic and the particle can give back energy to the surface, i.e., as they say, if partial accommodation occurs, then

$$w(v, v_1, r_S)=\frac{1}{2\pi} p(v, v_1) \quad \text{for} \quad nv_1<0, \tag{1.20}$$

where the function $p(\mathbf{v}, \mathbf{v}_1)$ describes the probability of partial accommodation. In the limiting case, when the particles (after collision with the surface) come into thermal equilibrium with it, i.e., for the case of total accommodation

$$w(v, v_1, r_S)=\frac{2}{\pi}\left(\frac{M}{2\varkappa T_0}\right)^2 (nv) \exp\left[-\frac{Mv^2}{2\varkappa T_0}\right], \tag{1.21}$$

where T_0 is the surface temperature of the body.

If the particles are completely absorbed by the surface, then $w = 0$ and

$$A(r_S, v)=\begin{cases} 0 & \text{for} \quad v\nabla F>0; \\ -(v\nabla F) f & \text{for} \quad v\nabla F<0. \end{cases} \tag{1.22}$$

For partial absorption, the right side of the equation has the form (1.14) as before. Only the normalization of the function w is changed. Indeed, the relation

$$\int w(v, v_1, r_S)\, d^3v = R(v_1, r_S),$$

holds, where R is the coefficient of reflection, i.e., the total probability for reflection of a particle from the body's surface ($0 < R \leq 1$ always). In the absence of absorption, $R \equiv 1$. The possibility of ejection of particles from the surface, due to its vaporization and destruction under the influence of collisions with other particles, can also be taken into account with the aid of the function w. In this connection, the integral in the last formula may be greater than unity.

c. Equations for the Distribution Functions of Electrons and Ions and for the Electric Field. The ion and electron distributions are described by the distribution functions $f_i(\mathbf{r}, \mathbf{v})$ and $f_e(\mathbf{r}, \mathbf{v})$. The kinetic equations for the functions f_i and f_e differ from the equation for the function f_n by the fact that the Lorentz force acting on a charged particle is taken into account:

$$v\frac{\partial f_i}{\partial r}+\frac{e}{M_i}\left(E+\frac{1}{c}[v+V_0, H]\right)\frac{\partial f_i}{\partial v}=A_i(r, v)\,\delta(F)+Y_i, \tag{1.23}$$

$$v\frac{\partial f_e}{\partial r}-\frac{e}{m}\left(E+\frac{1}{c}[v+V_0, H]\right)\frac{\partial f_e}{\partial v}=A_e(r, v)\,\delta(F)+Y_e. \tag{1.24}$$

Here e is the charge on an ion,* $-e$ is the charge of an electron, M_i is the mass of an ion, m is the electron's mass, Y_i and Y_e are the collision integrals for ions and electrons, $A_i(\mathbf{r}, \mathbf{v})$ and $A_e(\mathbf{r}, \mathbf{v})$ are functions characterizing the interaction of ions and electrons with the surface of the body. Of course, the functions $A_i(\mathbf{r}, \mathbf{v})$, $A_e(\mathbf{r}, \mathbf{v})$, and $A_n(\mathbf{r}, \mathbf{v})$ may significantly differ. In particular, an ion of not too high energy usually has an extremely high probability of capturing an electron upon collision with the surface, i.e., there is a high probability that it will turn into a neutral particle. Electrons may be absorbed by the surface with large probability. Therefore, below we shall often consider the case when all charged particles are absorbed upon collision with the surface of the body. In this case the functions $A_i(\mathbf{r}, \mathbf{v})$ and $A_e(\mathbf{r}, \mathbf{v})$ have the form (1.22).† Partial reflection of ions and electrons will sometimes be taken into account.

At large distances from the body – in the undisturbed plasma – the ion and electron distributions are, of course, Maxwellian:

$$f_i = f_{0i} = N_0 \left(\frac{M_i}{2\pi\varkappa T} \right)^{3/2} \exp\left\{ -\frac{M_i (v + V_0)^2}{2\varkappa T} \right\}, \tag{1.25}$$

$$f_e = f_{0e} = N_0 \left(\frac{m}{2\pi\varkappa T} \right)^{3/2} \exp\left\{ -\frac{m (v + V_0)^2}{2\varkappa T} \right\}, \tag{1.26}$$

where N_0 is the density of ions and electrons in the undisturbed plasma.

It is important to note, as indicated above in § 2, that only the case when the velocity of the body's motion is much smaller than the average thermal velocity of the electrons, i.e.,

$$V_0 \ll v_e = \sqrt{\frac{2\varkappa T}{m}}\,. \tag{1.27}$$

is of interest to us.

In addition, a body is usually negatively charged, so that the potential $\varphi_0(\mathbf{r}_S)$ of the electric field on the surface satisfies the condition

$$-e\varphi_0(r_S) \gg \varkappa T \tag{1.28}$$

(see § 14). If, in addition, the direction of motion of the body does not coincide with the direction of the magnetic field, then the electrons manage to reach thermal equilibrium, since the velocity of the moving body and the absorption of electrons on its surface is small. In this connection, the electron distribution function f_e, correct to terms of order (V_0/v_e) and $\exp(e\varphi_0/\varkappa T)$, is simply the Maxwell-Boltzmann distribution function

$$f_e(r, v) = N_0 \left(\frac{m}{2\pi\varkappa T} \right)^{3/2} \exp\left\{ \frac{e\varphi(r)}{\varkappa T} - \frac{m (v + V_0)^2}{2\varkappa T} \right\}, \tag{1.29}$$

where $\varphi(\mathbf{r})$ is the potential of the electric field at the point $\mathbf{r}$.

In the general case, the electric and magnetic fields in Eqs. (1.23), (1.24) are determined by Maxwell's equations. However, the earth's magnetic field in the vicinity of a moving body is only insignificantly disturbed. The influence of disturbances of the magnetic field on the motion of electrons and ions is also not great. Therefore, one can assume that the magnetic field **H** in Eqs. (1.23) and (1.24) is simply the earth's constant field $\mathbf{H}_0$. The problem of disturbances of the magnetic field and their influence on the motion of charged particles is considered in Chapter VI.

*It is assumed for simplicity that all ions are singly charged and of the same kind. The generalization to the case of several types of ions is obvious.

†In this connection, it is necessary to add to A_i a term describing the "creation" of neutral particles upon collision of ions with the surface.

The electric field in the vicinity of a body is created both by the electric charge of the body itself and by the volume charge which appears in the plasma due to the difference in densities of electrons and ions. Thus, the potential φ of the electric field is determined by Poisson's equation

$$\Delta\varphi = -4\pi e\left\{\int f_i(r, v)\,d^3v - \int f_e(r, v)\,d^3v\right\}. \tag{1.30}$$

At large distances from the body in the undisturbed plasma, the potential of the electric field, naturally, must vanish. The usual boundary conditions for φ must be satisfied on the surface of the body. In particular, $\varphi = \varphi_0$ on the surface of a metallic object, where φ_0 is the surface potential.

Equations (1.23), (1.24), and (1.30), under the condition that the magnetic field **H** is constant ($\mathbf{H} = \mathbf{H}_0$), form a complete system of equations determining the disturbance of charged particles in the vicinity of the body.

d. Collision Integral. It was indicated above that for conditions of interest to us the mean free path of the particles is large in comparison with the dimensions of the body. This means that the influence of collisions on the disturbances produced in the plasma by a moving body is unimportant in a large region in the vicinity of the body. Therefore, it is not necessary to consider particle collisions in order to solve many problems.

However, in the case when regions far away from a body ($r \gg l$) play a role in the phenomena investigated, for example, for an analysis of the problem of scattering of radio waves by a body's trail, collisions are important (see Chapter V). Furthermore, without account of collisions it is generally impossible to find the distribution of the ions or electrons which undergo finite motion, i.e., the particles which move in spatially restricted orbits in the vicinity of the body (see Chapter VIII, §35). In these cases, in Eqs. (1.23) and (1.24), it is necessary to introduce collision integrals, describing the change of the distribution function of the particles considered due to their collisions with each other and with all other particles in the plasma (with electrons, molecules, atoms).

In the general case, the Boltzmann collision integral has the form

$$Y = \int W(v, v'; v_1, v_1')\{f(v)F(v') - f(v_1)F(v_1')\}\,d^3v'\,d^3v_1\,d^3v_1', \tag{1.31}$$

where **v** is the velocity of the particle considered, **v'** is the velocity of that particle with which it collides (let us call it the primed particle) after the collision; $\mathbf{v}_1$ and $\mathbf{v}_1'$ are the velocities of the particle and the primed particle before the collision, F is the distribution function for the primed particles, $W(\mathbf{v}, \mathbf{v}'; \mathbf{v}_1, \mathbf{v}_1')$ is the probability of a collision in which the particle's velocity $\mathbf{v}_1$ is changed to **v** and $\mathbf{v}_1'$ is changed to **v'**.

Thus, the collision integral in general form is an extremely complicated integral term in the kinetic equation. If collisions between the particles themselves play a significant role, then $F(\mathbf{v}') = f(\mathbf{v}')$ and $F(\mathbf{v}_1') = f(\mathbf{v}_1')$, so that the kinetic equation also becomes nonlinear. Simplified expressions for the collision integral, which one can obtain in various specific cases, are therefore extremely important. A considerable literature is devoted to these questions. Below we present only those expressions which will be used in what follows.

If the particles interact according to Coulomb's law (collisions of ions with ions, electrons with ions, electrons with electrons), then the collision integral for them can, as shown by Landau [22], be represented in the form

$$Y = -\operatorname{div}_v j = \sum_k \frac{\partial j_k}{\partial v_k}, \tag{1.32}$$

where

$$j_k = \frac{\pi e^4 \Lambda}{m} \sum_l \int \left[\frac{f(v)}{m'}\frac{\partial F(v')}{\partial v_l'} - \frac{F(v')}{m}\frac{\partial f(v)}{\partial v_l}\right]\frac{w\delta_{lk} - w_l w_k}{w^3}\,d^3v'. \tag{1.33}$$

Here Λ is the Coulomb logarithm

$$\Lambda = \ln \frac{D\varkappa T}{e^2} = \frac{3}{2} \ln \frac{0.37\varkappa T}{e^2 N_0^{-1/3}} \tag{1.34}$$

(usually $\Lambda \sim 10$), m is the mass of the particle considered, m' is the mass of the primed particle, $\mathbf{w} = \mathbf{v} - \mathbf{v}'$.

The Landau collision integral (1.33) is significantly simpler than the Boltzmann collision integral (1.31), since the functions f and F in (1.33) only depend on the velocities $\mathbf{v}$ and $\mathbf{v}'$ of the particle after collision. This simplification is possible because, in the case of the Coulomb interaction, distant "collisions" play the major role. In such "collisions" the particles fly past each other at a comparatively large distance (of order D), and they interact weakly. A particle's velocity is changed only slightly in each such "collision." Thanks to this, $\mathbf{v}$ is close to $\mathbf{v}_1$, $\mathbf{v}'$ is close to $\mathbf{v}_1'$, and one can represent the integral (1.31) in integrodifferential from (1.32).

Also the following more general assertaion turns out to be valid (see, for example, [23]): if the distribution function only depends on those variables which change very little during each collision of the particle, the change of the distribution function caused by collisions is described by one equation, quite similar to the Fokker-Planck equation for the Brownian motion of a particle. In this case, therefore,

$$Y = -\operatorname{div} \boldsymbol{j},$$

where $\mathbf{j}$ is the current density of particles in phase space (in the space of velocities and coordinates) which is caused by the collisions. For example, if the distribution function f only depends on the energy of the particle $\mathcal{E}$, then

$$Y = -\operatorname{div} \mathcal{J}(\mathcal{E}) = -A(\mathcal{E}) \frac{\partial}{\partial \mathcal{E}} \mathcal{J}(\mathcal{E}), \tag{1.35}$$

where $A(\mathcal{E})$ is a normalization constant, and $\mathcal{J}(\mathcal{E})$ is the current of particles in energy space. In particular, in elastic collisions with neutral particles, electrons change their energies very little, because the electron's mass is much less than the molecular mass. This circumstance has been used and enabled one to obtain simple expressions of the form (1.35) for the integral associated with collisions of electrons with molecules (see [24, 25]).

CHAPTER II

DISTURBANCE OF THE CONCENTRATION OF PARTICLES, WITHOUT TAKING INTO ACCOUNT THE INFLUENCE OF ELECTRIC AND MAGNETIC FIELDS

§ 4. Qualitative Examination of the Disturbances Produced by a Rapidly Moving Body

Before going on to a rigorous calculation of the density of particles on the kinetic equation, let us consider purely qualitatively the nature of the disturbances to be expected in the vicinity of a body.

Let the body be moving with velocity $\mathbf{V}_0$. We shall carry out the analysis in a coordinate system fixed in the body. In this coordinate system, the body is at rest, and a beam of particles is incident on it with velocity $-\mathbf{V}_0$. Since, as already noted above, we have everywhere in mind the case in which the mean free path l of the particles is much larger than the linear dimensions of the body, it is not possible to take collisions between particles into account, in so far as the first approximation is concerned. Under these conditions, the disturbances of the particles' concentrations will have the following character:

a. Density of Neutral Particles. Neutral particles (molecules, atoms) interact only with the surface of the body itself. After collision with the surface, they are scattered in various directions. In this connection, an excess of particles is, of course, established in front of the body – a "region of compression"; conversely, a "region of rarefaction" appears behind the body (see Fig. 1).

The concentration in the "region of compression" has its maximum value near the surface of the body. Here, if the particles are specularly scattered upon collision with the surface, then the velocities of the particles incident on the body and of the scattered particles are identical; in this case the maximum value of the concentration of reflected particles, n_2, i.e., the excess concentration, is equal to the density of particles in the incoming beam, i.e., in the undisturbed gas. For diffuse scattering and especially in the presence of accommodation, the average velocity of the scattered particles in the direction normal to the surface of the body is decreased, and the excess concentration of particles increases.

At large distances from the body, the excess particles fly apart and their density must fall off in proportion to $1/r^2$, where r is the distance from the body. Indeed, the number of particle undergoing collisions with

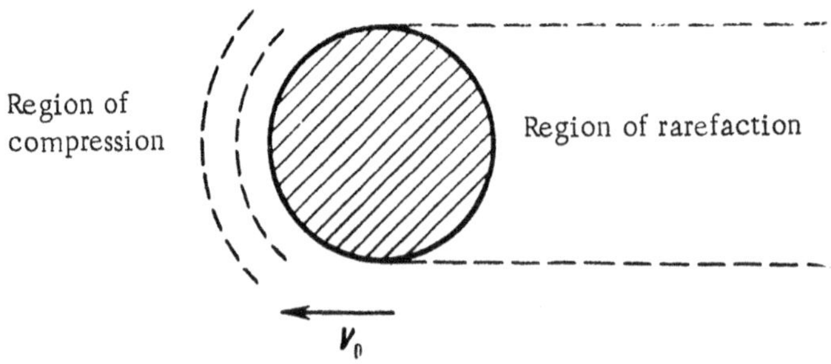

Fig. 1. Schematic representation of the disturbed regions in the vicinity of a body.

the body during a time interval Δt is equal to, in order of magnitude, $n_0R_0^2V_0\Delta t$. After collision, the particles are scattered in various directions with the same velocity V_0, and at a distance r from the body they occupy a volume of the order of magnitude of $r^2V_0\Delta t$. Therefore, the excess concentration is given by

$$n_2 \approx \frac{n_0R_0^2V_0\Delta t}{r^2V_0\Delta t} = n_0\left(\frac{R_0}{r}\right)^2. \tag{2.1}$$

We also note that the density of particles near the surface of the body may increase because of thermal evaporation from the surface and also due to its destruction (atomization) under the action of the incident particles.

A region of rarefaction appears behind the body. This region is occupied by particles of the incident beam because of the presence of thermal dispersion of the velocities. In fact, if all incident particles had the same velocity, $-\mathbf{V}_0$, then they in general would not fill the space behind the body. Filling of the rarefied region occurs because of the presence of particles whose velocities are arbitrary in direction. In this connection, it is clear that filling of this region basically occurs because of the motion of particles with thermal velocity v_n in a direction perpendicular to the direction of motion of the current, whose velocity is $-\mathbf{V}_0$. Therefore, the dimension of the strongly rarefied region is of order $R_0(V_0/v_n)$; $V_0 \gg v_n$ in the case of a rapidly moving body, and the rarefied region is much larger than the disturbed region in front of the body. At large distances from the body (for $r \gg R_0 V_0/v_n$), the density of molecules in the rarefied region is, of course, close to the density of the undisturbed gas, n_0. In this connection, the disturbed region spreads out so that in a direction ρ perpendicular to the direction of motion, its linear dimensions are of order $\Delta\rho \sim r(v_n/V_0) \gg R$. The disturbance of the density decreases in proportion to $1/r^2$ with increasing distance from the body, so that

$$|\delta n| = n_0 - n \approx \frac{R_0^2}{\Delta\varrho^2} \sim \left(\frac{V_0}{v_n}\right)^2\left(\frac{R_0}{r}\right)^2. \tag{2.2}$$

At small distances from the body's surface, where $r \ll R_0(V_0/v_n)$, the density of particles in the rarefied zone is very small. In fact, only a particle whose velocity $v \sim V_0(R_0/r)$ can travel to a point removed by a distance r from the maximum cross section of the body. In this connection, if $r \ll R_0(V_0/v_n)$, then $v \gg v_n$, i.e., the velocity of the particle must be much greater than the thermal velocity v_n. The number of such particles is, of course, small since it is proportional to $\exp[-Mv^2/2\varkappa T]$. Therefore, for $V_0 \gg v_n$, the density near the body's surface is also exponentially small:

$$n \sim n_0 \exp\left[-\frac{Mv^2}{2\varkappa T}\right] \sim n_0 \exp\left[-\left(\frac{V_0}{v_n}\right)^2\left(\frac{R_0}{r}\right)^2\right]. \tag{2.3}$$

Naturally, n cannot become significantly less than $n_0 \exp[-(V_0/v_n)^2]$ since the density of particles which are moving faster than the body is insignificantly disturbed by the body.

It is important to emphasize that the density of particles in the rarefied region is basically determined by the freely moving particles; it does not depend, therefore, on the characteristic properties of their interaction with the body's surface.

b. Density of Ions. In contrast to neutral particles, not only the interaction of charged particles with the body, but also the electric and magnetic fields existing in a plasma significantly influence the density of charged particles – ions and electrons. For ions it is important, however, that they, like molecules, are incident on the body with energies $M_iV_0^2/2$, i.e., in the case of a rapidly moving body, the ions are incident with energies many times exceeding the thermal energy $\varkappa T$. At the same time, the potential energy $e\varphi$ of an ion in an electric field, arising because of disturbances in the plasma, altogether usually is only on the order of the thermal energy. For this reason, the influence of an electric field on the motion of the ions must not be very strong. Therefore, if there is no magnetic field, then as a first approximation, one can assume that the ions move in the same way as neutral particles. In other words, the disturbance of the density of ions caused by their interaction with the body has the same character, in the absence of a magnetic field, as for the case of neutral particles.

On the other hand, a constant external magnetic field $\mathbf{H}_0$ has an important effect on the motion of ions, and consequently on their filling of the rarefied region behind the body. In contrast to neutral particles, ions in the presence of a magnetic field can move freely only in a direction parallel to $\mathbf{H}_0$. The ions revolve in a plane perpendicular to $\mathbf{H}_0$ with gyromagnetic frequency $\Omega = e\mathbf{H}_0/M_i c$. It is clear, however, that the magnetic field significantly changes the nature of an ion's motion only over a time interval $\Delta t \gtrsim 2\pi/\Omega$. During this time interval the flux of ions, incident on the body with velocity $\mathbf{V}_0$, moves over a distance $\Delta r \sim 2\pi V_0/\Omega$. Therefore, the influence of a magnetic field is practically absent in the zone nearest to the body, for distances $r \ll 2\pi V_0/\Omega$. In the ionosphere, this zone is of the order of ten meters or larger. Now for distances $r \gtrsim 2\pi V_0/\Omega$ the influence of a magnetic field is very essential. In fact, here the disturbance δN_i of the ions' density is usually not large. However δN_i now decreases with increasing distance r from the body not in proportion to $1/r^2$, but as $1/r$, as a consequence of the fact that the ions freely fill the region (disturbed by the body) only along one direction – the direction of the magnetic field. Indeed, during a time interval r/V_0, in the course of which the flux of ions incident on the body is displaced through a distance r, the disturbed (by the passage of the body) region spreads over a distance $R_0(v_i/V_0)\sin\alpha$, where α is the angle between the direction of the body's motion and the direction of the magnetic field, and v_i is the average thermal velocity of the ions. Therefore, in the far zone the disturbance in the ionic density is

$$|\delta N_i| = N_0 - N_i \sim N_0 \frac{R_0 V_0}{r v_i \sin\alpha} \sim \frac{1}{r \sin\alpha} \tag{2.4}$$

and decreases with increasing distance from the body in proportion to $1/r$, i.e., more slowly than the disturbance in the density of neutral particles.

Thus, the disturbed region – the "trail" of the body – increases as a result of the presence of a magnetic field. Together with this, the disturbance δN_i increases with decrease of the angle α. It goes without saying that estimate (2.4) is not applicable for very small values of α.

In the special case when $\alpha = 0$, i.e., when the body is moving exactly along the direction of the magnetic field, the ions in the far zone cannot move perpendicularly to the direction of the body's motion. In this case, without taking collisions into account, the trail must not be filled by particles. If, now, collisions are taken into account, then the trail is filled by particles, and the disturbance of the density decreases in proportion to $1/r$ with increasing distance from the body.

c. Density of Electrons and the Potential of the Electric Field. Under the conditions of interest to us, the velocity of the body, as already noted above, is always much smaller than the velocity of the electrons. In addition, a considerable negative potential is usually established on the surface of the body. Under these conditions, the body only very weakly disturbs the equilibrium distribution of electrons. Therefore, the density distribution of the electrons, as is evident from Eq. (1.29), is usually described by Boltzmann's formula

$$N(r) = N_0 \exp\left[\frac{e\varphi}{\varkappa T}\right],$$

where φ is the potential of the electric field. The value of the density of electrons in the disturbed region is completely determined by the potential of the electric field. Thus,

$$\varphi = -\frac{\varkappa T}{e} \ln \frac{N_0}{N(r)}, \tag{2.5}$$

i.e., the potential logarithmically depends on the ratio $N_0/N(\mathbf{r})$.

It is evident from this formula that, if the disturbance of the density of ions and electrons is of the same order as their unperturbed density, then the magnitude of the potential field, generally speaking, is of order $-\varkappa T/e$.

On the other hand, the potential is determined by the charge density in the medium, i.e., by the difference between the concentrations N of electrons and N_i of ions. It is easy to estimate this difference. In fact, the potential φ, arising because of the difference in concentrations, $N_i - N$, in a region of dimension r is approximately given by

$$\varphi \sim \frac{e(N_i - N)r^3}{r} \sim e(N_i - N)r^2.$$

If we now equate it (the above expression for φ) to the thermal potential $-\varkappa T/e$, then we find that

$$\frac{|N_i - N|}{N_0} \sim \frac{\varkappa T}{e^2 r^2 N_0} \sim \left(\frac{D}{r}\right)^2, \tag{2.6}$$

i.e., the relative difference in the electrons' and ions' densities is exceedingly small for $r \gg D$.

Thus, the density of electrons in the disturbed region is practically equal to the density of ions. The potential then of the electric field is of order $-\varkappa T/e$. However, in the most strongly disturbed region, i.e., in the region of maximum rarefaction near the body's surface, φ may attain a significant magnitude. Here the density of ions is negligibly small, so that $|N_i - N| \approx N$, and we obtain, using relations (2.5) and (2.6) presented above,

$$\varphi \sim -\frac{\varkappa T}{e}\ln\frac{N_0}{N} = -\frac{\varkappa T}{e}\ln\frac{N_0}{(N - N_i)} \approx -\frac{\varkappa T}{e}\ln\left(\frac{R_0}{D}\right)^2. \tag{2.7}$$

For a body with dimensions of the order of meters, in the ionosphere, the maximum value of φ may be an order of magnitude larger than the thermal potential.

Far from the body, where the disturbances in the density of charged particles are not large, $N - N_0 \ll N_0$, the potential φ is proportional to these disturbances

$$N = N_0 \exp\left[\frac{e\varphi}{\varkappa T}\right] \approx N_0\left(1 + \frac{e\varphi}{\varkappa T}\right), \quad \varphi = \frac{\varkappa T}{e}(N - N_0). \tag{2.8}$$

Therefore, at large distances from the body, the potential of the electric field falls off in the same manner as the disturbances in the concentration, i.e., proportional to $1/r^2$ in the absence of a magnetic field, or proportional to $1/r$ in the presence of a constant magnetic field.

Summing up the results of the analysis carried out above, one can conclude that the disturbances of the ions' density basically originate because of "shading" of the incident beam of ions by the body. Now electrons, because of their large thermal velocity, are very slightly "shaded," and the density of electrons is principally disturbed by the influence of the electric field. The latter (the electric field) is established such that the density of electrons is close to the density of ions. An electric field has very little effect on the motion of ions, so that in the absence of a constant magnetic field the disturbances in the density of ions, and therefore also in the density of electrons, do not differ appreciably from the disturbances in the density of neutral particles.*

By virtue of this, the simplest case is considered in this chapter, i.e., the case when the disturbances in the plasma are caused only by collisions of the particles with the surface of a moving body. Although such a notion is rigorously valid only for neutral particles, the results obtained in the absence of a magnetic field are also true, to the first approximation, for ions, and consequently also true for electrons. The influence of electric and magnetic fields on the motion of charged particles will be investigated in subsequent chapters.

*Of course, here it is necessary to consider the properties of the interaction of the particles with the body's surface; these, however, significantly manifest themselves only in the region of "compression."

a. The Distribution Function. The distribution function for neutral particles, in a coordinate system fixed in the moving body, is determined by Eqs. (1.5) and (1.11)

$$v\frac{\partial f_n}{\partial r} = A_n \delta(F), \tag{2.9}$$

where $F(\mathbf{r}) = 0$ is the equation of the body's surface, and $A_n(\mathbf{r}, \mathbf{v})$ is the function characterizing the interaction of neutral particles with the body.

A beam of particles with velocity $-\mathbf{V}_0$ is incident on the body; a rarefied region is formed behind the body. We shall assume that the body is moving rapidly, so that its velocity is much larger than the thermal velocity, v_n, of the molecules. In this case, collisions of molecules with the body's surface, affiliated with filling of the rarefied region, have a small probability. Therefore, these collisions are of little significance far from the surface. This means that the specific shape of the body has little influence on the properties of the rarefied zone for distances $z \gg R_0$; only the maximum cross section of the body in a plane perpendicular to the incident beam is important. Therefore, for an approximate calculation, one can replace the body itself by its transverse cross section. For example, one can replace a spherical surface by a circular disk of radius R_0, situated at the point $z = 0$. Then the right side of Eq. (2.9) differs from zero only for $z = 0$. The equilibrium distribution function (1.6) is altered at the point $z = 0$ because of the interaction with the body, namely

$$f(x_0, y_0, 0; v_x, v_y, v_z) = \begin{cases} n_0\left(\dfrac{M}{2\pi\varkappa T}\right)^{3/2} \exp\left\{-\dfrac{M\,[v_x^2+v_y^2+(v_z+V_0)^2]}{2\varkappa T}\right\} \\ \quad \text{for arbitrary } v_z, \text{ if the point } (x_0, y_0) \text{ lies outside} \\ \quad \text{the region } S_0, \text{ and for } v_z > 0, \text{ if the point } (x_0, y_0) \\ \quad \text{lies inside the region } S_0; \\ f_2(x_0, y_0, 0; v_x, v_y, v_z) \\ \quad \text{for } v_z < 0, \text{ if the point } (x_0, y_0) \text{ lies inside the} \\ \quad \text{region } S_0. \end{cases}$$

Here S_0 is the maximum cross section of the body. The z axis is directed along the direction of the body's motion; the incident beam of particles is moving in the $-z$ direction. Then f_2 is the distribution function for the particles reflected from the back side of the cross section S_0 and moving in the $-z$ direction. In the case of a rapidly moving body, when $V_0 \gg (2\varkappa T/M)^{1/2}$, the number n_2 of such particles is exponentially small, since before reflection they must be moving laterally (to the side), opposite to the flux:

$$n_2 \sim n_0 \exp\left\{-\frac{MV_0^2}{2\varkappa T}\right\}.$$

Since the specific shape of the body's surface is not considered here, one can neglect the reflected particles. Also the number of particles having velocity $v_z > 0$ is exponentially small. Neglecting them, we may write the distribution function at $z = 0$ in the following simple form:

$$f(x_0, y_0, 0; v_x, v_y, v_z) = \begin{cases} n_0\left(\dfrac{M}{2\pi\varkappa T}\right)^{3/2} \exp\left\{-\dfrac{M\,[v_x^2+v_y^2+(v_z+V_0)^2]}{2\varkappa T}\right\}, & \text{if } v_z < 0 \text{ and the point } (x_0, y_0) \text{ lies outside the region } S_0; \\ 0, & \text{if } v_z > 0 \text{ or the point } (x_0, y_0) \text{ lies inside the region } S_0. \end{cases} \tag{2.10}$$

The region of rarefaction corresponds to values of $z < 0$. For $z < 0$ the distribution function is determined by Eq. (1.5):

$$v_x \frac{\partial f_n}{\partial x} + v_y \frac{\partial f_n}{\partial y} + v_z \frac{\partial f_n}{\partial z} = 0. \tag{2.11}$$

It is easy to find the solution of this equation by the method of characteristics. The equations of the characteristics,

$$\frac{dx}{v_x} = \frac{dy}{v_y} = \frac{dz}{v_z},$$

obviously have the solution

$$\left.\begin{aligned} x - \frac{v_x}{v_z} z &= x_0, \\ y - \frac{v_y}{v_z} z &= y_0. \end{aligned}\right\} \tag{2.12}$$

Therefore, the general solution of Eq. (2.11) has the form

$$f_n(x, y, z; v_x, v_y, v_z) = F\left(x - \frac{v_x}{v_z} z,\ y - \frac{v_y}{v_z} z;\ v_x, v_y, v_z\right), \tag{2.13}$$

where F is an arbitrary function. The form of the function F is determined from the condition that at $z = 0$ it must coincide with the distribution function (2.10). Taking this into account, we find that the distribution function for neutral particles in the region of rarefaction has the form

$$f_n = f\left(x - \frac{v_x}{v_z} z,\ y - \frac{v_y}{v_z} z,\ 0;\ v_x, v_y, v_z\right) = \begin{cases} n_0 \left(\dfrac{M}{2\pi \varkappa T}\right)^{3/2} \exp\left\{-\dfrac{M\left[v_x^2 + v_y^2 + (v_z + V_0)^2\right]}{2\varkappa T}\right\}, \\ \text{if } v_z < 0 \text{ and the point } \left(x - \frac{v_x}{v_z} z;\ y - \frac{v_y}{v_z} z\right) \\ \text{lies outside the region } S_0; \\ 0, \\ \text{if } v_z > 0 \text{ or the point } \left(x - \frac{v_x}{v_z} z;\ y - \frac{v_y}{v_z} z\right) \\ \text{lies inside the region } S_0. \end{cases} \tag{2.14}$$

b. Density of Particles. In order to find the density of particles in the rarefied region, it is only necessary to integrate the distribution function (2.14) with respect to velocities, namely

$$n(x, y, z) = \int f\left(x - \frac{v_x}{v_z} z,\ y - \frac{v_y}{v_z} z,\ 0;\ v_x, v_y, v_z\right) dv_x\, dv_y\, dv_z. \tag{2.15}$$

In this expression, for the integration with respect to dv_x and dv_y it is convenient to change over to new variables

$$\left.\begin{aligned} x_0 &= x - \frac{v_x}{v_z} z, \\ y_0 &= y - \frac{v_y}{v_z} z, \end{aligned}\right\} \tag{2.16}$$

i.e.,

$$v_x = \frac{x - x_0}{z} v_z, \quad v_y = \frac{y - y_0}{z} v_z.$$

Here the Jacobian of the transformation equals

$$\begin{vmatrix} \frac{\partial v_x}{\partial x_0}, & \frac{\partial v_y}{\partial x_0}, & \frac{\partial v_z}{\partial x_0} \\ \frac{\partial v_x}{\partial y_0}, & \frac{\partial v_y}{\partial y_0}, & \frac{\partial v_z}{\partial y_0} \\ \frac{\partial v_x}{\partial v_z}, & \frac{\partial v_y}{\partial v_z}, & \frac{\partial v_z}{\partial v_z} \end{vmatrix} = \begin{vmatrix} -\frac{v_z}{z}, & 0, & \frac{x-x_0}{z} \\ 0, & -\frac{v_z}{z}, & \frac{y-y_0}{z} \\ 0, & 0, & 1 \end{vmatrix} = \frac{v_z^2}{z^2}. \tag{2.17}$$

Therefore

$$n(x, y, z) = \frac{1}{z^2} \int v_z^2 f\left(x_0, y_0, 0; \frac{x-x_0}{z} v_z, \frac{y-y_0}{z} v_z, v_z\right) dx_0\, dy_0\, dv_z. \tag{2.18}$$

By virtue of the properties of the function f for $z=0$ (see Eq. (2.10)), the integration with respect to dx_0dy_0 in (2.18) only goes over the region outside the cross section S_0 of the body. The analogous integral, taken over the cross section S_0, is evidently equal to the disturbance of the particle density

$$-\delta n(x, y, z) = n_0 - n(x, y, z),$$

since the integral over the entire plane is equal to n_0. Thus,

$$-\delta n(x, y, z) = \frac{n_0}{z^2}\left(\frac{M}{2\pi\varkappa T}\right)^{3/2} \int_{S_0} dx_0\, dy_0 \int_{-\infty}^{0} v_z^2 \exp\left\{-\frac{M}{2\varkappa T}\left[\frac{(x-x_0)^2+(y-y_0)^2}{z^2} v_z^2 + (v_z+V_0)^2\right]\right\} dv_z. \tag{2.19}$$

Integrating with respect to dv_z in Eq. (2.19), we obtain

$$\int_{-\infty}^{0} v_z^2 \exp\left\{-\frac{M}{2\varkappa T}\left[\frac{(x-x_0)^2+(y-y_0)^2}{z^2} v_z^2 + (v_z+V_0)^2\right]\right\} dv_z$$

$$= \left(\frac{2\varkappa T}{M}\right)^{3/2} \exp\left(-\frac{MV_0^2}{2\varkappa T}\right) \int_0^{\infty} y^2 \exp(-d_0 y^2 - b_0 y)\, dy$$

$$= \left(\frac{2\varkappa T}{M}\right)^{3/2} \exp\left(-\frac{MV_0^2}{2\varkappa T}\right) \left\{\frac{\sqrt{\pi}\, b_0^2}{8 d_0^{5/2}}\left(1+\frac{2d_0}{b_0^2}\right)\left[1+\Phi\left(\frac{b_0}{2\sqrt{d_0}}\right)\right] \exp\left(\frac{b_0^2}{4d_0}\right) + \frac{b_0}{4d_0^2}\right\}. \tag{2.20}$$

Here $\Phi(x) = \frac{2}{\sqrt{\pi}} \int_0^x e^{-t^2} dt$ is the probability integral,

$$d_0 = \frac{z^2+(x-x_0)^2+(y-y_0)^2}{z^2}, \quad b_0 = 2\left(\frac{MV_0^2}{2\varkappa T}\right)^{1/2}. \tag{2.21}$$

Further, one can consider that in the region of rarefaction, not too close to the cross section of the body, the ratio $b_0^2/2d_0$ has a large magnitude for $(MV_0^2/\varkappa T)z^2 \gg (x-x_0)^2+(y-y_0)^2$. Then, substituting (2.20) and (2.21) into (2.19) and neglecting exponentially small terms, we finally obtain

$$-\delta n(x, y, z) = \frac{n_0 MV_0^2}{2\pi\varkappa T} z^3 \int_{S_0} [z^2+(x-x_0)^2+(y-y_0)^2]^{-5/2} \exp\left[-\frac{MV_0^2}{2\varkappa T}\, \frac{(x-x_0)^2+(y-y_0)^2}{z^2+(x-x_0)^2+(y-y_0)^2}\right] dx_0\, dy_0. \tag{2.22}$$

This formula determines the density of particles in the rarefied region behind a body of arbitrary shape, whose maximum cross section is S_0. For its derivation, the body itself was replaced by the disk S_0, and the specific shape of the body's surface was not taken into consideration. Therefore, this formula may be valid only at considerable distances from the surface of a rapidly moving body. Moreover, far from the body for large values of z, if the condition

$$z^2 \gg (x - x_0)^2 + (y - y_0)^2$$

is satisfied (more precisely, if $z^2 > [(x - x_0)^2 + (y - y_0)^2](MV_0^2/2\varkappa T)^{1/2}$, where (x_0, y_0) is an arbitrary point of the cross section), expression (2.22) for δn simplifies considerably:

$$-\delta n(x, y, z) = \frac{n_0 M V_0^2}{2\pi\varkappa T z^2} \exp\left(-\frac{MV_0^2}{2\varkappa T}\frac{x^2+y^2}{z^2}\right) \int\limits_{S_0} \exp\left[-\frac{MV_0^2}{2\varkappa T}\frac{x_0^2+y_0^2-2xx_0-2yy_0}{z^2}\right] dx_0\, dy_0. \quad (2.23)$$

It is possible to obtain formula (2.23) directly from (2.19), if we consider that for a rapidly moving body $(v_z + V_0) \ll V_0$ almost always, and therefore one can replace v_z by $-V_0$ in Eq. (2.19), except inside the brackets $(v_z + V_0)$. In this connection, the thermal motion of the particles in the direction of the body's motion is neglected in comparison with the velocity V_0 of the body, since only the thermal motion of the particles in a plane perpendicular to the direction of the body's motion is effective for filling the rarefied zone. Therefore, the problem of an approximate determination of the concentration of particles in the rarefied zone behind a rapidly moving body is in fact a two-dimensional problem: it is only necessary to determine how the particles fill up in the course of time the empty region, equal to the maximum cross section of the body in a plane perpendicular to the direction of motion. Then, considering that during this same time interval t, all the particles are displaced as a whole along the z direction through a distance $-V_0 t$, with the aid of the simple substitution $t = -z/V_0$ one can go over to a system of coordinates moving together with the body. Such a simplified method of calculation leads to formula (2.23).

Expression (2.23) for the disturbances in the density of neutral particles has an extremely simple and convenient (for integration) form. Such integration is carried out below for the case when the body's cross section in a plane perpendicular to the direction of motion is circular or rectangular; in addition, a simple expression is obtained for the disturbance δn for a body of arbitrary shape, valid at sufficiently great distances from the body.

TABLE V. Ratio of the Concentration of Particles $n(\rho, z)$ Behind a Rapidly Moving Body to the Undisturbed Concentration of Particles, n_0

	$\frac{z}{R_0}\sqrt{\frac{2\varkappa T}{MV_0^2}}$									
ρ/R_0	0	−0.2	−0.4	−0.6	−0.8	−1.0	−1.2	−1.5	−2.0	−4.0	−8.0
0.0	0	0	0	0.06	0.21	0.37	0.50	0.64	0.78	0.94	0.98
0.2	0	0	0.01	0.09	0.24	0.38	0.51	0.65	0.78	0.94	0.98
0.4	0	0	0.06	0.14	0.28	0.42	0.53	0.67	0.78	0.94	0.98
0.6	0	0.04	0.11	0.23	0.36	0.48	0.57	0.69	0.79	0.94	0.98
0.8	0	0.09	0.27	0.39	0.48	0.56	0.63	0.71	0.80	0.94	0.98
1.0	0.50	0.51	0.54	0.58	0.62	0.65	0.68	0.75	0.82	0.94	0.98
1.2	1.00	0.92	0.78	0.76	0.75	0.74	0.75	0.78	0.84	0.94	0.98
1.5	1.00	1.00	0.96	0.88	0.87	0.86	0.84	0.84	0.86	0.94	0.98
2.0	1.00	1.00	1.00	0.97	0.95	0.95	0.93	0.91	0.90	0.94	0.98
4.0	1.00	1.00	1.00	1.00	1.00	1.00	0.99	0.99	0.99	0.97	0.98
8.0	1.00	1.00	1.00	1.00	1.00	1.00	1.00	1.00	1.00	0.99	0.99

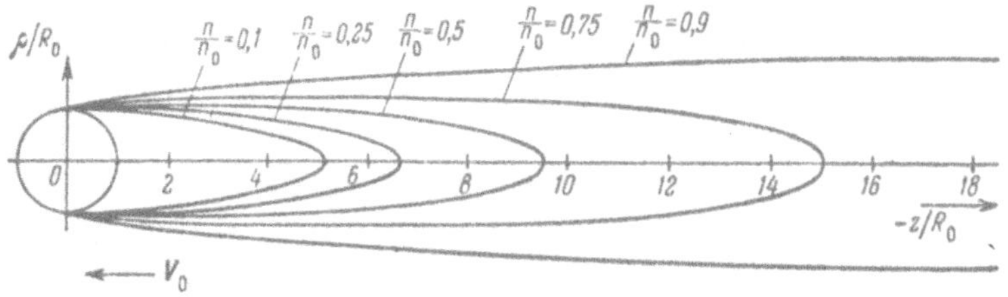

Fig. 2. Curves of constant particle density in the rarefaction region behind a body of circular cross section $[V_0/(2\varkappa T/M)^{1/2}] = 8$.

c. Body of Circular Cross Section. Let the body's cross section be a circle of radius R_0. Of course, the spherical body considered above also has, in particular, such a cross section.

For the calculation of δn, in the integral (2.23) it is natural to go over to polar coordinates φ, $\rho = (x^2 + y^2)^{1/2}$ and to consider that $z < 0$. If, in addition, the angle φ is measured from the direction (x, y) so that $x_0 x + y_0 y = \rho_0 \rho \cos\varphi$, then in place of Eq. (2.23) we obtain

$$-\delta n(x, y, z) = \frac{n_0 M V_0^2}{2\pi\varkappa T z^2} \exp\left[-\frac{MV_0^2}{2\varkappa T}\left(\frac{\varrho}{z}\right)^2\right] \int_0^{R_0}\int_0^{2\pi} \varrho_0\, d\varrho_0\, d\varphi \exp\left\{-\frac{MV_0^2}{2\varkappa T}\left[\left(\frac{\varrho_0}{z}\right)^2 - \frac{2\varrho\varrho_0\cos\varphi}{z^2}\right]\right\} =$$

$$= 2n_0 \exp\left[-\frac{MV_0^2}{2\varkappa T}\left(\frac{\varrho}{z}\right)^2\right] \int_0^{-\frac{R_0}{z}\sqrt{\frac{MV_0^2}{2\varkappa T}}} \tau e^{-\tau^2}\, I_0\left(-\frac{\varrho}{z}\sqrt{\frac{2MV_0^2}{\varkappa T}}\,\tau\right) d\tau, \qquad (2.24)$$

where $I_0(x) = \frac{1}{\pi}\int_0^{\pi} e^{-x\cos\theta}\, d\theta$ is the Bessel function of zero order having an imaginary argument.

This expression simplifies considerably on the z axis: in this case $\rho = 0$, $I_0 = 1$, so that

$$-\delta n(0, z) = n_0 - n(0, z),$$

where

$$n(0, z) = n_0 \exp\left(-\frac{MV_0^2 R_0^2}{2\varkappa T z^2}\right). \qquad (2.25)$$

For large values of z

$$-\delta n(0, z) = n_0 \frac{MV_0^2 R_0^2}{2\varkappa T z^2}. \qquad (2.25a)$$

Values of the ratio $n(\rho, z)/n_0$ for different values of ρ and z are presented in Table V, and curves indicating the density variation in the rarefied region are shown in Fig. 2. These curves, as well as the series of curves in the following figures were constructed under the assumption that the ratio $V_0/v_n = V_0/(2\varkappa T/M)^{1/2} = 8$. It is clear from Fig. 2 that the region where the disturbances are large is extremely elongated: $-\delta n \simeq n_0/2$ for $z \simeq 1.2R_0(MV_0^2/2\varkappa T)^{1/2} \simeq 10\,R_0$; $-\delta n \simeq 0.1n_0$ for $z \simeq 25R_0$. Thus, an extended, disturbed region is formed behind the body; with increasing z the disturbance of the particle density decreases like $1/z^2$.

d. Body of Rectangular Cross Section. Now let us consider a body of rectangular cross section with dimensions $2R_x$ and $2R_y$.

In this case, it is easy to carry out the integration of expression (2.23) at once, and we obtain

$$-\delta n(x, y, z)=\frac{n_0MV_0^2}{2\pi\varkappa Tz^2}\int\limits_{-R_x}^{R_x}\int\limits_{-R_y}^{R_y} dx_0\,dy_0 \exp\left[-\frac{MV_0^2}{2\varkappa T}\frac{(x-x_0)^2+(y-y_0)^2}{z^2}\right]=$$

$$=\frac{n_0}{\pi}\left|\int\limits_{u_1}^{u_2} e^{-u^2}du \,|\times|\int\limits_{u_3}^{u_4} e^{-u^2}du\right|=\frac{n_0}{4}\left|\Phi\left(\frac{x-R_x}{z}\sqrt{\frac{MV_0^2}{2\varkappa T}}\right)-\right.$$

$$\left.-\Phi\left(\frac{x+R_x}{z}\sqrt{\frac{MV_0^2}{2\varkappa T}}\right)\right|\left|\Phi\left(\frac{y-R_y}{z}\sqrt{\frac{MV_0^2}{2\varkappa T}}\right)-\Phi\left(\frac{y+R_y}{z}\sqrt{\frac{MV_0^2}{2\varkappa T}}\right)\right|, \tag{2.26}$$

where

$$u_1=\frac{x+R_x}{z}\sqrt{\frac{MV_0^2}{2\varkappa T}},$$

$$u_2=\frac{x-R_x}{z}\sqrt{\frac{MV_0^2}{2\varkappa T}},$$

$$u_3=\frac{y+R_y}{z}\sqrt{\frac{MV_0^2}{2\varkappa T}},$$

$$u_4=\frac{y-R_y}{z}\sqrt{\frac{MV_0^2}{2\varkappa T}}.$$

Here, as before, $z < 0$, and $\Phi(x)=\frac{2}{\sqrt{\pi}}\int\limits_0^x e^{-t^2}dt$ is the probability integral.

On the z axis, i.e., for $x=0$, $y=0$, expression (2.26) simplifies somewhat:

$$-\delta n(0, 0, z)=n_0\Phi\left(\frac{R_x}{z}\sqrt{\frac{MV_0^2}{2\varkappa T}}\right)\Phi\left(\frac{R_y}{z}\sqrt{\frac{MV_0^2}{2\varkappa T}}\right). \tag{2.27}$$

At large distances from the body

$$-\delta n(0, 0, z)=n_0\frac{4}{\pi}\frac{MV_0^2}{2\varkappa T}\frac{R_xR_y}{z^2}. \tag{2.28}$$

Thus, with increase of z, the disturbance $|\delta n|$ decreases like $1/z^2$. The variation of n/n_0 behind a body of rectangular cross section is shown in Fig. 3.

e. Disturbance of the Concentration of Particles at Large Distances from a Body of Arbitrary Cross Section. Formula (2.23) enables one to obtain a simple expression for the disturbance δn at large distances behind a body, for arbitrary cross section S_0. For large values of z, the exponent in the exponential term in integral (2.23) is always small, and this term is consequently close to unity. Therefore, in this case

$$-\delta n(x, y, z)=\frac{n_0S_0MV_0^2}{2\pi\varkappa Tz^2}\exp\left(-\frac{MV_0^2}{2\varkappa T}\frac{x^2+y^2}{z^2}\right), \tag{2.29}$$

provided the following conditions is satisfied:

$$z^2 \gg |\overline{x_0^2}+\overline{y_0^2}-2x\overline{x}_0-2y\overline{y}_0|\frac{MV_0^2}{2\varkappa T}, \tag{2.30}$$

where

$$\overline{x}_0=\frac{1}{S_0}\int\limits_{S_0} x_0\,dx_0\,dy_0,\quad \overline{y}_0=\frac{1}{S_0}\int\limits_{S_0} y_0\,dx_0\,dy_0$$

and so forth.

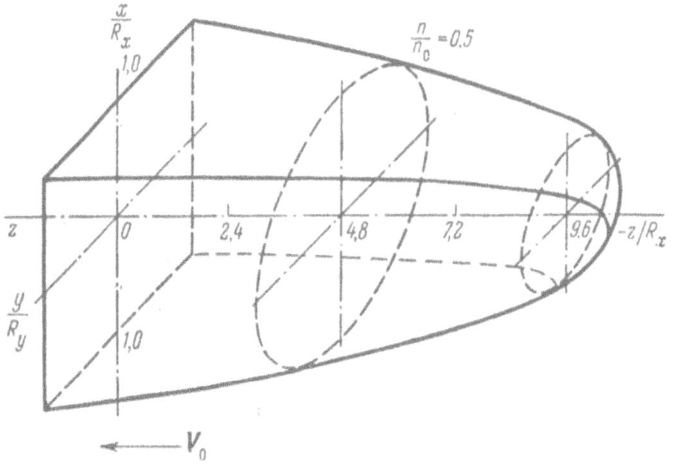

Fig. 3. Surface of constant density, $n/n_0 = 0.5$, in the region of rarefaction for a body of square cross section.

On the z axis, i.e., for $x = y = 0$, at large distances from the body when $z^2 \gg (MV_0^2/2\varkappa T)(\overline{x_0^2} + \overline{y_0^2})$, the disturbance of the particle density has the form:

$$-\delta n = n_0 \frac{S_0 MV_0^2}{2\pi \varkappa T z^2}. \tag{2.31}$$

It is easy to understand the physical meaning of this expression, using the argument presented on pages 22 and 23. The region disturbed by (the passage of) the body spreads out uniformly in a plane perpendicular to the direction of motion, with a velocity $v_n = (2\varkappa T/M)^{1/2}$, i.e., during a time interval $t = -z/V_0$ it spreads out over a circle of radius $(2\varkappa T/M)^{1/2} z/V_0$. Formula (2.31) coincides, of course, to the corresponding expressions (2.25a), (2.28), obtained for cross sections of particular shape. In the general case considered, the disturbance of the density also falls off like $1/z^2$ with increasing z.

It is clear from Eq. (2.29) that a weakly disturbed region behind the body may extend over extremely large distances. For example, $-\delta n = 0.1 n_0$ at a distance $|z| \approx 3.16 (MV_0^2/2\varkappa T)^{1/2} R_0$, where $R_0 = (S_0/\pi)^{1/2}$ is the effective radius of the body. In the ionosphere, $(MV_0^2/2\varkappa T)^{1/2} \sim 8$ to 10 and the corresponding distance $|z| \sim 25$ to 30 R_0. At a distance $|z| \approx 10 (MV_0^2/2\varkappa T)^{1/2} R_0$ (i.e., of the order of 100 R_0), $-\delta n = 0.01 n_0$, and for $|z| \approx 31.6 (MV_0^2/2\varkappa T)^{1/2} R_0$ (i.e., at a distance of the order of 300 R_0), $-\delta n = 10^{-3} n_0$.

§ 6. Region of Compression Ahead of a Rapidly Moving Sphere

A region of compression is formed in front of the body: here the concentration increases due to particles reflected from the body's surface. It is clear that thermal motion in the incoming beam has little influence on the concentration of excess particles in the case of a rapidly moving body, and (to the first approximation) one can disregard it, i.e., assume that all the particles incident on the body have the same velocity, $-\mathbf{V}_0$. On the other hand, the specific nature of the interaction of particles with the body's surface plays an important role, which has already been noted earlier. Expressions will be obtained below for the excess concentration of of particles for different properties of their interaction with the body's surface. For simplicity, the surface of the body is assumed to be a sphere of radius R_0.

a. Specular Reflection. Let us assume that upon collision with the body's surface, a particle is specularly scattered, i.e., the particle's angle of incidence on the surface (the angle between the direction of the particle's velocity and the normal to the surface) is equal to its angle of reflection, and the magnitude of the particle's velocity does not change. Such scattering is shown in Fig. 4a. In this case, the number of particles in the small (infinitesimal) volume $dV_{\rho z}$ near an arbitrary point $A(\rho, z)$ is equal to the number of

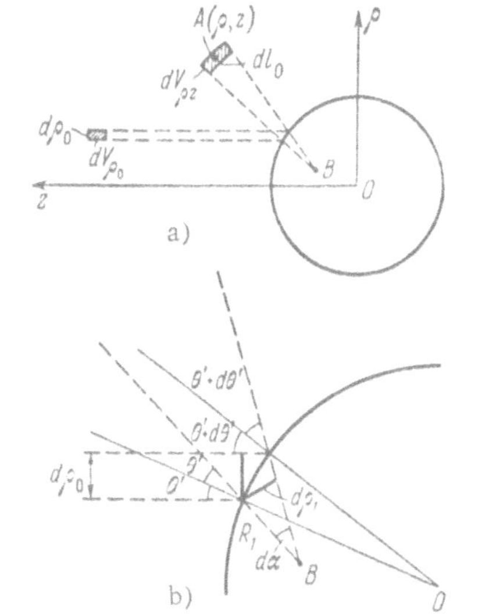

Fig. 4. Illustrating the derivation of the concentration of relfected particles.

particles in the corresponding small (infinitesimal) volume dV_{ρ_0} in the incoming beam (see Fig. 4a). Therefore, the excess concentration n_2 is

$$n_2(\varrho, z) = n_0 \frac{dV_{\rho_0}}{dV_{\rho z}} . \tag{2.32}$$

Considering, then, that the magnitude of the velocity does not change here, we find that the ratio of the volumes is equal to the ratio of their transverse cross sections,

$$\frac{dV_{\rho_0}}{dV_{\rho z}} = \frac{\varrho_0 d\varrho_0}{\varrho\, dl} , \tag{2.33}$$

where $d\rho_0$ and dl are shown in Fig. 4a. All of these quantities can be expressed in terms of ρ, z, R_0, and the angle θ' – the angle between the normal to the sphere at the point of impact and the z axis, which, as usual, is parallel to the velocity $\mathbf{V}_0$ (see Fig. 4b). Indeed,

$$\left.\begin{aligned} \varrho_0 &= R_0 \sin\theta', \\ d\varrho_0 &= R_0 \cos\theta'\, d\theta'. \end{aligned}\right\} \tag{2.34}$$

The angle θ' is obviously related to the coordinates of the point $A(\rho, z)$ by the equation

$$\frac{\varrho - R_0 \sin\theta'}{z - R_0 \cos\theta'} = \tan 2\theta'$$

or

$$R_0 = 2z\cos\theta' + 2\varrho\sin\theta' - \frac{\varrho}{\sin\theta'} . \tag{2.35}$$

Further, let the reflected rays for the angles of reflection θ' and $\theta' + d\theta'$ converge to the point B (see Fig. 4b). The angle $d\alpha$ formed by these rays is obviously equal to

$$d\alpha = 2(\theta' + d\theta') - 2\theta' = 2d\theta',$$

and therefore

$$dl = 2R_{AB}\, d\theta', \tag{2.36}$$

where R_{AB} is the distance between the points A and B. This distance is made up of the segment R_1 from the point B to the body's surface and the segment from the body's surface to the point $A(\rho, z)$. Thus,

$$R_{AB} = R_1 + \sqrt{(\varrho - R_0 \sin\theta')^2 + (z - R_0\cos\theta')^2} = R_1 + (\varrho - R_0 \sin\theta')\sqrt{1 + \mathrm{ctg}^2 2\theta'} = R_1 + \frac{\varrho - R_0 \sin\theta'}{\sin 2\theta'} . \tag{2.37}$$

Finally, using equality of the angles upon reflection, we find that

$$d\varrho_0 = d\varrho_1 = R_1 \cdot 2d\theta'.$$

Hence

$$R_1 = \frac{1}{2}\frac{d\varrho_0}{d\theta'} = \frac{R_0 \cos\theta'}{2} . \tag{2.38}$$

TABLE VI. Values of the Ratio $n(\rho, z)/n_0$ ahead of a Rapidly Moving Body, for Specular Reflection of the Particles

	$\frac{z}{R_0}$																
ρ/R_0	−1,0	−0,4	−0,2	0	0,2	0,4	0,6	0,8	1,0	1,2	1,4	1,6	1,8	2,0	2,5	3,0	4,0
0	—	—	—	—	—	—	—	—	2.00	1.51	1.31	1.21	1.15	1.11	1.06	1.04	1.02
0.2	—	—	—	—	—	—	—	—	1.92	1.49	1.30	1.20	1.14	1.11	1.06	1.04	1.02
0.4	—	—	—	—	—	—	—	—	1.72	1.41	1.26	1.19	1.13	1.10	1.06	1.04	1.02
0.6	—	—	—	—	—	—	—	2.00	1.53	1.33	1.22	1.16	1.12	1.10	1.06	1.04	1.02
0.8	—	—	—	—	—	—	2.00	1.56	1.38	1.25	1.19	1.14	1.11	1.09	1.05	1.04	1.02
1.0	1.0	1.0	1.0	1.33	1.67	1.55	1.45	1.35	1.26	1.20	1.15	1.12	1.10	1.08	1.05	1.04	1.02
1.2	1.02	1.11	1.16	1.21	1.31	1.32	1.28	1.22	1.18	1.15	1.13	1.11	1.09	1.07	1.04	1.03	1.02
1.4	1.03	1.10	1.13	1.16	1.19	1.20	1.19	1.17	1.15	1.13	1.11	1.09	1.08	1.06	1.04	1.03	1.02
1.6	1.04	1.08	1.10	1.12	1.14	1.15	1.14	1.13	1.12	1.10	1.09	1.08	1.07	1.06	1.04	1.03	1.02
1.8	1.05	1.07	1.08	1.09	1.10	1.11	1.11	1.10	1.09	1.08	1.07	1.06	1.06	1.05	1.03	1.03	1.02
2.0	1.05	1.06	1.07	1.07	1.08	1.08	1.08	1.08	1.07	1.07	1.06	1.06	1.05	1.04	1.03	1.03	1.02
2.5	1.04	1.05	1.05	1.05	1.05	1.05	1.05	1.05	1.05	1.04	1.04	1.04	1.04	1.03	1.03	1.02	1.01

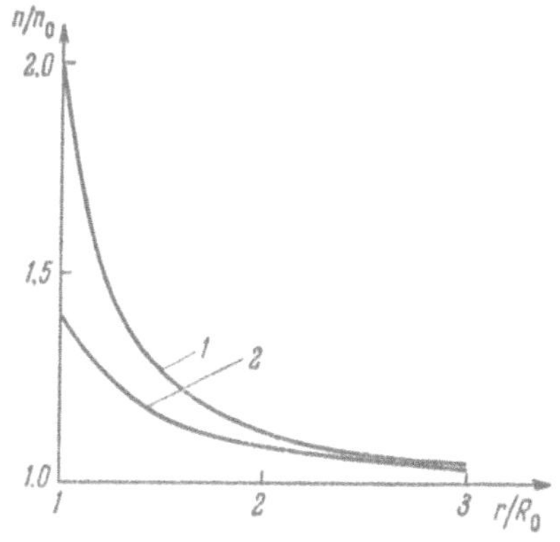

Fig. 5. Dependence of the concentration in the region of compression on the distance from the center of the body, in the case of specular reflection. Curve 1 refers to the distance along the z axis; curve 2 refers to distance measured from the center of the body in the plane z = 0.

Thus,

$$dl = 2\left(R_1 + \frac{\varrho - R_0 \sin\theta'}{\sin 2\theta'}\right) d\theta' = \frac{\varrho - R_0 \sin^3\theta'}{\sin\theta' \cos\theta'} d\theta'. \quad (2.39)$$

Now substituting (2.39) and (2.34) into (2.33) and (2.32), finally we have

$$n_2(\varrho, z) = n_0 \frac{R_0^2}{\varrho^2} \frac{\sin^2\theta' \cos^2\theta'}{1 - \frac{R_0}{\varrho}\sin^3\theta'}, \quad (2.40)$$

where the angle θ' can be expressed in terms of ρ, z, and R_0 with the aid of Eq. (2.35).

In front of the body for $\rho \to 0$, i.e., on its axis, it follows from Eq. (2.35) that

$$\sin\theta' \approx \frac{\varrho}{2z - R_0} \to 0, \quad \cos\theta' \to 1,$$

therefore

$$\left.\begin{aligned} n_2(0, z) &= n_0 \frac{R_0^2}{(2z - R_0)^2}, \\ n(0, z) &= n_0\left[1 + \frac{R_0^2}{(2z - R_0)^2}\right]. \end{aligned}\right\} \quad (2.41)$$

The dependence of the ratio n/n_0 on the distance from the body, in the direction of its motion along the z axis, and in a direction along the ρ axis perpendicular to its motion, is shown in Fig. 5. The distribution of the concentration of particles in the compressed region is given in Table VI and in Fig. 6 for various values of ρ and z, calculated according to Eq. (2.40). It is clear from the table and the figure that the concentration of reflected particles is extremely high near the body's surface. Close to the surface, $n_2/n_0 \approx 1$, $n/n_0 \approx 2$; for $z \approx 1.2\ R_0$; $n_2/n_0 \approx 0.5$; for $z = 1.5\ R_0$, $n_2/n_0 \approx 0.2$; and for $z = 2\ R_0$, the value of $n_2/n_0 \approx 0.1$. As the distance from the body's surface increases, the disturbance (i.e., the deviation from the unperturbed density) of the concentration in the compressed region falls off more rapidly than in the rarefied region.

b. Elastic Diffuse Scattering. Now let us assume that the particles are scattered diffusely upon impact with the body's surface, so that a particle may be reflected at any angle with the same probability. In this case, the surface element dS near a point $\mathbf{r}_S$ on the surface scatters dn_S particles during a time interval dt, where

$$dn_S = -(\mathbf{V}_0 \mathbf{n})\, n_0\, dS\, dt = n_0 V_0 \cos\theta'\, dS\, dt. \quad (2.42)$$

Here $\mathbf{n}$ is the outward normal to the body's surface at the point $\mathbf{r}_S$, θ' is, as before, the angle between the velocity $\mathbf{V}_0$ and the normal $\mathbf{n}$ (see Fig. 7).

For elastic diffuse scattering, the velocity of a reflected particle is directed, with equal probability, at any angle to the surface, but the absolute magnitude (of the velocity) is conserved. Therefore, all particles scattered at the moment t_0 during the time interval dt by a surface element dS are, after a lapse of time $(t - t_0)$ uniformly distributed in a hemispherical shell, whose center is located at the point $\mathbf{r}_S$. The radius of this shell is equal to $V_0(t - t_0)$, and its thickness is equal to $V_0 dt$. Therefore,

$$dn_S = 2\pi V_0^2 (t - t_0)^2 V_0\, dt\, dn_2(A) = 2\pi R_{SA}^2 V_0\, dt\, dn_2(A), \quad (2.43)$$

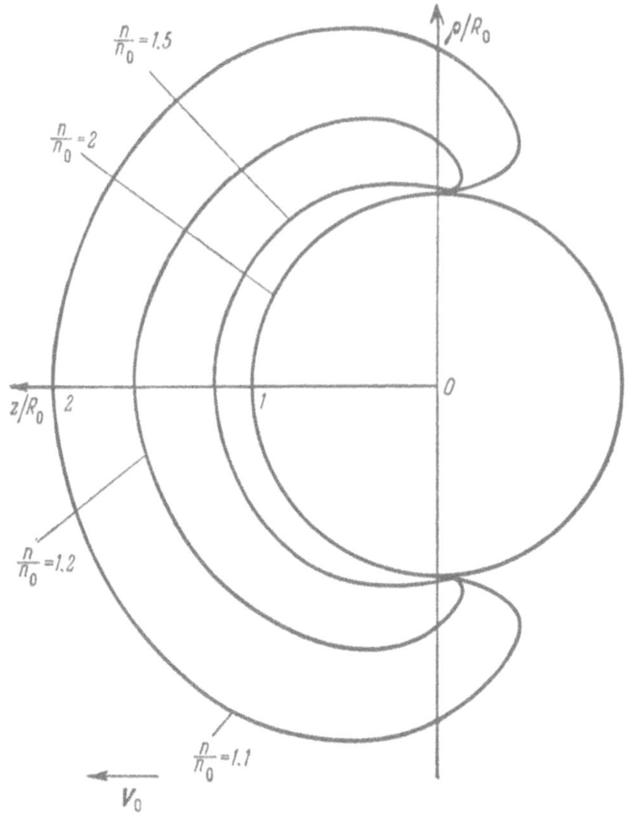

Fig. 6. Curves of constant density in the region of compression in the case of specular reflection.

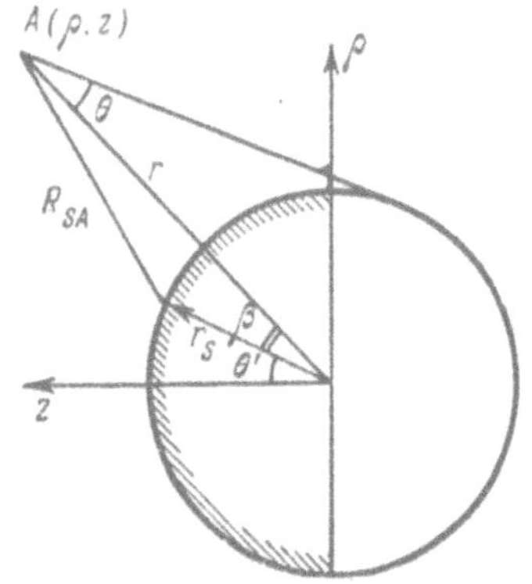

Fig. 7. Illustration of the derivation of the concentration of particles for diffuse scattering.

where $dn_2(A)$ is the density of particles, scattered by the surface element dS, near the point A(ρ, z), and $R_{SA} = V_0(t - t_0)$ is the distance from the point $\mathbf{r}_S$ to the point A. Now equating (2.42) and (2.43), we find that

$$dn_2(A) = n_Q \frac{\cos\theta'\, dS}{2\pi R_{SA}^2}. \tag{2.44}$$

The total density of scattered particles near the point A is obtained, of course, by integrating $dn_2(A)$ over the entire surface S_A of the body which can be seen from the point A (taking into account, of course, that only the front part of the surface scatters the beam of incoming particles).

For a spherical surface $dS = R_0^2 \sin\beta\, d\beta\, d\varphi$ and

$$R_{SA}^2 = r^2 + R_0^2 - 2R_0 r\cos\beta = R_0^2\left(1 + \frac{1}{\sin^2\theta} - 2\frac{\cos\beta}{\sin\theta}\right),$$

where $r = (\rho^2 + z^2)^{1/2}$, β is the angle between the vector $\mathbf{r}$ directed to the point A and $\mathbf{r}_S$, and 2θ is the angle which the sphere subtends as seen from the point A (see Fig. 10)

$$\sin\theta = \frac{R_0}{r} = \frac{R_0}{\sqrt{\rho^2 + z^2}}.$$

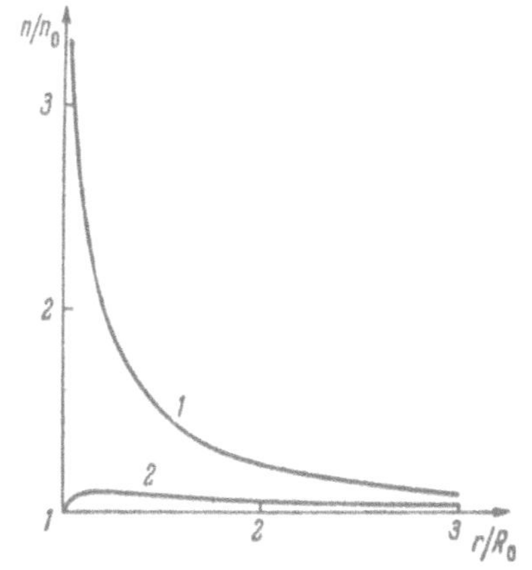

Fig. 8. Dependence of the concentration in the region of compression on the distance to the center of the body, in the case of diffuse scattering. Curve 1 is for distance measured along the z axis; curve 2 is for distance measured along the ρ axis (for z = 0).

TABLE VII

$(z-R_0)/R_0$	n/n_0	$(z-R_0)/R_0$	n/n_0
0.0001	5.95	0.01	3.56
0.001	4.80	0.02	3.26
0.002	4.46	0.05	2.74
0.005	4.00	0.1	2.34

Therefore, the concentration of particles scattered by the sphere is given by

$$n_2(\varrho, z)=\frac{1}{2\pi}\int_{-\pi}^{\pi} d\varphi \int_0^{\frac{\pi}{2}-\theta} \frac{n_0(\theta')\cos\theta' \sin\beta\, d\beta}{1+\frac{1}{\sin^2\theta}-2\frac{\cos\beta}{\sin\theta}}, \quad (2.45)$$

where $n_0(\theta')$ is the density of particles in the incident (on the body's surface) beam – it differs from zero only on the front surface of the body, which is crosshatched in Fig. 7:

$$n_0(\theta')=\begin{cases} n_0 & \text{for} \quad \theta'\leqslant\frac{\pi}{2}, \\ 0 & \text{for} \quad \theta'>\frac{\pi}{2}. \end{cases} \quad (2.46)$$

The angle θ', as is clear from Fig. 7, is related to the angle β by the equation

$$\cos\theta'=\frac{z}{r}\cos\beta+\frac{\varrho}{r}\sin\beta\cos\varphi. \quad (2.47)$$

In the region directly in front of the body (for $\rho \leq R_0$) the density $n_0(\theta')=n_0$, since from any point lying in this region, only the front surface of the body is seen. In this case, the integration in Eq. (2.45) is elementary to perform:

$$n_2(\varrho, z)=\frac{n_0}{2\pi}\int_{-\pi}^{\pi} d\varphi \int_0^{\frac{\pi}{2}-\theta} \frac{\left(\frac{z}{r}\cos\beta+\frac{\varrho}{r}\sin\beta\cos\varphi\right)\sin\beta\, d\beta}{1+\frac{1}{\sin^2\theta}-\frac{2\cos\beta}{\sin\theta}}=\frac{n_0 z}{2r}\sin\theta\int_{\sin\theta}^{1}\frac{\cos\beta\, d(\cos\beta)}{\frac{\sin\theta}{2}+\frac{1}{2\sin\theta}-\cos\beta}=$$

$$=\frac{n_0 z}{4r}\sin\theta\left[\left(\sin\theta+\frac{1}{\sin\theta}\right)\ln\left(\frac{1+\sin\theta}{1-\sin\theta}\right)+2\sin\theta-2\right].$$

Hence, considering that $\sin\theta = R_0/r$, for $\rho \leq R_0$ we finally obtain

$$n_2(\varrho, z)=\frac{n_0 z R_0^2}{4r^3}\left[\left(1+\frac{r^2}{R_0^2}\right)\ln\left(\frac{r+R_0}{r-R_0}\right)+2-2\frac{r}{R_0}\right], \quad (2.48)$$

where $r=(\rho^2+z^2)^{1/2}$. In particular, on the z axis

$$n_2(0, z)=\frac{n_0 R_0^2}{4z^2}\left[\left(1+\frac{z^2}{R_0^2}\right)\ln\left(\frac{z+R_0}{z-R_0}\right)+2-2\frac{z}{R_0}\right]. \quad (2.49)$$

Expression (2.45) can also be integrated in an analogous manner for points lying on the ρ axis (i.e., for z = 0). In this case

$$n_2(\varrho, 0)=\frac{n_0}{2\pi}\left\{\left(1+\frac{R_0^2}{\varrho^2}\right)\left(\frac{\pi}{4}-\frac{1}{2}\arcsin\frac{R_0}{\varrho}\right)+\right.$$

$$\left.+\frac{R_0}{\varrho}\sqrt{1-\frac{R_0^2}{\varrho^2}}-\left(1-\frac{R_0^2}{\varrho^2}\right)\arctan\left[\frac{R_0+\varrho}{\varrho-R_0}\tan\left(\frac{\pi}{4}-\frac{1}{2}\arcsin\frac{R_0}{\varrho}\right)\right]\right\}.$$

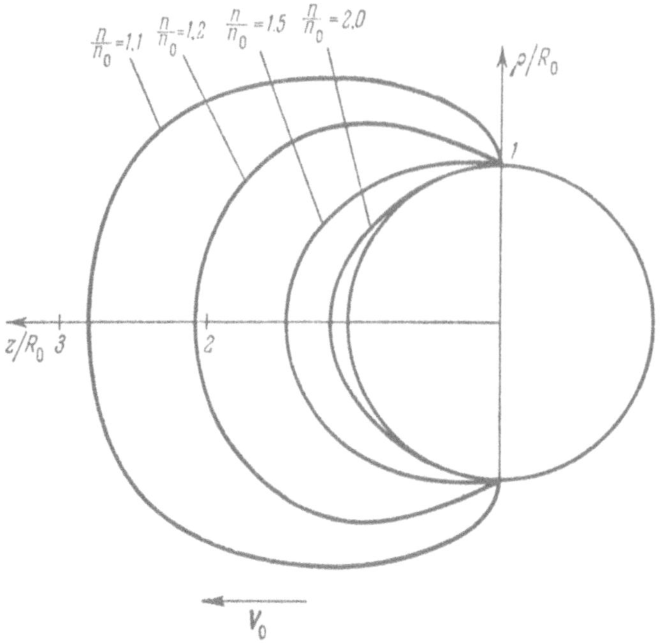

Fig. 9. Curves of constant particle density in the region of compression for diffuse scattering.

The ratio of the total density of particles, $n = n_0 + n_2$, to n_0 in front of the body, as a function of distance along the z and ρ axes, is shown in Fig. 8 for the case of diffuse scattering, and the particle density distribution in the vicinity of a sphere is shown in Fig. 9. It is clear from the figures that, for the case of diffuse scattering, the flux of reflected particles turns out to be more directional than for specular scattering (see Figs. 5 and 6). Therefore, the particle density directly in front of a body is higher for the case of diffuse scattering than it is for specular scattering. For example, on the z axis at large distances from the body, the density of reflected particles $n_2 = (n_0/2)(R_0/z)^2$ as follows from Eq. (2.49), whereas for specular scattering $n_2 = (n_0/4)(R_0/z)^2$. However, n_2 increases most strongly near the body's surface. Thus, for $(r - R_0) \ll R_0$

$$n_2 \approx \frac{n_0 z}{2R_0} \ln \frac{2R_0}{r - R_0},$$

so that as $r \to R_0$ the density of reflected particles increases logarithmically, and may become considerably larger then the density of particles in the incoming beam. Values of n/n_0 on the z axis, close to the body's surface, are presented in Table VII.

The reason for the increase in the density of reflected particles near the body's surface in the case of elastic, diffuse scattering is clear from simple considerations. Actually, the flux of reflected particles is constant. At the same time, the average velocity, directed along the normal, of the particles decreases as one approaches the surface for elastic, diffuse scattering. Hence, the density of particles naturally increases.

c. Accommodation, Absorption, and the Emission of Particles. Only elastic collisions of particles with a body's surface were considered above. At the same time, a particle is usually reflected from the surface, retaining only part of its initial energy. This inelastic scattering process is called accommodation. In the general case, as indicated in §3b, it is necessary to characterize this by a scattering function $w_n(\mathbf{v}, \mathbf{v}_0, \mathbf{r}_S)$, i.e., the probability that a particle, colliding with the surface at the point $\mathbf{r}_S$, changes its initial velocity $\mathbf{v}_0$ to the velocity $\mathbf{v}$. However, at the present time, very little is known concerning the function w_n. An inelastic scattering process is usually characterized by an effective coefficient of accommodation – the average value of the ratio of the energy, transmitted to the wall, to the particle's initial kinetic energy:

$$p_a = \frac{\mathscr{E}_0 - \mathscr{E}}{\mathscr{E}_0} = 1 - \frac{v^2}{v_0^2},$$

where $\mathscr{E}_0$ and $\mathscr{E}$ are, respectively, the values of the particle's kinetic energy before and after reflection. The value of the accommodation coefficient essentially depends on the ratio of the mass M of the incident particle to the mass M_S of the particles which the wall is composed of, on the body's temperature, and on other parameters (see [19]). On the average, apparently one can assume $p_a \sim (1/2)(M/M_S)$; we note that in the ionosphere, and even more in interplanetary space, M is less than, or much less than M_S, and the coefficient of accommodation, p_a, is apparently small, i.e., the scattering is nearly elastic. (Elastic reflection of a particle from the wall corresponds to the value $p_a = 0$, the complete loss of a particle's energy upon collision gives $p_a = 1$.)

The influence of the average accommodation on the density of particles reflected from the body can be taken into account very simply. It is sufficient to only consider that the average velocity of the reflected particle is now equal to

$$v = v_0 (1 - p_a)^{1/2}.$$

Therefore, with account of accommodation, the concentration of reflected particles, $n_2(\mathbf{r})$, in expressions (2.40), (2.45), and so forth, is increased on the average by a factor of $1/(1 - p_a)^{1/2}$ times.

The concentration of reflected particles essentially depends on their absorption by the body's surface, and also on the inverse process, i.e., the emission of these particles. As already indicated above, these processes are extremely complicated and have been investigated very little, so that one is not able to take detailed account of them in an investigation of the structure of the disturbed region. One can only remark that emission processes are apparently not very essential, because of the low temperature of the body and the comparatively small energy of the incoming particles in the upper ionosphere. Conversely, the probability for the (charge) neutralization of ions incident on a body with small energies, 3 to 5 eV, is usually large: for example, the coefficient for the reflection of ions from a pure metal is only of the order of a percent (see [19]). Consequently, the number of ions reflected from a body may be extremely small, which essentially manifests itself in their concentration in the "compressed" region. On the average, one can take account of this process of ion neutralization by introducing a coefficient R_i for the reflection of ions from the body. In this case, the concentration of ions $N_{2i}(\mathbf{r})$, reflected from the body is determined by the relation

$$N_{2i}(r) = R_i n_2(r) \frac{N_0}{n_0}, \tag{2.50}$$

where $n_2(\mathbf{r})$ is the concentration of reflected neutral particles, determined by formulas (2.40), (2.45), and others, N_0 is the ion concentration in the undisturbed plasma. In a number of cases below, we shall neglect, in general, the ions which are reflected from the body, assuming the ionic reflection coefficient R_i to be small.

We note here that the concentration of neutral particles around a sphere is also calculated in articles [77] and [78] for $V_0/v_n = 5$. Behind the body, the results [77] exactly coincide with the results obtained above, which one can, in particular, directly verify with the aid of Eq. (2.25) on the z axis of the body's motion. In front of the body, the concentration of particles obtained in [77] is larger than we obtained, since the authors considered the case of diffuse reflection in the presence of accommodation. In article [78], the concentration of particles in front of the body is calculated for diffuse and specular reflection by the method used in [77]. The results obtained in [78] exactly agree with our results. It should also be noted that the authors [77], in the evaluation of (the concentration of) reflected particles in front of the body, incorrectly evaluate integrals which, however, leads to mistakes which are smaller than the deviation indicated above.

§7. The Concentration of Particles Around a Sphere Moving with Arbitrary Velocity

a. The Distribution Function. Effects (produced) in the vicinity of a rapidly moving body have been analyzed in the preceding paragraphs. Here we shall consider the case of a sphere moving with arbitrary velocity.

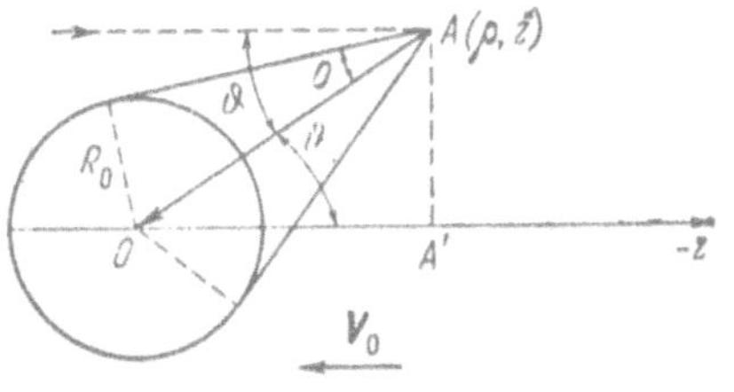

Fig. 10

Let us suppose, as usual, that a sphere with center at the point 0 is moving with a velocity $\mathbf{V}_0$. In the coordinate system in which the sphere is at rest, the entire picture is symmetrical with respect to the z axis, which passes through the point 0 in the direction of $\mathbf{V}_0$. We select a point $A(\rho, z)$ and we introduce, as usual, the angle ϑ between (the line) OA and the z axis and the angle 2θ, which the sphere subtends as seen from the point A (see Fig. 10), i.e.,

$$\sin\vartheta = \frac{\varrho}{\sqrt{\varrho^2+z^2}} = \frac{\varrho}{r}, \quad \sin\theta = \frac{R_0}{\sqrt{\varrho^2+z^2}} = \frac{R_0}{r}. \tag{2.51}$$

Let OA be an axis of a spherical coordinate system in velocity space. Then particles occurring at the point A, whose velocities form with OA an angle β larger than θ obviously arrive at the point A, by-passing the sphere, and particles whose velocities form with OA an angle $\beta < \theta$ arrive at the point A after colliding with the sphere's surface. Then Eqs. (2.9) and (1.14) for the neutral particles' distribution function is written in the form

$$v\frac{\partial f_n}{\partial r} = \frac{r}{r}\delta(r-R_0)\begin{cases} \int w_n(v, v_1, r_S)v_1 f d^3v_1, & \beta<\theta, \\ -vf, & \beta>\pi-\theta, \\ 0, & \pi-\theta>\beta>\theta. \end{cases} \tag{2.52}$$

Here it is assumed that the equation for the surface of the sphere is $F = r - R_0 = 0$, and that the condition $\beta < \theta$ simultaneously denotes $\mathbf{v}\mathbf{r}_S > 0$, but the condition $\beta > \pi - \theta$ means that $\mathbf{v}\mathbf{r}_S < 0$. The function w_n is the scattering probability for neutral particles (see §3b). If the surface is homogeneous, then w_n does not depend on $\mathbf{r}_S$. In this case the solution of Eq. (2.52), as one can easily verify by direct substitution, has the form

$$f_n = \begin{cases} f_0(v+V_0) = \left(\frac{M}{2\pi\varkappa T}\right)^{3/2} n_0 \exp\left[-\frac{M(v+V_0)^2}{2\varkappa T}\right], & \beta>\theta, \\ -\frac{1}{(nv)}\int f_0(v_1+V_0)(nv_1)w_n(v, v_1)d^3v_1, & \beta<\theta, \end{cases} \tag{2.53}$$

where f_0 is the undisturbed (Maxwellian) distribution function (given by Eq. (1.6)), $\mathbf{n}$ is the outward normal to the sphere at the point of impact of the particle with the surface. In fact, the distribution function (2.53) does not depend on the spatial coordinates anywhere, except at points on the surface defined by the condition $\theta = \pi/2$, i.e., on the surface of the sphere $r = R_0$. The distribution function (2.53) has a discontinuity on this surface; consequently $\partial f/\partial r = n\delta(r - R_0)\Delta f$, where Δf is the jump in the distribution function. For $\beta > \pi/2$, as is clear from Eq. (2.53), this jump is simply equal to $-f_0$, and for $\beta < \pi/2$ it is equal to

$$\frac{1}{nv}\int f_0\, nv_1 w_n d^3v_1,$$

and therefore the distribution function (2.53) satisfies Eq. (2.52).

The concentration of particles at an arbitrary point A, which one can characterize by the coordinates ρ and z or by the angles ϑ and θ [see Eq. (2.51) and Fig. 10], is made up of the concentration of "direct" particles n_1 (not undergoing collisions with the body) and the concentration of reflected particles n_2, i.e., $n = n_1 + n_2$, where

$$n_1 = \int_0^{2\pi} d\varphi \int_\theta^{\pi} \sin\beta\, d\beta \int_0^{\infty} v^2 f_0(v+V_0)\, dv, \tag{2.54}$$

$$n_2 = -\int_0^{2\pi} d\varphi \int_0^{\theta} \sin\beta\, d\beta \int_0^{\infty} \frac{v^2\, dv}{(nv)} \int f_0(v_1 + V_0)(nv_1)\, w_n(v, v_1)\, d^3v_1, \tag{2.55}$$

and φ and β are the azimuthal and polar angles with the axis OA in the spherical coordinate system.

b. Concentration of "Direct" (Unscattered) Particles. First of all, let us write down an expression for the concentration of "direct" particles n_1. Considering that

$$(v + V_0)^2 = v^2 + V_0^2 + 2vV_0(-\cos\vartheta\cos\beta + \sin\vartheta\sin\beta\cos\varphi),$$

we have

$$n_1(\vartheta, \theta) = \int_0^{2\pi} d\varphi \int_{\theta}^{\pi} \sin\beta\, d\beta \int_0^{\infty} v^2 n_0 \left(\frac{M}{2\pi\varkappa T}\right)^{3/2} \times$$

$$\times \exp\left\{-\frac{M}{2\varkappa T}\left[V_0^2 + v^2 + 2vV_0(-\cos\vartheta\cos\beta + \sin\vartheta\sin\beta\cos\varphi)\right]\right\} dv. \tag{2.54a}$$

Now expressing the integral with respect to $d\varphi$ in terms of a Bessel function of imaginary argument, we obtain

$$n_1 = n_0\left(\frac{M}{2\pi\varkappa T}\right)^{3/2} \int_{\theta}^{\pi} \sin\beta\, d\beta \int_0^{\infty} v^2 \exp\left[-\frac{M}{2\varkappa T}(V_0^2 + v^2 - 2vV_0\cos\vartheta\cos\beta)\right] I_0\left(\frac{MvV_0}{\varkappa T}\sin\vartheta\sin\beta\right) dv. \tag{2.56}$$

At points on the z axis, the value of $\sin\vartheta$ is equal to zero, and the integral with respect to $d\beta$ is elementary to evaluate, namely

$$n_1(0, \theta) = n_0\left(\frac{M}{2\pi\varkappa T}\right)^{3/2} \frac{\varkappa T}{MV_0} \int_0^{\infty} v \exp\left[-\frac{M(V_0^2 + v^2)}{2\varkappa T}\right] \left\{\exp\left(\frac{MvV_0}{\varkappa T}\cos\theta\right) - \exp\left(-\frac{MvV_0}{\varkappa T}\right)\right\} dv =$$

$$= n_0\left[\frac{1}{2} - \frac{1}{\sqrt{\pi}} \int_0^{\sqrt{\frac{MV_0^2}{2\varkappa T}}} e^{-x^2} dx + \cos\theta\, e^{-\frac{MV_0^2}{2\varkappa T}\sin^2\theta} \left(\frac{1}{2} + \frac{1}{\sqrt{\pi}} \int_0^{\sqrt{\frac{MV_0^2}{2\varkappa T}}\cos\theta} e^{-x^2} dx\right)\right] =$$

$$= \frac{n_0}{2}\left\{1 - \Phi\left(\sqrt{\frac{MV_0^2}{2\varkappa T}}\right) + \cos\theta\, e^{-\frac{MV_0^2}{2\varkappa T}\sin^2\theta} \left[1 + \Phi\left(\sqrt{\frac{MV_0^2}{2\varkappa T}}\cos\theta\right)\right]\right\}, \tag{2.56a}$$

where, as usual, $\Phi(x) = \frac{2}{\sqrt{\pi}} \int_0^x e^{-t^2} dt$ is the probability integral.

We consider the case of large distances from the body, where

$$\theta \ll 1 \text{ and } \theta\sqrt{\frac{MV_0^2}{2\varkappa T}} \ll 1. \tag{2.57}$$

Then it is more convenient to represent $\int_{\theta}^{\pi}\{\ldots\}\,d\beta$ in the form

$$\int_{\theta}^{\pi}\{\ldots\}\,d\beta=\int_{0}^{\pi}\{\ldots\}\,d\beta-\int_{0}^{\theta}\{\ldots\}\,d\beta.$$

It is obvious that the term with $\int_{0}^{\pi}\{\ldots\}\,d\beta$ is equal to n_0, since the integration in it is carried out over the entire velocity space. Thus

$$n_1=n_0-n_0\left(\frac{M}{2\pi\varkappa T}\right)^{3/2}2\pi\int_{0}^{\theta}\sin\beta\,d\beta\,\times$$

$$\times\int_{0}^{\infty}v^2\exp\left[-\frac{M}{2\varkappa T}(V_0^2+v^2-2vV_0\cos\vartheta\cos\beta)\right]\times I_0\left(\frac{MvV_0}{\varkappa T}\sin\vartheta\sin\beta\right)dv. \qquad (2.58)$$

By virtue of condition (2.57), in the integrand we may now set $\beta=0$ everywhere, except in the factor $\sin\beta$, and we replace $\sin\beta$ by β. Then

$$n_1=n_0-n_0\left(\frac{M}{2\pi\varkappa T}\right)^{3/2}\pi\theta^2\int_{0}^{\infty}v^2\exp\left[-\frac{M}{2\varkappa T}(V_0^2+v^2-2vV_0\cos\vartheta)\right]dv=$$

$$=n_0-n_0\left(\frac{M}{2\pi\varkappa T}\right)^{3/2}\frac{\pi R_0^2}{r^2}\int_{0}^{\infty}v^2\exp\left[-\frac{M}{2\varkappa T}(V_0^2+v^2-2vV_0\cos\vartheta)\right]dv, \qquad (2.59)$$

where $r=(\rho^2+z^2)^{1/2}$. It is evident from here that, at large distances from the body, the disturbances of the concentration of "direct" particles always decreases like $1/r^2$.

c. Concentration of Reflected Particles. As is clear from Eq. (2.55), the concentration of reflected particles essentially depends on the law describing the scattering of particles by the surface. For specular reflection, the scattering function w_n is determined by expression (1.15):

$$w_n=\delta(v_1-v+2n(nv)),$$

where $\mathbf{n}$ is the normal to the surface at the scattering point. In this case

$$n_2(r)=-\int_{0}^{2\pi}d\varphi\int_{0}^{\theta}\sin\beta\,d\beta\int_{0}^{\infty}\frac{v^2\,dv}{(nv)}\int f_0(v_1+V_0)(nv_1)\,w_n\,d^3v_1=$$

$$=\int_{0}^{2\pi}d\varphi\int_{0}^{\theta}\sin\beta\,d\beta\int_{0}^{\infty}v^2 f_0(v+V_0)\exp\left[\frac{2M\,(nv)\,(nv_0)}{\varkappa T}\right]dv.$$

One can also obtain a somewhat different expression for $n_2(\mathbf{r})$ by first carrying out the integration with respect to d^3v. In this connection, we consider that

$$\int_{0}^{2\pi}d\varphi\int_{0}^{\theta}\sin\beta\,d\beta\int_{0}^{\infty}v^2\delta(v_1-v+2n(nv))\,dv=\begin{cases}\dfrac{d^3r_1}{d^3r}, & \theta<\beta_1<\pi,\\ 0, & 0<\beta_1<\theta,\end{cases} \qquad (2.60)$$

where d^3r is the (infinitesimal) volume element near the point $\mathbf{r}$ which the particles, occupying the volume d^3r_1 before collision with the surface and having velocity $\mathbf{v}_1$, occupy. Therefore,

$$n_2(r) = \int_0^{2\pi} d\varphi_1 \int_\theta^\pi d\beta_1 \sin\beta_1 \int_0^\infty v_1^2 f_0(v_1 + V_0) \frac{d^3r_1}{d^3r} dv. \tag{2.61}$$

This expression has a simple physical meaning. One can obtain it immediately, not solving the kinetic equation, but using Liouville's theorem. In fact, according to Liouville's theorem, the concentration of particles in the six-dimensional space of $\mathbf{v}$ and $\mathbf{r}$ does not change upon scattering, i.e.,

$$f_2(v, r)\, d^3v\, d^3r = f_0(v_1 + V_0)\, d^3v_1\, d^3r_1,$$

where f_2 is the distribution function for the reflected particles, f_0 is the distribution function for the particles incident on the surface. Consequently,

$$n_2 = \int f_2\, d^3v = \int f_0(v_1 + V_0) \frac{d^3r_1}{d^3r} d^3v_1.$$

Now considering that, in the chosen (by us) coordinate system, the polar angle β for velocity $\mathbf{v}$ of the reflected particles only changes from 0 to θ and correspondingly the polar angle for the velocity $\mathbf{v}_1$ varies from θ to π, we immediately arrive at expression (2.61) for the concentration of reflected particles.

Thus, in order to find n_2, it is sufficient to determine the ratio d^3r_1/d^3r, i.e., to find out how the volumes before and after the collision are related. But such a problem was already solved above in connection with the determination of the concentration of particles in the "compressed" region for a rapidly moving body (see §6a). The only difference consists in the fact that now the particles incident on the body are moving at arbitrary angles to the z axis. Using Eqs. (2.40) and (2.35), we have

$$\frac{d^3r_1}{d^3r} = \frac{R_0^2 \cos^2\theta' \sin^2\theta'}{\varrho'(\varrho' - R_0 \sin^3\theta')}, \tag{2.40a}$$

where ρ' and z' are the coordinates of the point $\mathbf{r}$ in the system of coordinates whose z' axis is parallel to the velocity $\mathbf{v}_1$ of the incoming particle. The angle θ' is related to ρ' and z' by the equation

$$R_0 = 2z'\cos\theta' + 2\varrho'\sin\theta' - \frac{\varrho'}{\sin\theta'}. \tag{2.35a}$$

Considering now that

$$\frac{\varrho'}{\sqrt{\varrho'^2 + z'^2}} = \sin\beta_1, \qquad \frac{z'}{\sqrt{\varrho'^2 + z'^2}} = -\cos\beta_1,$$

we may finally write expression (2.40a) for d^3r_1/d^3r in the form

$$\frac{d^3r_1}{d^3r} = \frac{\sin^2\theta \cos^2\theta' \sin^2\theta'}{\sin^2\beta_1 - \sin\beta_1 \sin\theta \sin^3\theta'}, \tag{2.62}$$

where the angle θ is defined by Eq. (2.51), and θ' is related to θ and β_1 by the equation

$$\sin\beta_1 = -2\sin\theta' \cos(\beta_1 + \theta') - \sin\theta \sin\theta'. \tag{2.63}$$

For diffuse scattering, if the scattering probability is given by the general expression (1.20)

$$w_n = \frac{1}{2\pi} p(v, v_1),$$

then

$$n_2(r) = -\frac{1}{2\pi} \int d\Omega \int_0^\infty \frac{v^2\, dv}{(nv)} \int f_0(v_1 + V_0)(nv_1)\, p(v, v_1)\, d^3v_1. \tag{2.64}$$

Here the integration in the space of velocity **v** is carried out inside the angle which the scattering surface S subtends, as seen from the point **r**. In the general case, it is convenient to change to an integration over the surface S. We consider that $d\Omega = (\mathbf{nv}/R_{Sr}^2 v)dS$, where R_{Sr} is the distance from the point **r** to the scattering point $\mathbf{r}_S$, and **n** is the normal to the surface at the scattering point. Then

$$n_2(r) = -\frac{1}{2\pi} \int \frac{dS}{R_{Sr}^2} \int_0^\infty v\, dv \int f_0(v_1 + V_0)(nv_1)\, p(v, v_1)\, d^3v_1, \tag{2.65}$$

where the integration with respect to dS is carried out over the entire surface region which is seen from the point **r**.

In the case of elastic diffuse scattering (in the absence of accommodation)

$$p(v, v_1) = \frac{1}{v^2} \delta(v - v_1)$$

(see Eq. (1.18)). In addition, for a spherical surface,

$$dS = R_0^2 \sin\beta\, d\beta\, d\varphi, \quad R_{Sr}^2 = R_0^2\left(1 + \frac{1}{\sin^2\theta} - \frac{2\cos\beta}{\sin\theta}\right), \tag{2.66}$$

where β is the polar angle measured from the direction to the point **r** (see Fig. 7); consequently, in this case

$$n_2(r) = \frac{1}{2\pi} \int_0^{2\pi} d\varphi \int_0^{\frac{\pi}{2}-\theta} \frac{\sin\beta\, d\beta}{1 + \frac{1}{\sin^2\theta} - \frac{2\cos\beta}{\sin\theta}} \int_{(v_1 n) < 0} \cos(v_1 n)\, f_0(v_1 + V_0)\, d^3v_1. \tag{2.67}$$

Here the angle θ is, as before, defined by Eq. (2.51).

d. Rapidly Moving Sphere. In the case when the velocity of the sphere is much larger than the mean thermal velocity of the particles, i.e., when

$$\sqrt{\frac{MV_0^2}{2\varkappa T}} \gg 1,$$

one can significantly simplify the expressions obtained above.

First let us consider the rarefied region at large distances behind the body, so that

$$\frac{r}{z}, \ \frac{R_0}{z} \leqslant \sqrt[4]{\frac{\varkappa T}{MV_0^2}} \ll 1. \tag{2.68}$$

In this region the contribution of the reflected particles, as indicated in §5, is exponentially small, since a particle being reflected backwards from the body must have a velocity $\sim V_0$ before reflection, and the number of such particles is very small. Therefore, $n \approx n_1$. Provided that conditions (2.68) are satisfied, i.e., when

$$\vartheta,\ \theta \lesssim \sqrt[4]{\frac{\varkappa T}{MV_0^2}} \ll 1, \tag{2.69}$$

in Eq. (2.56) one can set

$$\sin\vartheta \approx \vartheta,\ \sin\beta \approx \beta,\quad \cos\vartheta \approx 1-\frac{\vartheta^2}{2},\ \cos\beta \approx 1-\frac{\beta^2}{2}.$$

As a result, we obtain

$$n = n_0 - n_0\left(\frac{M}{2\pi\varkappa T}\right)^{3/2} 2\pi \int_0^\theta \beta\, d\beta \int_0^\infty v^2 \exp\left[-\frac{M}{2\varkappa T}[(v-V_0)^2 + \right.$$

$$\left. + v_0 V_0(\beta^2+\vartheta^2)]\right] I_0\left(\frac{MvV_0\beta\vartheta}{\varkappa T}\right) dv. \tag{2.70}$$

Further, the fulfillment of the condition $MV_0^2/2\varkappa T \gg 1$ means that the factor $\exp[-(M/2\varkappa T)(v-V_0)^2]$ has a sharp maximum for $v \approx V_0$. This allows one to replace v by V_0 everywhere, except in this factor, and to replace the lower limit on the integration with respect to dv by $-\infty$, i.e.,

$$n = n_0 - 2n_0\left(\frac{MV_0^2}{2\varkappa T}\right) e^{-\frac{MV_0^2}{2\varkappa T}\vartheta^2} \int_0^\theta e^{-\frac{MV_0^2}{2\varkappa T}\beta^2} I_0\left(\frac{MV_0^2}{\varkappa T}\vartheta\beta\right)\beta\, d\beta. \tag{2.71}$$

Considering that $\vartheta \approx \rho/z$, $\theta \approx R_0/z$, and letting the integration variable $\beta = (2\varkappa T/MV_0^2)^{1/2}\tau$, we find that Eq. (2.71) coincides with Eq. (2.24).

Now let us consider the region behind the body, at small distances from its surface, but along the z axis. That is to say, we shall assume that

$$\vartheta < \theta,\ \frac{2\varkappa T}{MV_0^2} \ll \vartheta \sim 1. \tag{2.72}$$

In this region, the number of both "direct" (unscattered) as well as reflected particles is exponentially small. For example, let us calculate n_1. Here it is more convenient to start directly from Eq. (2.54). Under conditions (2.72), as one can easily see, the exponent in the integrand of (2.54a) has its maximum value for $\varphi = \pi$. One can evaluate the integral with respect to φ by the saddle-point method. Carrying out an expansion in powers of $x = \varphi - \pi$ and replacing the limits of the x integration by $x = \infty$, we obtain

$$n_1 = n_0\left(\frac{M}{2\pi\varkappa T}\right)^{3/2} \int_0^\infty v^2\, dv \int_{-\infty}^\infty dx\ \exp\left[-\frac{MvV_0}{\varkappa T}\sin\beta\sin\vartheta x^2\right] \int_\theta^\pi \exp\left[-\frac{M}{2\varkappa T}(v^2+V_0^2) + \right.$$

$$\left. + \frac{MvV_0}{\varkappa T}(\cos\beta\cos\vartheta + \sin\beta\sin\vartheta)\right] \sin\beta\, d\beta \approx$$

$$\approx \frac{n_0}{\sqrt{V_0\sin\vartheta}}\frac{M}{2\pi\varkappa T}\int_0^\infty v^{3/2}\, dv \int_\theta^\pi \sqrt{\sin\beta}\exp\left[\frac{MvV_0}{\varkappa T}\cos(\beta-\vartheta)\right] d\beta. \tag{2.73}$$

Values of β close to θ give the major contribution to the integral over dβ. Assuming

$$\cos(\beta-\vartheta)\approx\cos(\theta-\vartheta)-\sin(\theta-\vartheta)(\beta-\theta),$$

and replacing the upper limit on the integral with respect to d(β − θ) by ∞, we obtain

$$n_1=n_0\frac{M}{2\pi\varkappa T}\sqrt{\frac{\sin\theta}{V_0\sin\vartheta}}\int\limits_0^{\infty}v^{3/2}\exp\left[-\frac{M}{2\varkappa T}\left(V_0^2+v^2-\right.\right.$$

$$\left.\left.-2vV_0\cos(\theta-\vartheta)\right)\right]dv\int\limits_0^{\infty}\exp\left[-\frac{MvV_0\sin(\theta-\vartheta)}{\varkappa T}\beta\right]d\beta=$$

$$=\frac{n_0}{2\pi}\sqrt{\frac{\sin\theta}{V_0^3\sin\vartheta}}\frac{1}{\sin(\theta-\vartheta)}\int\limits_0^{\infty}v^{1/2}\exp\left\{-\frac{M}{2\varkappa T}\left[V_0^2+v^2-2vV_0\cos(\theta-\vartheta)\right]\right\}dv. \tag{2.74}$$

In the integral with respect to dv, the exponent has a sharp maximum at the value $v=V_0\cos(\theta-\vartheta)$; taking this into consideration, we finally have

$$n_1=n_0\left(\frac{\varkappa T}{2\pi MV_0^2}\right)^{1/2}\frac{\sin\theta\cos(\theta-\vartheta)}{\sin\vartheta\sin^2(\theta-\vartheta)}\exp\left[-\frac{MV_0^2}{2\varkappa T}\sin^2(\theta-\vartheta)\right]. \tag{2.75}$$

Finally, on the z axis close to the surface of the sphere ($(z-R_0)\ll R_0$), as follows directly from Eq. (2.56a),

$$n_1=n_0\left(\frac{\varkappa T}{2\pi MV_0^2}\right)^{1/2}e^{-\frac{MV_0^2}{2\varkappa T}}\left\{1+\left(\frac{\pi MV_0^2}{\varkappa T}\frac{z-R_0}{R_0}\right)^{1/2}\exp\left[\frac{MV_0^2(z-R_0)}{\varkappa TR_0}\right]\left[1+\Phi\left(\sqrt{\frac{MV_0^2(z-R_0)}{\varkappa TR_0}}\right)\right]\right\}. \tag{2.76}$$

In this connection, if $(MV_0^2/2\varkappa T)\,[(z-R_0)/R_0]\gg 1$, then

$$n_1(0,z)=n_0\frac{\sqrt{z^2-R_0^2}}{z}\exp\left[-\frac{MV_0^2R_0^2}{2\varkappa Tz^2}\right]. \tag{2.77}$$

Now let us determine the concentration of reflected particles, n_2, behind the body in the rarefied region for the case of specular reflection [26]. We consider n_2 on the z axis. In this case, expression (2.61) is rewritten in the form

$$n_2=\frac{n_0}{\sqrt{2\pi}}\left(\frac{M}{\varkappa T}\right)^{3/2}e^{-\frac{MV_0^2}{2\varkappa T}}\int\limits_0^{\infty}v_1^2e^{-\frac{Mv_1^2}{2\varkappa T}}dv_1\int\limits_{\theta}^{\pi}\frac{d^3r_1}{d^3r}\sin\beta_1e^{\frac{Mv_1V_0}{\varkappa T}\cos\beta_1}d\beta_1. \tag{2.78}$$

It will be evident from what follows that particles having velocities $v_1\approx V_0\cos\theta$ introduce the major contribution to n_2. Therefore, one can rewrite the coefficient associated with $\cos\beta_1$ in the exponent of the exponential in the second integral (we shall denote this coefficient by b) in the form

$$b=\frac{Mv_1V_0}{\varkappa T}\approx\frac{MV_0^2}{\varkappa T}\cos\theta=\frac{MV_0^2}{\varkappa T}\sqrt{1-\frac{R_0^2}{z^2}}. \tag{2.79}$$

Not overly close to the body's surface, the value of $b\gg 1$. For $b\gg 1$, the second integral in Eq. (2.78) is for the most part determined by values of $\beta_1=\theta+x$, where $x\ll 1$. In this case, it follows from Eq. (2.63) that $\theta'=(\pi/2)-(x/2)$. Hence, substituting into Eq. (2.62), we find that

$$\frac{d^3r_1}{d^3r} = \frac{\sin^2\theta\cos^2\theta'\sin^2\theta'}{\sin^2\beta_1 - \sin\beta_1\sin\theta\sin^3\theta'} \approx \frac{1}{4}\frac{\sin\theta}{\cos\theta}x. \quad (2.80)$$

The region, not very far removed from the body, is defined by the following values of the angle θ:

$$\sin\theta = \frac{R_0}{z} \gg \sqrt{\frac{2\varkappa T}{MV_0^2}}. \quad (2.81)$$

Substituting (2.80) and β_1 into the second integral in Eq. (2.78) and denoting it by d, we have

$$d = \int_{\theta}^{\pi} \frac{d^3r_1}{d^3r}\sin\beta_1 e^{b\cos\beta_1}\,d\beta_1 = \frac{1}{4}\frac{\sin^2\theta}{\cos\theta}e^{b\cos\theta}\int_0^{\infty} xe^{-b\sin\theta x}\,dx = \frac{1}{4}\frac{e^{b\cos\theta}}{b^2\cos\theta}.$$

Considering that, upon fulfillment of condition (2.80), the exponent of the exponential has a sharp maximum for $v_1 = V_0\cos\theta$, we finally discover

$$n_2 = \frac{n_0}{4}\frac{\varkappa T}{MV_0^2\cos\theta}e^{-\frac{MV_0^2}{2\varkappa T}\sin^2\theta} = \frac{n_0}{4}\frac{\varkappa T}{MV_0^2}\frac{z}{\sqrt{z^2 - R_0^2}}\exp\left[-\frac{MV_0^2R_0^2}{2\varkappa Tz^2}\right]. \quad (2.82)$$

The concentration of "direct" (unscattered) particles under the same conditions is given by Eq. (2.77). Comparing Eqs. (2.77) and (2.82), we convince ourselves that $n_2 \ll n_1$. In general, the concentration of reflected particles is much smaller than the concentration of "direct" particles everywhere behind the body, except for the region immediately adjacent to its surface, $(z - R_0) \ll R_0(2\varkappa T/MV_0^2)^{1/2}$. We observe that the principal part of the reflected particles is composed of particles whose velocities are in a direction nearly tangential to the sphere, drawn through the point of observation.

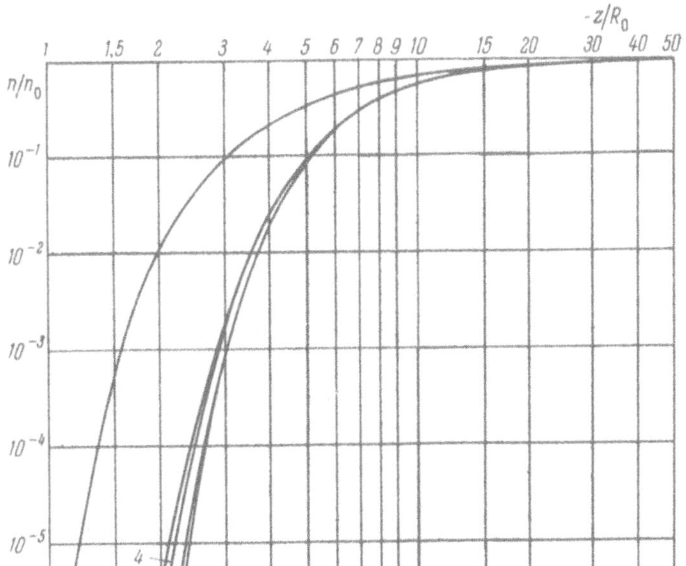

Fig. 11. The concentration of particles on the z axis behind a rapidly moving body $[V_0/(2\varkappa T/M)^{1/2} = 8]$ for the same cross section $S = \pi R_0^2$ of the body: 1) sphere; 2) disk; 3) body of circular cross section; 4) body of square cross section; 5) body of rectangular cross section ($R_x = 4R_y$).

TABLE VIII. Value of the Ratios of Particle Concentration n/n_0 in the Rarefied Zone for Bodies of Different Cross Sectional Shape

z/R_0	Sphere	Disk	Circular cross section	Square cross section	Rectangular cross section $R_y = 4R_x$
1	$5.65\cdot10^{-30}$	$8.98\cdot10^{-15}$	$1.60\cdot10^{-28}$	$2.28\cdot10^{-23}$	$5.40\cdot10^{-7}$
2	$9.75\cdot10^{-8}$	$2.44\cdot10^{-6}$	$1.12\cdot10^{-7}$	$1.07\cdot10^{-6}$	0.012
3	$7.77\cdot10^{-4}$	$1.57\cdot10^{-3}$	$8.25\cdot10^{-4}$	$1.74\cdot10^{-3}$	0.095
4	0.0177	0.0226	0.0183	0.0247	0.211
5	0.0757	0.0836	0.0773	0.0894	0.316
6	0.167	0.176	0.169	0.188	0.404
7	0.269	0.273	0.270	0.282	0.477
8	0.365	0.370	0.368	0.378	0.537
9	0.452	0.456	0.455	0.463	0.589
10	0.524	0.527	0.527	0.532	0.634
12	0.640	0.642	0.642	0.644	0.708
14	0.720	0.721	0.722	0.724	0.764
16	0.778	0.779	0.779	0.781	0.807
18	0.820	0.821	0.821	0.822	0.839
20	0.852	0.852	0.852	0.853	0.865
25	0.903	0.903	0.903	0.903	0.908
30	0.931	0.931	0.931	0.931	0.934
35	0.949	0.949	0.949	0.949	0.951
40	0.961	0.961	0.961	0.961	0.962

Now let us consider the region ahead of the body. Here the number of "direct" particles differs from n_0 only by an exponentially small amount, since only those particles which have a velocity greater than the velocity of the body are "shaded" by the body. Hence $n = n_1 + n_2 \approx n_0 + n_2$. We shall find n_2 for the case of a specularly reflecting surface. Provided $MV_0^2/2\varkappa T \gg 1$, the exponent of the exponential in the expression for f_0 in (2.61) has a maximum at $\beta_1 = \vartheta$. For $\beta_1 = \vartheta$, the z' axis coincides with the z axis, and the angle θ' once again is determined by Eq. (2.35), since $z' = z$ and $\rho' = \rho$, i.e.,

$$R_0 = 2z\cos\theta' + 2\varrho\sin\theta' - \frac{\varrho}{\sin\theta'}. \tag{2.35}$$

Therefore one can take d^3r_1/d^3r out from under the integrals. In addition, one can replace the lower limit on the integral over $d\beta_1$ by 0, since for $0 < \beta_1 < \theta$ the integrand is very small anyway. Then the double integral turns out to equal simply n_0, and we obtain formula (2.40) from (2.61):

$$n_2 = n_0\left(\frac{R_0}{\varrho}\right)^2 \frac{\sin^2\theta'\cos^2\theta'}{1-\frac{R_0}{\varrho}\sin^3\theta'}.$$

For diffuse scattering, considering that the directions of motion of the particles incident on the body are close to $-\mathbf{V}_0$, i.e., $\cos(\mathbf{v}_1\mathbf{n}) \approx \cos(\mathbf{V}_0, \mathbf{n}) = \cos\theta'$, in formula (2.67), we immediately obtain expression (2.45) from (2.67).

e. The Influence of the Shape of the Surface on the Structure of the Rarefied Region in the Case of a Rapidly Moving Body. Approximate general expressions for the concentration of particles in the rarefied zone far from a rapidly moving body were obtained in § 5. We shall compare the results of an exact solution of the problem for a rapidly moving sphere with them. On the z axis, behind a sphere of radius R_0, the concentration of particles is given by Eq. (2.77):

$$n(0, z) = n_0\frac{\sqrt{z^2-R_0^2}}{z}\exp\left[-\frac{MV_0^2R_0^2}{2\varkappa Tz^2}\right]. \tag{2.77}$$

A similar calculation of $n_1(0, z)$, with the aid of Eq. (2.22), behind a disk of the same radius R_0 (also see [26] and [82]) leads to the expression

$$n(0, z) = n_0 \frac{z}{\sqrt{z^2+R_0^2}} \exp\left[-\frac{MV_0^2}{2\varkappa T}\frac{R_0^2}{z^2+R_0^2}\right]. \tag{2.83}$$

The approximate formula (2.23) for a body whose maximum cross section is a circle of radius R_0 has the form (see Eq. (2.25))

$$n(0, z) = n_0 \exp\left[-\frac{MV_0^2R_0^2}{2\varkappa T z^2}\right]. \tag{2.25}$$

Values of the particle concentration along the z axis behind a sphere, disk, and a body of arbitrary shape, whose maximum cross sections are $S_0 = \pi R_0^2$ and which have the shape of circle, square, or rectangle are given in Table VIII and in Fig. 11 for $V_0/(2\varkappa T/M)^{1/2} = 8$. It is clear from the figure and the table that, near the body's surface, in the region where the concentration of particles is small, the influence of the shape of a body's cross section is essential; far away from the body's surface, disturbance of the concentration is determined only by the magnitude of the cross section.

Let R_0 be the effective radius of the maximum cross section of the body, $R_0 = (S_0/\pi)^{1/2}$. Depending on the value of the ratio z/R_0, one can distinguish four regions.

1) The far region: $z/R_0 > V_0/(2\varkappa T/M)^{1/2}$. Here, only the magnitude of the cross section S_0 is important. The disturbed concentrations in this region are small; they are determined by formula (2.29).

2) The intermediate region: $V_0/(2\varkappa T/M)^{1/2} > z\ R_0 > [V_0/(2\varkappa T/M)^{1/2}]^{1/2}$. Here, not only the magnitude but also the shape of a body's maximum cross section are important. The concentration disturbances in this region are already extremely large, so that the concentration of particles n is much smaller than the undisturbed concentration n_0. The disturbance of the concentration is determined by the general formula (2.23), and for bodies having specifically circular or rectangular cross sections – it is determined by Eqs. (2.24) and (2.26).

3) The region near the body: $z/R_0 < [V_0/(2\varkappa T/M)^{1/2}]^{1/2}$. Here the influence of the shape of a body's surface is appreciable, and not only the shape of the cross section.* The particle concentration in this region is small:

$$n < n_0 \exp\left[-\sqrt{\frac{MV_0^2}{2\varkappa T}}\right].$$

For a spherical surface, it is determined by formula (2.77); for a disk, by formula (2.83).

4) The region immediately adjoining the surface of a body, $(z - R_0) < R_0(\varkappa T/MV_0^2)^{1/2}$. Here both the "direct" as well as the reflected particles are essential. The particle concentration in this region is insignificantly small:

$$n \sim n_0 \exp\left[-\frac{MV_0^2}{2\varkappa T}\right].$$

§ 8. Fourier Components of the Disturbances of the Concentration of Particles for a Rapidly Moving Sphere

In order to solve the problem of the scattering of radiowaves, it is necessary to determine the disturbance of the electron-concentration in the "trail" of the body. In this connection, as we shall see below, not the concentration disturbance itself, as a function of the coordinates, but the Fourier components of the disturbance enter into the calculated formulas.

*We note that the boundary between regions 2 and 3 varies, depending on the longitudinal dimension of the body.

The electron concentration in the vicinity of a large body ($R_0 \gg D$), as indicated above in §4, coincides with the ion concentration, and the ion concentration, in the absence of a magnetic field and neglecting the influence of an electric field, is determined by the same formulas which also determine the concentration of neutral particles. In other words, in this approximation the ion and electron concentrations are described by the expressions obtained in the preceding paragraphs – it is only necessary to replace n_0 by N_0 and M by M_i, where N_0 denotes the concentration of ions in the undisturbed plasma, and M_i denotes the mass of an ion. For the calculation of the Fourier components of the electron density, it is sufficient to take the Fourier transform of $N(\mathbf{r})$. However, it is more convenient to calculate $N_{\mathbf{q}}$, solving Eq. (2.9) directly for the distribution function, which in this case one can write in the form

$$v\frac{\partial f_i}{\partial r} = \begin{cases} -\frac{rv}{r} f_i(v)\,\delta(r-R_0) & \text{for} \quad rv < 0, \\ 0 & \text{for} \quad rv > 0. \end{cases} \tag{2.84}$$

Here it is assumed that the ions are completely neutralized upon collision with the surface (see Eq. (1.22)). On the right side of Eq. (2.84), one can replace $f_i(\mathbf{v})$ by

$$f_{0i} = \left(\frac{M_i}{2\pi\varkappa T}\right)^{3/2} N_0 \exp\left[-\frac{M_i(v+V_0)^2}{2\varkappa T}\right],$$

since the presence of the body obviously has no influence on the incoming ions. In addition, for a large velocity of the sphere, as we saw above in §§5 and 7, one can replace $\mathbf{v}$ by $-\mathbf{V}_0$ on the right side of the equation; hence, in place of (2.84) we have

$$v\frac{\partial f_i}{\partial r} = \begin{cases} -\frac{r}{r} V_0 f_{0i}(v)\,\delta(r-R_0), & rV_0 < 0, \\ 0, & rV_0 > 0. \end{cases} \tag{2.85}$$

In order to calculate the Fourier components of the distribution function f_i, we multiply (2.85) by $e^{i\mathbf{qr}}$ and integrate with respect to d^3r. We obtain

$$iqvf_q = -\int \frac{r}{r} V_0 f_{0i}(v)\,\delta(r-R_0)\,e^{-iqr}\,d^3r, \tag{2.86}$$

where

$$f_q = \int f_i'(r)\,e^{-iqr}\,d^3r,$$
$$f_i' = f_i - f_{0i}.$$

We mention here that as we shall see in Chapter VII, in the problem of the scattering of radiowaves by a body's "trail," the vector $\mathbf{q}$ turns out to equal the change of the radiowave's wave vector upon scattering, i.e., $\mathbf{q} = \mathbf{k}' - \mathbf{k}$, where $\mathbf{k}$ is the wave vector of the incident wave, and $\mathbf{k}'$ is the wave vector of the scattered wave. Now if we are interested in the disturbance of the distribution function $f'_i(\mathbf{r})$ in coordinate space, then, knowing f_q, it is, of course, possible to find it according to the formula for the inverse Fourier transform

$$f'(r) = \frac{1}{(2\pi)^3}\int f_q e^{iqr}\,d^3q.$$

Here now $\mathbf{q}$ runs over all possible values. It is convenient to take the integral to the right in Eq. (2.86) in spherical coordinates with the axis along $\mathbf{V}_0$. Let χ denote the angle between $\mathbf{q}$ and $\mathbf{V}_0$. Then

$$qr = qr(\cos\vartheta\cos\chi + \sin\vartheta\sin\chi\cos\varphi)$$

and

$$f_q = i\frac{R_0^2 V_0 f_{0i}}{qv}\int_0^{\pi/2}\cos\vartheta\sin\vartheta\, d\vartheta\int_0^{2\pi} d\varphi\exp[iqR_0(\cos\vartheta\cos\chi +$$

$$+\sin\vartheta\sin\chi\cos\varphi)] = i\frac{V_0 f_{0i}}{qv}\pi R_0^2 2\int_0^{\pi/2}\cos\vartheta\sin\vartheta\times$$

$$\times\exp[iqR_0\cos\vartheta\cos\chi]\, J_0(qR_0\sin\vartheta\sin\chi)\, d\vartheta = i\frac{V_0 f_{0i}}{qv}\pi R_0^2 G(qR_0\chi), \quad (2.87)$$

where

$$G(qR_0, \chi) = 2\int_0^{\pi/2}\cos\vartheta\sin\vartheta\exp[iqR_0\cos\vartheta\cos\chi]\quad J_0(qR_0\sin\vartheta\sin\chi)\, d\vartheta, \quad (2.88)$$

and J_0 is the Bessel function of real argument.

From Eq. (2.87) we find $N_{\mathbf{q}}$, the Fourier component of the ion or electron concentration:

$$N_q = \int \delta N(r)e^{-iqr}d^3r = \int f_q d^3v = iV_0\pi R_0 G\int\frac{f_{0i}}{qv}d^3v.$$

In order to evaluate the integral on the right side of Eq. (2.89), we change from an integration with respect to d^3v to an integration with respect to d^3u, where $\mathbf{u} = \mathbf{v} + \mathbf{V}_0$. Then

$$\int\frac{f_{0i}}{qv}d^3v = \int\frac{f_{0i}\, d^3u}{q(u - V_0)} = N_0\left(\frac{M_i}{2\pi\varkappa T}\right)^{3/2}\int\frac{\exp\left[-\frac{M_i}{2\varkappa T}(u_x^2 + u_y^2 + u_z^2)\right]}{qu_x - qV_0}du_x\, du_y\, du_z =$$

$$= N_0\left(\frac{M_i}{2\pi\varkappa T}\right)^{1/2}\int_{-\infty}^{\infty}\frac{\exp\left[-\frac{M_i u_x^2}{2\varkappa T}\right]}{qu_x - qV_0}du_x. \quad (2.90)$$

In (2.90), the x axis is directed along $\mathbf{q}$.

The integral (2.90) is an improper integral, since the integrand has a pole at the point

$$u_x = \frac{qV_0}{q}.$$

Therefore, in order for expression (2.90) to have meaning, it is necessary to indicate the method of evaluating the integral. This method is chosen, starting from physical considerations. It is obvious that the theory we are considering, in which collisions are neglected, is a limiting case of the theory which takes collisions into account. In Chapter V we shall see that, when taking account of collisions, the quantity $\mathbf{qV}_0$ enters into all formulas in the combination $\mathbf{qV}_0 + i\nu$, where ν is the effective number of collisions [see formulas (5.6) and (5.12)]. Here all integrals are convergent. Then, in order to go over to the case of no collisions, it is necessary to let ν go to zero. After this, the answer will no longer depend on ν. The arguments cited indicate that in order to "take" (evaluate) the integral correctly, we must replace $\mathbf{qV}_0$ in it by $\mathbf{qV}_0 + i\delta$, where $\delta \to +0$ is an infinitesimal positive quantity. In other words, it is necessary to evaluate the improper integral (2.90) by going around the pole with respect to u_x in the complex plane from below. This rule for going around poles was established by Landau [22]. It is possible to carry out further transformations of expression (2.90) with the aid of the formula

$$\int_{-\infty}^{\infty} \frac{e^{-u^2}}{u-a-i\delta} du = i\sqrt{\pi}\, Q(a), \tag{2.91}$$

where

$$Q(a) = \left(\sqrt{\pi} + 2i \int_0^a e^{x^2} dx\right) e^{-a^2}. \tag{2.92}$$

Then

$$N_q = -N_0 \frac{\pi R_0^2}{q} G(qR_0, \chi) \left(\frac{M_i V_0^2}{2\varkappa T}\right)^{1/2} Q(a), \tag{2.93}$$

where

$$a = \frac{(qV_0)}{q\sqrt{2\varkappa T/M_i}} = \frac{q_z V_0}{q\sqrt{2\varkappa T/M_i}}.$$

For $qR_0 = 0$, $G(0, \chi) = 1$ and consequently

$$N_q = \frac{N_0 \pi R_0^2}{q} \left(\frac{M_i V_0^2}{2\varkappa T}\right)^{1/2} Q(a). \tag{2.94}$$

This expression for $N_{\mathbf{q}}$ is the Fourier transform of the concentration disturbance $\delta N(\mathbf{r})$ [see formula (2.29)]. In fact, using expression (2.29) for the disturbance of the concentration of particles, we have

$$N_q = \int \delta N(r) e^{-iqr} d^3r = -\frac{N_0 R_0^2 M_i V_0^2}{2\varkappa T} \int_{z<0} \frac{1}{z^2} \exp\left\{-i(q_x x + q_y y + q_z z) - \frac{M_i V_0^2}{2\varkappa T} \frac{x^2+y^2}{z^2}\right\} dx\, dy\, dz.$$

Here it is assumed that $S_0 = \pi R_0^2$. Having used the formula

$$\int_{-\infty}^{\infty} e^{-d_0 x^2 - i c_0 x} dx = \sqrt{\frac{\pi}{d_0}}\, e^{-\frac{c_0^2}{4d_0}},$$

after integration with respect to x and y, we find

$$N_q = -N_0 \pi R_0^2 \int_{-\infty}^{0} dz \exp\left\{-\frac{(q_x^2+q_y^2)\varkappa T}{2M_i V_0^2} z^2 - i q_z z\right\} = -\frac{N_0 \pi R_0^2}{\sqrt{q_x^2+q_y^2}} \left(\frac{M_i V_0^2}{2\varkappa T}\right)^{1/2} Q(a_1), \tag{2.95}$$

where $a_1^2 = M_i V_0^2 q_z^2 / 2\varkappa T\,(q_x^2 + q_y^2)$. We see that Eq. (2.95) coincides with Eq. (2.94), if the fact that formula (2.29) is valid only at large distances behind the body, when

$$z^2 \gg x^2 + y^2, \text{ i.e., } q_z^2 \ll q_x^2 + q_y^2.$$

is taken into consideration. In this case, when $\mathbf{q} \perp \mathbf{V}_0$ ($\chi = \pi/2$),

$$G\left(qR_0, \frac{\pi}{2}\right) = 2\int_0^{\pi/2} \cos\vartheta \sin\vartheta J_0(qR_0 \sin\vartheta)\, d\vartheta = \int_0^1 x J_0(qR_0 x)\, dx = 2\frac{J_1(qR_0)}{qR_0}. \tag{2.96}$$

For $\mathbf{q} \parallel \mathbf{V}_0$ ($\chi = 0$)

$$G(qR_0, 0) = 2\left[\frac{e^{iqR_0} - 1}{(qR_0)^2} - i\frac{e^{iqR_0}}{qR_0}\right]. \tag{2.97}$$

CHAPTER III

INFLUENCE OF A CONSTANT MAGNETIC FIELD ON THE DISTURBANCE OF THE ION AND ELECTRON CONCENTRATIONS

§ 9. Distribution Function of the Ions

Since a constant magnetic field $\mathbf{H}_0$ strongly influences the motion of ions in a plasma, it is natural that the presence of $\mathbf{H}_0$ has a significant effect on the distribution of ions in the region "shaded" by the body.* In this connection, one can neglect the collisions of particles with the body's surface in the rarefied region in the case of a rapidly moving body, considered in this chapter, as was indicated above in §§5 and 7, and one can replace the body by its transverse cross section in a plane perpendicular to its direction of motion. Starting from this equation for the ions' distribution function (1.23) in the rarefied region, i.e., for $z < 0$, in the presence of a constant magnetic field one can write (it) in the form

$$v \frac{\partial f}{\partial r} + \frac{e}{M_i c} [\boldsymbol{H}_0, v + V_0] \frac{\partial f}{\partial v} = 0, \tag{3.1}$$

if the influence of an electric field on the motion of the ions is ignored. The boundary condition for Eq. (3.1) for $z = 0$ has the form (2.10)

$$f(x_0, y_0, 0; v_x, v_y, v_z) = \begin{cases} N_0 \left(\frac{M_i}{2\pi\varkappa T} \right)^{3/2} \exp\left[-\frac{M_i (v + V_0)^2}{2\varkappa T} \right], \\ \text{if } v_z < 0 \text{ and the point } (x_0, y_0) \text{ is located outside } S_0; \\ 0, \\ \text{if } v_z > 0 \text{ or the point } (x_0, y_0) \text{ is located inside } S_0, \end{cases} \tag{3.2}$$

where S_0 is the transverse cross section of the body.

We shall solve the linear equation (3.1) by the method of characteristics. The equations of the characteristics are, naturally, the equations of motion of an ion; they have the form

$$\frac{dr}{dt} = v, \quad \frac{dv}{dt} = \frac{e}{M_i c} [\boldsymbol{H}_0, v + V_0],$$

where the time t is a free parameter.

We choose the z axis, as usual, along the direction of the body's motion, the y axis perpendicular to the $(\mathbf{H}_0, \mathbf{V}_0)$ plane, and the x axis in the $(\mathbf{H}_0, \mathbf{V}_0)$, plane perpendiular to the direction of $\mathbf{V}_0$. In this coordinate system, the equations of the characteristics can be written in the form

*At distances $r \lessdot 2\pi V_0/\Omega$ from the body in the near zone, the influence of a magnetic field is unimportant, as was already noted (see §4). In the ionosphere, this distance r is extremely large: thus, for oxygen and nitrogen ions it exceeds 200 m, for hydrogen it is of the order of 20 m. Therefore, the influence of a magnetic field on the concentration of ions in the region of "compression," whose dimensions are comparatively small (see §6), is for simplicity not considered here.

$$\left.\begin{aligned}&\frac{dx}{dt}=v_x,\quad \frac{dy}{dt}=v_y,\quad \frac{dz}{dt}=v_z,\\&\frac{dv_x}{dt}=-\Omega v_y\cos\alpha,\quad \frac{dv_y}{dt}=\Omega v_x\cos\alpha-\Omega(v_z+V_0)\sin\alpha,\quad \frac{dv_z}{dt}=\Omega v_y\sin\alpha.\end{aligned}\right\} \tag{3.3}$$

Here α is the angle between the direction $\mathbf{V}_0$ of the body's motion and the direction of the magnetic field $\mathbf{H}_0$, $\Omega = eH_0/M_ic$ is the gyromagnetic frequency.

The solution of the characteristic system of equations (3.3) has the form

$$\left.\begin{aligned}&x=x_0+[v_{x0}\sin\alpha-(v_{z0}+V_0)\cos\alpha]\,t\sin\alpha-\frac{u_\perp}{\Omega}\cos\alpha\,[\sin(\Omega t-\varphi_0)+\sin\varphi_0],\\&y=y_0+\frac{u_\perp}{\Omega}[\cos(\Omega t-\varphi_0)-\cos\varphi_0],\\&z=z_0-V_0t+[v_{x0}\sin\alpha-(v_{z0}+V_0)\cos\alpha]\,t\cos\alpha+\frac{u_\perp}{\Omega}\sin\alpha\,[\sin(\Omega t-\varphi_0)+\sin\varphi_0],\\&v_x=v_{x0}\sin^2\alpha-(v_{z0}+V_0)\cos\alpha\sin\alpha-u_\perp\cos\alpha\cos(\Omega t-\varphi_0),\\&v_y=-u_\perp\sin(\Omega t-\varphi_0),\\&v_z=-V_0+v_{x0}\sin\alpha\cos\alpha-(v_{z0}+V_0)\cos^2\alpha+u_\perp\sin\alpha\cos(\Omega t-\varphi_0).\end{aligned}\right\} \tag{3.4}$$

In (3.4) x_0, y_0, and z_0 are the initial coordinates; v_{x0}, v_{y0}, and v_{z0} are the initial velocities; $u_\perp = \{v_{y0}^2 + [v_{x0}\cos\alpha + (v_{z0} + V_0)\sin\alpha]^2\}^{1/2}$ is the projection of the velocity $\mathbf{u} = \mathbf{v}_{i0} + \mathbf{V}_0$ on a plane perpendiular to $\mathbf{H}_0$; and $\sin\varphi_0 = v_{y0}/u_\perp$.

Knowing the equations of the characteristics and the form of the distribution function on the boundary of the rarefied region (given by Eq. (3.2)), it is not difficult to determine it over the entire region of rarefaction, i.e., for arbitrary values of $z < 0$. For this, it is only necessary to express the initial values of the coordinates and velocities in terms of their values at the moment t. In this connection, we shall consider that the ions are freely revolving in a plane perpendicular to $\mathbf{H}_0$. Therefore, in the system of coordinates connected with $\mathbf{H}_0$, the total energy of an ion, and also the projection of its velocity on the plane orthogonal to $\mathbf{H}_0$, are constants of the motion, i.e.,

$$(v_z+V_0)^2+v_x^2+v_y^2=(v_{z0}+V_0)^2+v_{x0}^2+v_{y0}^2, \tag{3.5a}$$

$$u_\perp^2=v_y^2+[v_x\cos\alpha+(v_z+V_0)\sin\alpha]^2=v_{y0}^2+[v_{x0}\cos\alpha+(v_{z0}+V_0)\sin\alpha]^2. \tag{3.5b}$$

Considering relations (3.5) and the fact that during the revolution of an ion one always has

$$\sin\varphi=\frac{v_y}{u_\perp}, \tag{3.6}$$

where $\varphi = \varphi_0 - \Omega t$, from (3.4) we find that

$$\left.\begin{aligned}&x_0=x-[u_\perp\cos\alpha\cos\varphi+v_x]\,t+\frac{u_\perp}{\Omega}\cos\alpha\,[\sin(\Omega t+\varphi)-\sin\varphi],\\&y_0=y+\frac{u_\perp}{\Omega}[\cos(\varphi+\Omega t)-\cos\varphi],\\&z_0=z-v_zt+(u_\perp\sin\alpha\cos\varphi)\,t-\frac{u_\perp\sin\alpha}{\Omega}[\sin(\Omega t+\varphi)-\sin\varphi].\end{aligned}\right\} \tag{3.7}$$

Now it is possible to write down the solution of Eq. (3.1) which satisfies the boundary condition (3.2):

$$\left.\begin{aligned} f(x, y, z; v_x, v_y, v_z) = f\Big\{x - (u_\perp \cos\alpha\cos\varphi + v_x)\,t + \frac{u_\perp}{\Omega}\cos\alpha\,[\sin(\Omega t + \varphi) - \sin\varphi], \\ y + \frac{u_\perp}{\Omega}[\cos(\Omega t + \varphi) - \cos\varphi],\ 0;\ v_x^2 + v_y^2 + (v_z + V_0)^2\Big\}, \end{aligned}\right\} \qquad (3.8)$$

where the function on the right side of (3.8) is defined by formula (3.2). The parameter t is determined from the relation $z_0 = 0$, i.e., from the equation

$$z - v_z t + (u_\perp \sin\alpha\cos\varphi)\,t - \frac{u_\perp \sin\alpha}{\Omega}[\sin(\Omega t + \varphi) - \sin\varphi] = 0. \qquad (3.9)$$

In formulas (3.7)-(3.9) the velocity $u_\perp$ and the angle φ are expressed in terms of the velocity components v_x, v_y, and v_z with the aid of formulas (3.5b) and (3.6).

In the absence of a magnetic field, the distribution function (3.8) coincides, of course, with (2.14). Integrating the obtained expression for the distribution function with respect to the velocities, one can determine the ion concentration in the "shaded" zone. This concentration, of course, depends essentially on the angle α between the direction of the magnetic field and the direction of the body's motion.

§ 10. Motion of Body Along the Magnetic Field

We consider first the simple case when the body moves along the magnetic field $\mathbf{H}_0$, i.e., $\alpha = 0$. Here we shall assume that the body's cross section is a circle of radius R_0. Expression (3.8) for the distribution function then assumes the form

$$f(x, y, z; v_x, v_y, v_z) = \begin{cases} N_0\left(\dfrac{M_i}{2\pi\varkappa T}\right)^{3/2}\exp\left[-\dfrac{M_i(v+V_0)^2}{2\varkappa T}\right] \\ \text{for } \left\{x + \dfrac{u_\perp}{\Omega}\left[\sin\left(\Omega\dfrac{z}{v_z} + \varphi\right) - \sin\varphi\right]\right\}^2 + \\ \quad + \left\{y + \dfrac{u_\perp}{\Omega}\left[\cos\left(\Omega\dfrac{z}{v_z} + \varphi\right) - \cos\varphi\right]\right\}^2 \geqslant R_0^2, \\ 0 \\ \text{for } \left\{x + \dfrac{u_\perp}{\Omega}\left[\sin\left(\Omega\dfrac{z}{v_z} + \varphi\right) - \sin\varphi\right]\right\}^2 + \\ \quad + \left\{y + \dfrac{u_\perp}{\Omega}\left[\cos\left(\Omega\dfrac{z}{v_z} + \varphi\right) - \cos\varphi\right]\right\}^2 < R_0^2. \end{cases} \qquad (3.10)$$

Here it is considered that $t = z/v_z$ for $\alpha = 0$ and

$$u_\perp = v_\perp = \sqrt{v_x^2 + v_y^2}. \qquad (3.11)$$

In order to evaluate the ion concentration at the point $\mathbf{r}$, it is necessary to integrate the distribution function (3.10) with respect to the velocities dv_x, dv_y, and dv_z. For the integration with respect to dv_x and dv_y, it is convenient to change over to new variables u and φ_1, which are defined by the relations

$$\left.\begin{aligned} u_\perp\cos\left(\frac{\Omega z}{2v_z} + \varphi\right) &= \frac{u\cos\varphi_1 - \varrho\Omega}{2\sin\left(\dfrac{\Omega z}{2v_z}\right)}, \\ u_\perp\sin\left(\frac{\Omega z}{2v_z} + \varphi\right) &= \frac{u\sin\varphi_1}{2\sin\left(\dfrac{\Omega z}{2v_z}\right)}, \end{aligned}\right\} \qquad (3.12)$$

where, as usual, $\rho = (x^2 + y^2)^{1/2}$. Since $dv_x dv_y = u\,du\,d\varphi_1/4\sin^2(\Omega z/2v_z)$, then, introducing the variable $v' = v/(2\varkappa T/M_i)^{1/2}$, we obtain $[V'_0 = V_0/(2\varkappa T/M_i)^{1/2}]$.

$$N_i(\varrho, z) = \frac{N_0}{\sqrt{\pi}} \int_{-\infty}^{0} dv'_z \exp[-(V'_0 + v'_z)^2] \frac{1}{\pi} \int_{0}^{2\pi} d\varphi_1 \times$$

$$\times \int_{\frac{R_0}{2\rho_H \sin \frac{\Omega z}{2v_z}}}^{\infty} \frac{du'u'}{4\sin^2\left(\frac{\Omega z}{2v_z}\right)} \exp\left[-\frac{\left(u'\cos\varphi_1 - \frac{\varrho}{\varrho_H}\right)^2 + u'^2 \sin^2\varphi_1}{\left(2\sin\frac{\Omega z}{2v_z}\right)^2}\right],$$

or after integration with respect to φ_1:

$$N_i(\varrho, z) = \frac{N_0}{\sqrt{\pi}} \int_{-\infty}^{0} dv'_z \exp[-(V'_0 + v'_z)^2] \times$$

$$\times \left\{2\exp\left[-\frac{\varrho^2}{\left(2\varrho_H \sin\left(\frac{\Omega z}{2v_z}\right)\right)^2}\right] \int_{\frac{R_0}{2\rho_H \sin\left(\frac{\Omega z}{2v_z}\right)}}^{\infty} u' e^{-u'^2} I_0\left(\frac{\varrho u'}{\varrho_H \sin\left(\frac{\Omega z}{2v_z}\right)}\right) du'\right\}. \tag{3.13}$$

Here $\rho_H = (2\varkappa T/M_i)^{1/2}/\Omega = c(2\varkappa T M_i)^{1/2}/eH_0$ has the meaning of the average Larmor radius of an ion, I_0 is the zero order Bessel function of imaginary argument. Considering then that $V_0 \gg (\varkappa T/M_i)^{1/2}$, i.e., $V_0'^2 \gg 1$, for not very large values of $|z| \ll 2\pi V_0^2/\Omega\,(\varkappa T/M_i)^{1/2}$, one can replace v_z by $-V_0$ everywhere inside the curly brackets. Now integrating with respect to dv'_z, we obtain

$$N_i(\varrho, z) = 2N_0 \exp\left[-\frac{\varrho^2}{\left[2\varrho_H \sin\left(\frac{\Omega z}{2V_0}\right)\right]^2}\right] \int_{\frac{-R_0}{2\rho_H \sin\left(\frac{\Omega z}{2V_0}\right)}}^{\infty} u e^{-u^2} I_0\left(\frac{-\varrho u}{\varrho_H \sin\frac{\Omega z}{2V_0}}\right) du. \tag{3.14}$$

At small distances from the body, for $-z \ll V_0/\Omega$,

$$\sin\frac{\Omega z}{2V_0} \approx \frac{\Omega z}{2V_0},$$

and formula (3.14) coincides with (2.24). This confirms the assumption made above that the influence of the magnetic field is insignificant in the near zone (when $-z \ll V_0/\Omega$). Conversely, the influence of a magnetic field is very large when $-z \gtrsim V_0/\Omega$. It is obvious from Eq. (3.14) that $N_i(\rho, z)$ is a periodic function of z with period equal to $2\pi V_0/\Omega$.

The character of the variation of the ion concentration in the "shaded" zone depends essentially on the ratio between the radius R_0 of the body's cross section and the average Larmor radius ρ_H for the ions. If the dimensions of the body are very large, i.e., $R_0 \gg \rho_H$, then the disturbed zone is a semi-infinite cylinder of radius R_0, inside of which the ion concentration vanishes – only near the boundaries of the cylinder does the ion concentration vary somewhat over distances on the order of ρ_H.

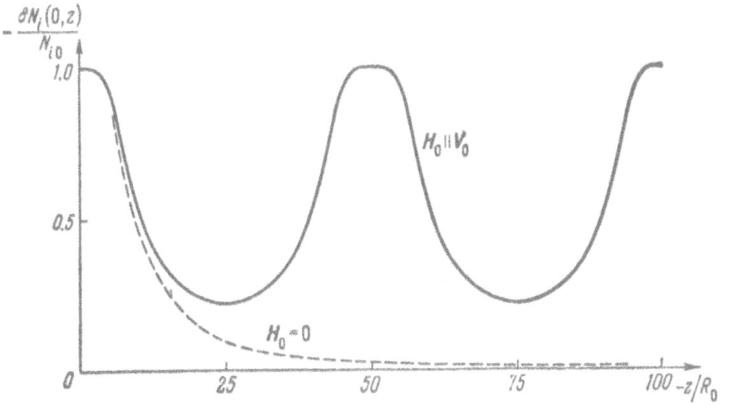

Fig. 12. Ion concentration disturbance $\delta N_i/N_0$ along the z axis in the rarefied region for the case when a spherical body moves parallel to a magnetic field: $V_0 \parallel \mathbf{H}_0$, $\rho_H/R_0 = 1$, $V_0/(2\varkappa T/M)^{1/2} = 8$. Dashed curve—the same for $\mathbf{H}_0 = 0$.

For bodies of small dimensions, when $R_0 < \rho_H$, the ion concentration in the "shaded" zone, on the contrary, varies very strongly over the length of the period $2\pi V_0/\Omega$.

On the z axis (for $\rho = 0$) expression (3.14) assumes a particularly simple form:

$$N_i(0, z) = N_0 \exp\left[-\frac{R_0^2}{4\varrho_H^2 \sin^2\left(\frac{\Omega z}{2V_0}\right)}\right]. \tag{3.15}$$

For the case, $\mathbf{V}_0 \parallel \mathbf{H}_0$, considered here the character of the variation of the ion concentration disturbance, $-\delta N_i/N_0 = [N_0 - N_i(\rho, z)]/N_0$, along the z azis for $\rho = 0$ is evident from Figs. 12 and 13, where curves for different values of $N_i(\rho, z)/N_0$ are shown. The corresponding curves are calculated for $R_0 = \rho_H$ and $(M_i V_0^2/2\varkappa T)^{1/2} = 8$. In this case, the period of variation of $\delta N_i/N_0$, expressed in (units of) the sphere's radius R_0 or ρ_H, is equal to 50.24.

It should be noted that a strictly periodic structure of the rarefied region behind a moving body was obtained here because of the fact that, in approximately integrating Eq. (3.13) with respect to dv'_z, we neglected the thermal motion of the ions along the z axis. With account of the thermal motion, the periodic structure is smeared out. It is strongly manifested only up to distances $|z|$ on the order of $2\pi V_0^2/\Omega(\varkappa T/M_i)^{1/2}$.

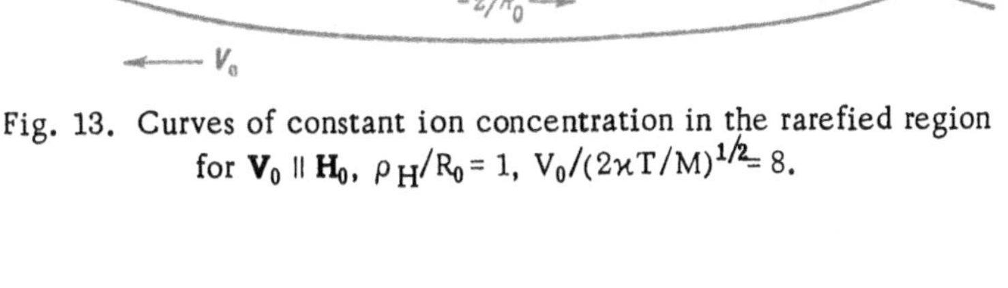

Fig. 13. Curves of constant ion concentration in the rarefied region for $\mathbf{V}_0 \parallel \mathbf{H}_0$, $\rho_H/R_0 = 1$, $V_0/(2\varkappa T/M)^{1/2} = 8$.

§11. Motion of Body Transverse to the Magnetic Field

Now let us assume that the body moves in a direction perpendicular to $\mathbf{H}_0$, i.e., $\alpha = \pi/2$. In this case, the problem has no axial symmetry, and the expression for the ion density in the case when the body's cross section is circular has a very complicated form. Simpler and more intuitive formulas are obtained if the body's cross section, in a plane perpendicular to the direction of motion, is rectangular with dimensions $2R_x$, $2R_y$.

In this case, expression (3.8) for the distribution function assumes the form

$$f(x, y, z; v_x, v_y, v_z) = \begin{cases} N_0\left(\frac{M_i}{2\pi\varkappa T}\right)^{3/2}\exp\left[-\frac{M_i(v+V_0)^2}{2\varkappa T}\right], \\ \quad \text{if} \quad R_x \geqslant |x - v_x t| \\ \quad \text{and} \quad R_y \geqslant \left| y - \frac{u_\perp}{\Omega}[\cos(\Omega t + \varphi) - \cos\varphi]\right|; \\ 0, \\ \quad \text{if} \quad R_x < |x - v_x t| \\ \quad \text{and} \quad R_y < \left| y - \frac{u_\perp}{\Omega}[\cos(\Omega t + \varphi) - \cos\varphi]\right|, \end{cases} \tag{3.16}$$

where the parameter t is determined by relation (3.9):

$$z - v_z t + u_\perp \cos\varphi t - \frac{u_\perp}{\Omega}[\sin(\Omega t + \varphi) - \sin\varphi] = 0, \tag{3.17}$$

where

$$u_\perp = \sqrt{v_y^2 + (v_z + V_0)^2}, \quad \sin\varphi = \frac{v_y}{u_\perp}.$$

The desired (sought for) ion density in the "shaded" zone is determined after integrating the distribution function (3.16) with respect to velocities. We shall, as is customary, investigate the disturbance of the particle concentration:

$$-\delta N_i(x, y, z) = N_0 - N_i(x, y, z), \tag{3.18}$$

where

$$\delta N_i(x, y, z) = \int \left\{ f(x, y, z; v_x, v_y, v_z) - N_0\left(\frac{M_i}{2\pi\varkappa T}\right)^{3/2}\exp\left[-\frac{M_i(v+V_0)^2}{2\varkappa T}\right]\right\} dv_x\, dv_y\, dv_z. \tag{3.19}$$

The integration with respect to dv_x is carried out in the same way as in the absence of a magnetic field (see (2.26)). Taking this into account, we have

$$-\delta N_i(x, y, z) = \frac{N_0}{2}\int dv_y\, dv_z \frac{M_i}{2\pi\varkappa T}\exp\left|-\frac{M_i[v_y^2 + (v_z + V_0)^2]}{2\varkappa T}\right| \left|\Phi\left(\frac{x - R_x}{t}\sqrt{\frac{M_i}{2\varkappa T}}\right) - \Phi\left(\frac{x + R_x}{t}\sqrt{\frac{M_i}{2\varkappa T}}\right)\right| \times$$

$$\times \begin{cases} 1 \ \text{for} \ R_y \leqslant \left| y - \frac{u_\perp}{\Omega}[\cos(\Omega t + \varphi) - \cos\varphi]\right|, \\ 0 \ \text{for} \ R_y > \left| y - \frac{u_\perp}{\Omega}[\cos(\Omega t + \varphi) - \cos\varphi]\right|. \end{cases} \tag{3.20}$$

Here, as before, the parameter t is determined by relation (3.17). Representing it in the form

$$t = -\frac{z}{V_0} + \frac{v_z + V_0 - u_\perp \cos\varphi}{V_0}t + \frac{u_\perp}{V_0}\frac{\sin(\Omega t + \varphi) - \sin\varphi}{\Omega}$$

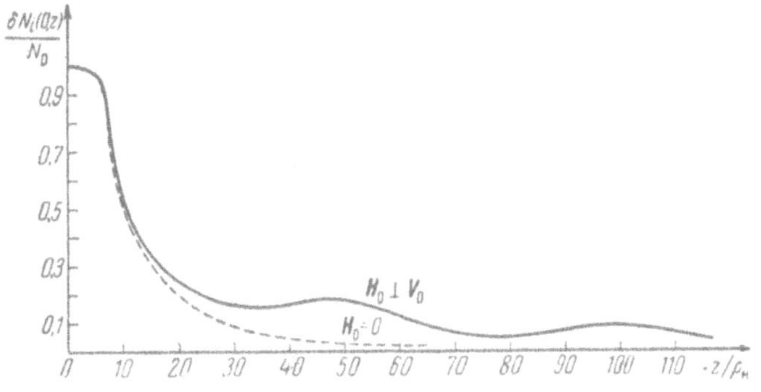

Fig. 14. Disturbances on the ion concentration $\delta N_i/N_0$ on the z axis in the rarefied region for a body of square cross section; $\mathbf{V}_0 \perp \mathbf{H}_0$, $R_x = R_y = \rho_H$. Dash curve—the same for $\mathbf{H}_0 = 0$.

and considering that $v_z + V_0 \sim (\varkappa T/M_i)^{1/2}$ and $u_\perp \sim (\varkappa T/M_i)^{1/2}$ but $V_0 \gg (\varkappa T/M_i)^{1/2}$, we find an expression for t by the method of successive approximations:

$$t = -\frac{z}{V_0} + \left[\frac{u_\perp}{V_0}\frac{-\sin\varphi + \sin\left(-\frac{\Omega z}{V_0} + \varphi\right)}{\Omega} - \frac{z}{V_0}\frac{v_z + V_0 - u_\perp\cos\varphi}{V_0}\right] + \ldots, \tag{3.21}$$

i.e., actually in the form of a series in powers of $(\varkappa T/M_i V_0^2)^{1/2}$. Below, for an evaluation of the ion concentration, we shall use the first approximation $t = -z/V_0$ which is equivalent to neglecting the thermal motion of the particles in the z direction.

It is more convenient to carry out further integration in (3.20) in the new variables (u, φ), where

$$u = \sqrt{\frac{M_i}{2\varkappa T}}u_\perp = \sqrt{\frac{M_i}{2\varkappa T}[v_y^2 + (v_z + V_0)^2]}. \tag{3.22}$$

Then we obtain

$$-\delta N_i(x, y, z) = \frac{N_0}{2}\left|\Phi\left(\frac{x - R_x}{z}\sqrt{\frac{M_i V_0^2}{2\varkappa T}}\right) - \Phi\left(\frac{x + R_x}{z}\sqrt{\frac{M_i V_0^2}{2\varkappa T}}\right)\right| \times$$

$$\times \frac{1}{\pi}\int_0^{2\pi}\int_0^{\infty} ue^{-u^2}du \begin{cases} 1, & R_y \geqslant |y - \varrho_H u[\cos(\Omega t + \varphi) - \cos\varphi]|, \\ 0, & R_y < |y - \varrho_H u[\cos(\Omega t + \varphi) - \cos\varphi]| \end{cases} =$$

$$= \frac{N_0}{2}\left|\Phi\left(\frac{x - R_x}{z}\sqrt{\frac{M_i V_0^2}{2\varkappa T}}\right) - \Phi\left(\frac{x + R_x}{z}\sqrt{\frac{M_i V_0^2}{2\varkappa T}}\right)\right| \frac{1}{\pi}\int_0^{2\pi} d\varphi\,[e^{-u_-^2(\varphi)} - e^{-u_+^2(\varphi)}], \tag{3.23}$$

where, as before, $\rho_H = c(2M_i\varkappa T)^{1/2}/eH$, and u_- and u_+ are the boundary values for u, determined by the conditions:

$$\left.\begin{aligned} -R_y \leqslant y - \varrho_H u\left[\cos\left(-\Omega\frac{z}{V_0} + \varphi\right) - \cos\varphi\right] \leqslant R_y, \\ u \geqslant 0. \end{aligned}\right\} \tag{3.24}$$

Considering that $[-\Omega(z/V_0) + \varphi] - \cos\varphi = 2\sin(-\Omega z/2V_0) \times \sin[-\Omega z/2V_0) + \varphi]$, we find that $u_\pm^2(\varphi) \sim 1/\sin^2[-\Omega z/2V_0) + \varphi]$. And so, in the last integral it is natural to change over to a new variable $x = \sin^{-2}[-\Omega z/2V_0 + \varphi]$. Then $d\varphi = -1/2x(x-1)^{1/2}$ and if the equality

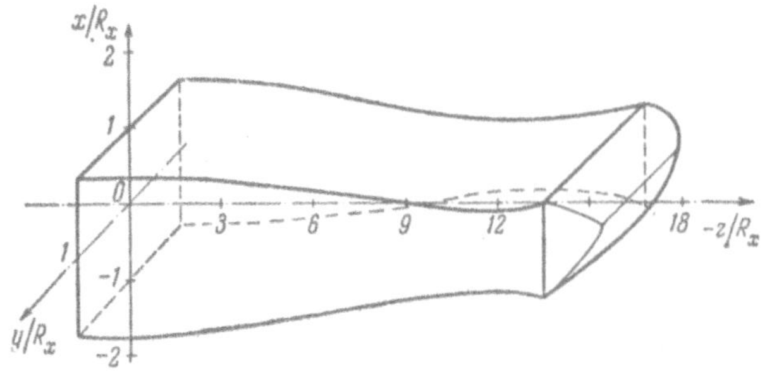

Fig. 15. General shape of the surface of constant ion concentration, $N_i\ (x, y, z)/N_0 = 0.5$, in the rarefied region for a body of square cross section; $\rho_H = 0.3R_x$, $\mathbf{V}_0 \perp \mathbf{H}_0$, $\mathbf{H}_0 \parallel OX$.

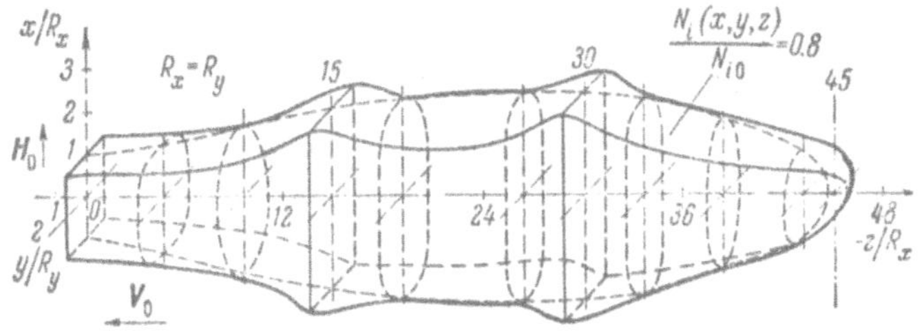

Fig. 16. The general shape of the surface of constant ion concentration, $N_i\ (x, y, z)/N_0 = 0.8$, in the rarefied region for a body of square cross section; $\rho_H = 0.3R_x$.

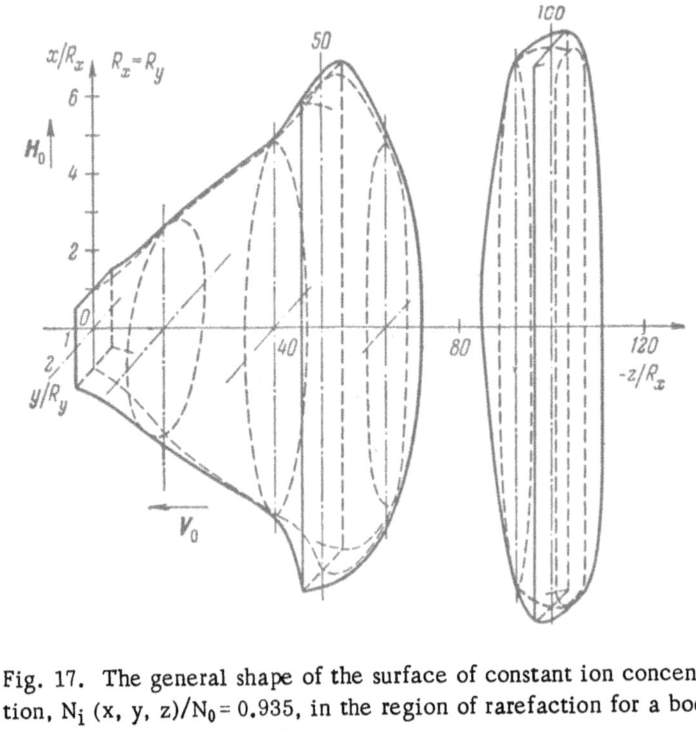

Fig. 17. The general shape of the surface of constant ion concentration, $N_i\ (x, y, z)/N_0 = 0.935$, in the region of rarefaction for a body of square cross section; $\rho_H = R_x$.

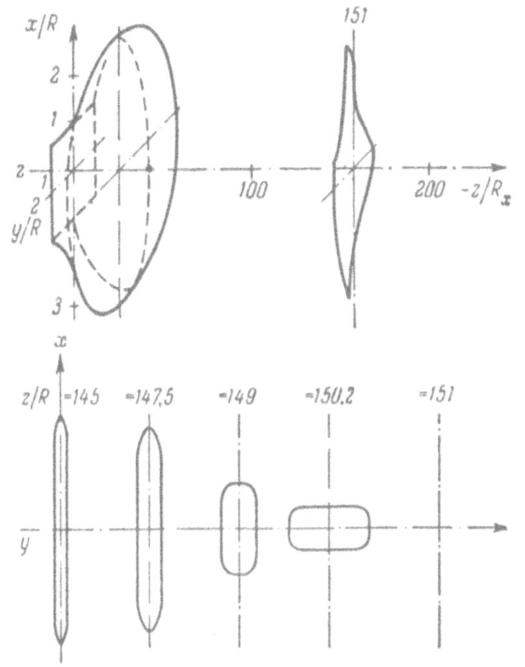

Fig. 18. General shape of the surface of constant ion concentration, $N_i(x, y, z)/N_0 = 0.940$, in the rarefied region for a body with square cross section: $\rho_H = 3R_x$. At the bottom are shown cross sections in the xy plane of this same surface for the various values of z/R_x indicated in the figure.

$$\frac{1}{2\pi}\int_1^\infty \frac{dx e^{-bx}}{x\sqrt{x-1}} = \frac{1}{2}[1-\Phi(|\sqrt{b}|)], \tag{3.25}$$

is taken into account, we obtain

$$\frac{1}{\pi}\int_0^{2\pi} d\varphi \, [e^{-u_-^2(\varphi)} - e^{-u_+^2(\varphi)}] = \frac{1}{2}\left|\Phi\left(\frac{y-R_y}{2\varrho_H \sin\left(\frac{\Omega z}{2V_0}\right)}\right) - \Phi\left(\frac{y+R_y}{2\varrho_H \sin\left(\frac{\Omega z}{2V_0}\right)}\right)\right|. \tag{3.26}$$

Thus, the final expression (3.23) for the ion concentration in the "shaded" zone takes the form

$$-\delta N_i(x, y, z) = N_0 - N_i(x, y, z) =$$

$$\frac{N_0}{4}\left|\Phi\left(\frac{x-R_x}{z}\sqrt{\frac{M_i V_0^2}{2\varkappa T}}\right) - \Phi\left(\frac{x+R_x}{z}\sqrt{\frac{M_i V_0^2}{2\varkappa T}}\right)\right| \left|\Phi\left(\frac{y-R_y}{2\varrho_H \sin\left(\frac{\Omega z}{2V_0}\right)}\right) - \Phi\left(\frac{y+R_y}{2\varrho_H \sin\left(\frac{\Omega z}{2V_0}\right)}\right)\right|. \tag{3.27}$$

From here it is clear that in the case, $\mathbf{V}_0 \perp \mathbf{H}_0$, under consideration the dependence of δN_i on x has the same form as in the case when $\mathbf{H}_0 = 0$ (see Eq. (2.26)), while the dependence on y is significantly altered because of the influence of the magnetic field. At small distances from the body, when $z \ll 2\pi V_0/\Omega$, formula (3.27) coincides with formula (2.26), which was obtained in the absence of a magnetic field, as of course must

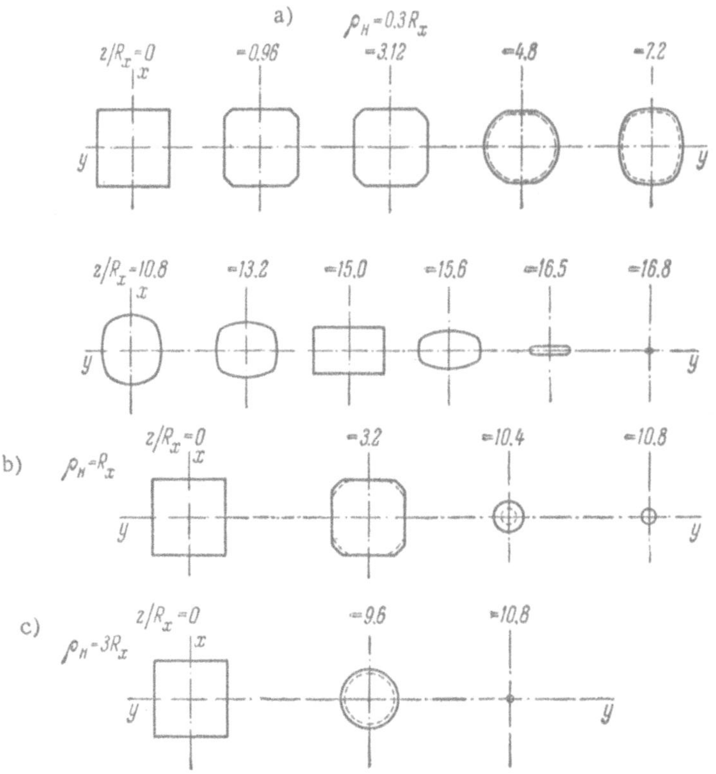

Fig. 19. Cross sections in the xy plane through the surface of constant ion concentration $N_i(x, y, z)/N_0 = 0.5$ in the rarefied region for different values of z/R_x, as indicated in the figure. The body has a square cross section, $R_x = R_y$. The dash curve–the same in the absence of a magnetic field. a) $\rho_H = 0.3R_x$; b) $\rho_H = R_x$; c) $\rho_H = 3R_x$.

happen. If $z \gtrsim 2\pi V_0/\Omega$, the influence of the magnetic field is very large; the second factor in (3.27) becomes, as in the case $\mathbf{V}_0 \parallel \mathbf{H}_0$, a periodic function of z with period $2\pi V_0/\Omega$.

Of course, in this case, with account of the thermal motion of the ions along the z axis, the periodic structure is smeared out over distances $z > (2\pi V_0/\Omega)(MV_0^2/\varkappa T)^{1/2}$.

On the z axis, the expression for N_i takes the simple form

$$N_i(0, 0, z) = N_0\left[1 - \Phi\left(\frac{R_x}{z}\sqrt{\frac{M_iV_0^2}{2\varkappa T}}\right)\Phi\left(\frac{R_y}{2\varrho_H \sin\left(\frac{\Omega z}{2V_0}\right)}\right)\right]. \tag{3.28}$$

At a sufficiently large distance, the average value of the disturbance δN_i, as already noted above, decreases like $1/z$ with increasing z, while in the absence of a magnetic field, the same quantity decreases like $1/z^2$. In the case of motion of the body along the field ($\mathbf{V}_0 \parallel \mathbf{H}_0$), the disturbance on the average generally does not decrease with increasing $|z|$.

The distribution of the ion concentration in the "shaded" zone has a rather complicated form, due to the influence of the magnetic field. One can understand the character of the ion concentration disturbance from Figs. 14-20.

The variation of $-\delta N_i/N_0$ along the z axis (formula (3.28)) in the case when the body's cross section is a square with $R_x = R_y = \rho_H$ and $(M_iV_0^2/2\varkappa T)^{1/2} = 8$ is shown in Fig. 14. In contrast to the case of longitudinal motion of a body ($\mathbf{V}_0 \parallel \mathbf{H}_0$), the disturbance does not remain constant on the average, but decreases with distance as $1/z$.

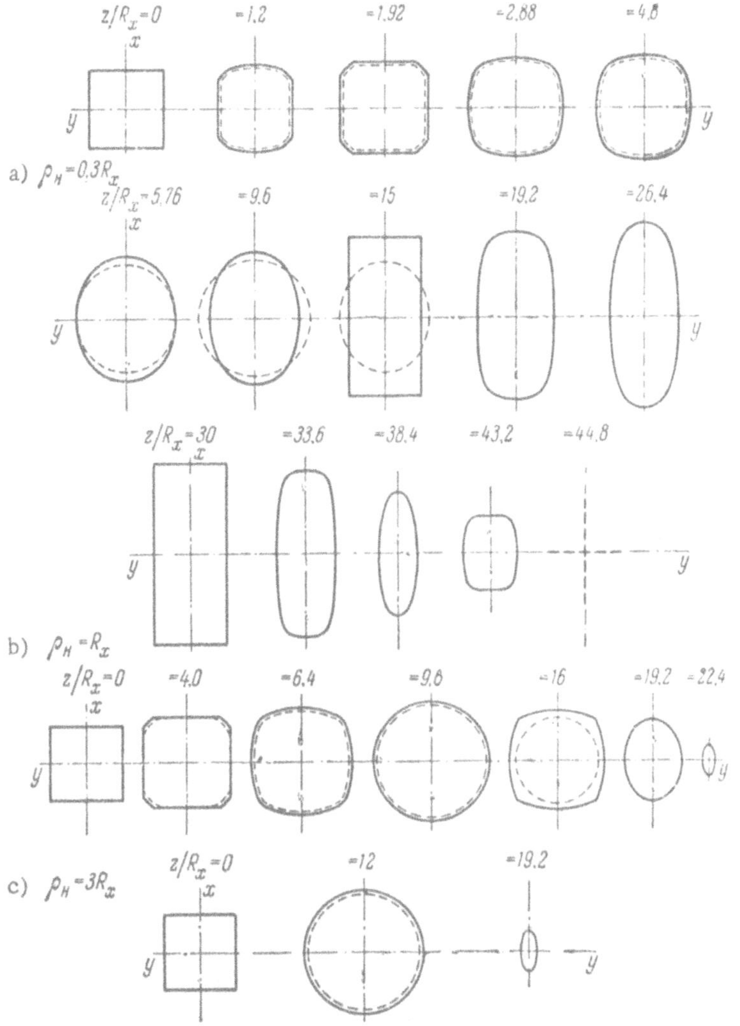

Fig. 20. Cross section in the xy plane through the surface of constant ion concentration N_i (x, y, z)/N_0 = 0.8 in the rarefied region for different values of z/R_x, as indicated in the figure. The body's cross section is square, $R_x = R_y$. The dashed curves–the same in the absence of a magnetic field. a) $\rho_H = 0.3R_x$; b) $\rho_H = R_x$; c) $\rho_H = 3R_x$.

Surfaces of different values of the ratio N_i/N_0 in the rarefied zone behind a moving body of square cross section are shown in Figs. 15-18. The shape of the corresponding surfaces N_i/N_0 = const is, in general, extremely complicated; the surfaces are not even related with each other (see Figs. 17 and 18). The complicated structure of these surfaces reflects the presence of periodic "compressions" and "rarefactions" in the particle concentrations behind a moving body. In contrast to the case of longitudinal motion of a body, the concentration disturbances for $\mathbf{V}_0 \perp \mathbf{H}_0$ decrease on the average with increasing distance from the body. The magnetic field exerts a strong influence on the structure of the disturbed region only at large distances. Therefore, with increasing values of the ratio N_i (x, y, z)/N_0 the shape of the surfaces of constant density becomes complicated, which is clearly evident from the figures presented. One can also discern the influence of the body's dimensions, i.e., the value of the ratio R/ρ_H, on the ion concentration N_i from Figs. 19 and 20, in which cross sections of the surfaces $N_i/N_0 = 0.5$ (Fig. 19) and $N_i/N_0 = 0.8$ (Fig. 20) at various distances behind the body are shown, respectively, for (a) $\rho_H = 0.3R_x = 0.3R_y$, (b) $\rho_H = R_x = R_y$, and (c) $\rho_H = 3R_x = 3R_y$. The dashed curves in these same figures show, for purposes of comparison, the corresponding cross sections in the

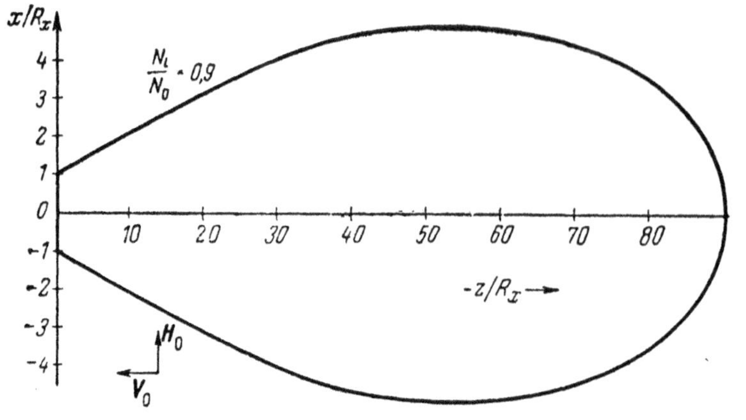

Fig. 21. Cross section in the xy plane through the surface of constant ion concentration N_i (x, y, z)/N_0 = 0.9 in the rarefied region, when $R_y \gg \rho_H$, $\mathbf{V}_0 \perp \mathbf{H}_0$, $\mathbf{H}_0 \parallel Ox$.

case when $\mathbf{H}_0 = 0$. It is clear from the figures that, with increasing values of the ratio R/ρ_H, the influence of a magnetic field on the structure of the disturbed region becomes stronger. In particular, when the linear dimension of the body is large in comparison with the Larmor radius of the ions ($R_y \gg \rho_H$), the disturbed zone has the shape of a thin "petal" lying in the xz plane, i.e., in the plane $\mathbf{V}_0\mathbf{H}_0$, namely:

$$N_i(x, y, z) \approx N_0\left[1-\frac{1}{2}\left|\Phi\left(\frac{x-R_x}{z}\sqrt{\frac{M_iV_0^2}{2\varkappa T}}\right)-\Phi\left(\frac{x+R_x}{z}\sqrt{\frac{M_iV_0^2}{2\varkappa T}}\right)\right|\left\{\begin{matrix}0, & |y|>R_y\\ 1, & |y|<R_y\end{matrix}\right.\right]. \quad (3.29)$$

The thickness of the "petal" (leaf) varies periodically with amplitude of order ρ_H and with period $2\pi V_0/\Omega$. The shape of the "petal" in the xz plane is shown in Fig. 21. We also note that it is clear from the figures presented above that, due to the influence of the magnetic field, the disturbance of the particle concentrations is still significant (even) at a large distance from the body.

§ 12. Arbitrary Direction of Motion

In this general case, for the calculation of the ion concentration, i.e., for the integration of the distribution function (3.8) with respect to velocities, it is convenient to change from the velocities v_x, v_y, and v_z to the velocity $u_\perp$ [see Eq. (3.5b)], the angle φ, and the velocity $v_\parallel$, parallel to the magnetic field. Here the following simple expression is obtained for the ion concentration at large distances from the body ($-z \gg V_0/\Omega$), where the influence of the magnetic field is most important:

$$N_i(x, y, z) = N_0\left\{1-\frac{1}{4}\left|\Phi\left(\frac{x-R_x}{z\sin\alpha+x\cos\alpha}\sqrt{\frac{M_iV_0^2}{2\varkappa T}}\right)-\Phi\left(\frac{x+R_x}{z\sin\alpha+x\cos\alpha}\sqrt{\frac{M_iV_0^2}{2\varkappa T}}\right)\right|\times\right.$$

$$\left.\times\left|\Phi\left(\frac{y-R_y}{2\varrho_H\sin\left(\frac{\Omega z}{2V_0}\right)}\right)-\Phi\left(\frac{y+R_y}{2\varrho_H\sin\left(\frac{\Omega z}{2V_0}\right)}\right)\right|\right\} \quad (3.30)$$

provided that the angle α is not very small, that is, $\sin\alpha > 2R_x/z$. In this case, of course, the shape of the disturbed region is extremely complicated. The curve $N_i/N_0 = 0.9$ in the xz plane is shown in Fig. 22 for comparison with Fig. 21; here the angle α between $\mathbf{V}_0$ and $\mathbf{H}_0$ equals 45° and $R_y \gg \rho_H$. In this case the disturbed region is asymmetric with respect to the z axis. In addition, the elongation of the region with $N_i/N_0 = 0.9$ is increased by approximately 1.5 times, which is a consequence of the decrease of the angle α between the direction of the magnetic field and the direction of the body's motion. Altogether, the length of the disturbed

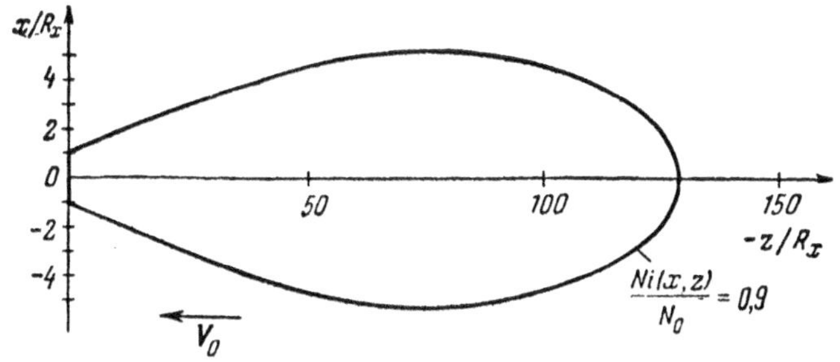

Fig. 22. Cross section in the xz plane through the surface of constant ion concentration N_i (x, y, z)/N_0 = 0.9 in the rarefied region, when $R_y \gg \rho_H$ and the body is moving at an angle of 45° to the direction of the magnetic field.

region increases with decreasing angle α in proportion to $1/\sin\alpha$. The average disturbance of the ion concentration, for arbitrary values of $\alpha \neq 0$, decreases with increasing $|z|$ in proportion to $1/|z|\sin\alpha$. Consequently, the case of strictly longitudinal motion is essentially an isolated case.

§ 13. Fourier Components of the Electron Concentration

Now we go on to a calculation of the Fourier components $N_{\mathbf{q}}$ of the electron concentration. The corresponding formulas for $N_{\mathbf{q}}$ are used in Chapter VII for a calculation of the scattering of electromagnetic waves by the "trail" of a body. As before in the case when the dimensions of the body are much larger than the Debye radius, the disturbance of the electron concentration is equal to the disturbance of the ion concentration, and in order to determine $N_{\mathbf{q}}$, it is sufficient to calculate the disturbance of the ions' distribution function. As in the preceding chapter, we consider the case of a rapidly moving metallic sphere which neutralizes all ions incident on its surface. In this case, the distribution function for the ions satisfies Eq. (2.85), to the left side of which is added, as in § 9, a term containing the Lorentz force (e/c) $[\mathbf{u}\mathbf{H}_0]$ which acts on the ions (here $\mathbf{u} = \mathbf{v} + \mathbf{V}_0$). Thus, in the case of an absorbing sphere, Eq. (3.1) has the form

$$(u - V_0)\frac{\partial f_i'}{\partial r} + \frac{e}{M_i c}[uH_0]\frac{\partial f_i'}{\partial u} = \begin{cases} -\frac{rV_0}{r} f_{0i}(u)\,\delta(r - R_0), & \text{if } rV_0 < 0; \\ 0, & \text{if } rV_0 > 0, \end{cases} \tag{3.31}$$

and $f_i' = f_i - f_{0i}$.

We obtain the following equation for the Fourier components:

$$\frac{e}{M_i c}[uH_0]\frac{\partial f_q}{\partial u} + iq(u - V_0) f_q = \pi R_0^2 V_0 f_{0i} G(qR_0, \chi), \tag{3.32}$$

where $f_q = \int f_i' e^{-iqr} d^3r$ (see the beginning of § 8), and G(qR_0, χ) is, as before, determined by formula (2.88).

We rewrite Eq. (3.31) in cylindrical coordinates in velocity space with the axis $u_\parallel$ along the direction of the magnetic field. We introduce the vectors $\mathbf{q}_\perp$ and $\mathbf{u}_\perp$, i.e., the projections of the vectors $\mathbf{q}$ and $\mathbf{u}$ on a plane perpendicular to the magnetic field, and the angle β between $\mathbf{q}_\perp$ and $\mathbf{u}_\perp$. Then Eq. (3.31) takes the form

$$\frac{\partial f_q}{\partial \beta} - i(\mu + \gamma\cos\beta) f_q = \frac{\pi R_0^2 V_0 f_{0i} G}{\Omega}. \tag{3.33}$$

Here the following notation is being used:

$$\Omega = \frac{eH_0}{M_i c}, \quad \mu = \frac{q_{\|}u_{\|} - qV_0}{\Omega}, \quad \gamma = \frac{q_{\perp}u_{\perp}}{\Omega}, \quad q_{\|} = \frac{qH_0}{H_0}, \quad u_{\|} = \frac{uH_0}{H_0}.$$

The general solution of Eq. (3.33) is given by

$$f_q = \exp\left[i(\mu\beta + \gamma \sin\beta)\right] \int_C^\beta \exp\left[i(\mu t + \gamma \sin t)\right] dt \, \frac{\pi R_0^2 V_0 f_{0i} G}{\Omega}. \tag{3.34}$$

In this connection, it is necessary to select the constant C in Eq. (3.34) in order to obtain a solution which is periodic in β with period 2π, since a change of the angle β by 2π means a rotation of the vector $\mathbf{u}_\perp$ by 2π, i.e., it leads to the same value of the vector $\mathbf{u}$. Thus, the periodicity in β expresses the single-valuedness of the function $f_q(\mathbf{u})$. It is easy to see that $C = \infty$. In fact, letting $C = \infty$ and $t = x + \beta$, we obtain

$$f_q = -\frac{\pi R_0^2 V_0 f_{0i} G}{\Omega} \int_0^\infty \exp\left[-i\{\mu x + \gamma\left[\sin(\beta + x) - \sin\beta\right]\}\right] dx, \tag{3.35}$$

and the periodicity of expression (3.35) with respect to β is obvious. Thus, expression (3.35) determines the desired Fourier components of the distribution function. Integrating (3.35) with respect to d^3u and substituting their values in place of μ and γ, we have

$$N_q = \int f_q d^3u = -\frac{\pi R_0^2 V_0 G}{\Omega} \int f_{0i}(u) \exp\left[-i\left\{\frac{q_{\|}u_{\|} - qV_0}{\Omega} x + \right.\right.$$
$$\left.\left. + \frac{q_\perp u_\perp}{\Omega}\left[\sin(\beta + x) - \sin\beta\right]\right\}\right] dx u_\perp du_\perp du_{\|} d\beta. \tag{3.36}$$

Integration with respect to β in Eq. (3.36) is carried out from 0 to 2π, and by virtue of the periodicity of the function f_q we may replace β by $\beta - (x/2)$. Then

$$N_q = -\frac{N_0 \pi R_0^2 V_0 G}{\Omega} \left(\frac{M_i}{2\pi\varkappa T}\right)^{3/2} \times$$
$$\times \int \exp\left[-i\left(\frac{q_{\|}u_{\|} - qV_0}{\Omega} x + \frac{2q_\perp u_\perp}{\Omega} \cos\beta \sin\frac{x}{2}\right) - \frac{M_i}{2\varkappa T}(u_\perp^2 + u_{\|}^2)\right] dx u_\perp du_\perp du_{\|} d\beta. \tag{3.37}$$

One can evaluate the integrals with respect to $du_{\|}$ and $u_\perp \, du_\perp \, d\beta$ in terms of elementary functions. Indeed, in the plane of the vector $\mathbf{u}_\perp$, since $q_\perp u_\perp \cos\beta = \mathbf{q}_\perp \mathbf{u}_\perp$, we have

$$\left(\frac{M_i}{2\pi\varkappa T}\right) \int_0^\infty \int_0^{2\pi} \exp\left(-\frac{M_i}{2\varkappa T} u_\perp^2 - 2i \frac{q_\perp u_\perp}{\Omega} \cos\beta \sin\frac{x}{2}\right) u_\perp du_\perp d\beta =$$
$$= \left(\frac{M_i}{2\pi\varkappa T}\right) \int \exp\left(-\frac{M_i}{2\varkappa T} u_\perp^2 - 2i \frac{\mathbf{q}_\perp \mathbf{u}_\perp}{\Omega} \sin\frac{x}{2}\right) d^2u_\perp.$$

Now, as in §11, changing from integration with respect to $d^2u_\perp$ to integration with respect to d^2w, where $\mathbf{w} = \mathbf{u}_\perp + 2\mathbf{q}_\perp \left[\varkappa T \sin(x/2)/M_i\Omega\right]$, we obtain

$$\left(\frac{M_i}{2\pi\varkappa T}\right) \exp\left(-\frac{2\varkappa T}{M_i\Omega^2} q_\perp^2 \sin^2\frac{x}{2}\right) \int \exp\left[-\frac{M_i w^2}{2\varkappa T}\right] d^2w = \exp\left(-\frac{2\varkappa T}{M_i\Omega^2} q_\perp^2 \sin^2\frac{x}{2}\right). \tag{3.38}$$

At last, taking into account that

$$\left(\frac{M_i}{2\pi\varkappa T}\right)^{1/2}\int\limits_{-\infty}^{\infty}\exp\left(-i\frac{q_{\|}u_{\|}}{\Omega}x-\frac{M_i}{2\varkappa T}u_{\|}^2\right)du_{\|}=\exp\left(-\frac{\varkappa T}{2M_i\Omega^2}q_{\|}^2x^2\right), \tag{3.39}$$

we finally obtain

$$N_q=-\frac{\pi R_0^2N_0V_0G}{\Omega}\int\limits_0^{\infty}\exp\left[i\frac{qV_0}{\Omega}x-\frac{\varkappa T}{2M_i\Omega^2}\left(q_{\|}^2x^2+4q_{\perp}^2\sin^2\frac{x}{2}\right)\right]dx. \tag{3.40}$$

The magnetic field in general turns out to have an important influence on the quantity $N_{\mathbf{q}}$. However, as was indicated in §10, the role of the magnetic field is small for fulfillment of the condition $z(\Omega/V_0) \ll 1$. In terms of the Fourier components, this condition is rewritten in the form

$$q_{\|}\sqrt{\frac{\varkappa T}{M_i}}\gg\Omega=\frac{eH_0}{M_ic}. \tag{3.41}$$

If condition (3.41) is satisfied, then the integrand in (3.40) is rapidly damped out for $x \gtrsim 1$. In this case, values of $x \ll 1$ are important in the integral, and one can set $\sin^2(x/2) \approx x^2/4$. Then

$$N_q=-\frac{\pi R_0^2N_0V_0G}{\Omega}\int\limits_0^{\infty}\exp\left(i\frac{qV_0}{\Omega}x-\frac{\varkappa T}{2M_i\Omega^2}q^2x^2\right)dx. \tag{3.42}$$

Now using the identity

$$\int\limits_0^{\infty}\exp\left(ita-\frac{t^2}{4}\right)dt=\left(\sqrt{\pi}+2i\int\limits_0^{a}e^{t^2}dt\right)e^{-a^2}=Q(a), \tag{3.43}$$

we directly bring (3.42) to the form (2.93). Thus, upon fulfillment of condition (3.41), the magnetic field is actually unimportant. It plays the greatest role in the opposite limiting case, when

$$q_{\|}\sqrt{\frac{\varkappa T}{M_i}}\ll\Omega. \tag{3.44}$$

We shall also consider this case. By virtue of the periodicity with respect to z of the concentration disturbance δN_i (x, y, z) with period equal to $2\pi V_0/\Omega$ (see §§ 10 and 11), the Fourier components $N_{\mathbf{q}}$ must have sharp maxima when the values of $\mathbf{qV}_0/\Omega$ are close to integers. Let us investigate the behavior of $N_{\mathbf{q}}$ in the vicinity of these maxima. In other words, we shall assume that

$$\frac{qV_0}{\Omega}=p+\varepsilon,\ |\varepsilon|\ll 1,\qquad p=0,\ \pm 1,\ \pm 2,\ldots \tag{3.45}$$

For this (purpose), we rewrite Eq. (3.40), introducing the quantity ε. Then

$$N_q=-\frac{\pi R_0^2N_0V_0G}{\Omega}\int\limits_0^{\infty}\exp\left(i\varepsilon x-\frac{\varkappa T}{2M_i\Omega^2}q_{\|}^2x^2\right)\exp\left(ipx-\frac{2\varkappa T}{M_i\Omega^2}q_{\perp}^2\sin^2\frac{x}{2}\right)dx. \tag{3.46}$$

If conditions (3.44) and (3.45) are satisfied, then the periodic function $\exp[ipx - (2\varkappa T/M_i\Omega^2)\, q_\perp^2 \sin^2(x/2)]$ is rapidly oscillating in comparison with the other factors in the integrand. Therefore, we may replace this function by its average value over one period, i.e., we may carry out the following substitution:

$$\exp\left(ipx - \frac{2\varkappa T}{M_i\Omega^2} q_\perp^2 \sin^2\frac{x}{2}\right) \rightarrow \frac{1}{2\pi}\int\limits_0^{2\pi} \exp\left(ipx - \frac{2\varkappa T}{M_i\Omega^2} q_\perp^2 \sin^2\frac{x}{2}\right) dx =$$

$$= \frac{1}{2\pi}\exp\left(-\frac{q_\perp^2\varkappa T}{M_i\Omega^2}\right)\int\limits_0^{2\pi}\exp\left(ipx + \frac{q_\perp^2\varkappa T}{M_i\Omega^2}\cos x\right) dx = \exp\left(-\frac{q_\perp^2\varkappa T}{M_i\Omega^2}\right) I_p\left(\frac{q_\perp^2\varkappa T}{M_i\Omega^2}\right), \tag{3.47}$$

where I_p denotes a Bessel function of imaginary argument. Afterwards transforming the integral (3.46) with the aid of formula (3.43), we obtain

$$N_q = -\frac{\pi R_0^2 N_0 V_0 G}{q_{\parallel}}\sqrt{\frac{M_i}{2\varkappa T}}\, Q(a_1)\exp\left(-\frac{q_\perp^2\varkappa T}{M_i\Omega^2}\right) I_p\left(\frac{q_\perp^2\varkappa T}{M_i\Omega^2}\right), \tag{3.48}$$

where

$$a_1 = \frac{\varepsilon\Omega}{q_{\parallel}}\sqrt{\frac{M_i}{\varkappa T}} = \frac{qV_0 - p\Omega}{q_{\parallel}}\sqrt{\frac{M_i}{2\varkappa T}}. \tag{3.49}$$

It follows from Eq. (3.48) that, if $q_{\parallel} \rightarrow 0$ and $\mathbf{q}\mathbf{V}_0 \rightarrow p\Omega$ ($\varepsilon \rightarrow 0$) in such a way that a_1 remains finite, then N_q tends to infinity $\sim 1/q_{\parallel}$.

As indicated above, the case of a body moving strictly along the magnetic field vector, when $\mathbf{V}_{0\perp} = 0$, $\mathbf{q}\mathbf{V}_0 = q_{\parallel} V_0$ and for $p = 0$, $a_1 = V_0(M_i/2\varkappa T)^{1/2}$, is a special case. Using the fact that $(M_i V_0^2/2\varkappa T)^{1/2} \gg 1$ and taking into account that

$$Q(a_1) \approx 2ie^{-a_1^2}\int\limits_a^{a_1} e^{t^2} dt \approx \frac{i}{a_1} \quad \text{for} \quad a_1 \gg 1, \tag{3.50}$$

we obtain

$$N_q = -\frac{\pi R_0^2 N_0 G}{q_{\parallel}}\, i \exp\left(-\frac{q_\perp^2\varkappa T}{M_i\Omega^2}\right) I_0\left(\frac{q_\perp^2\varkappa T}{M_i\Omega^2}\right). \tag{3.51}$$

Thus, in the case when the body moves strictly along the direction of the field, $N_q \rightarrow \infty$ as $q_{\parallel} \rightarrow 0$, independently of the value of $\mathbf{q}_\perp$. This result has a simple physical meaning. Indeed, for motion of the body along the magnetic field, as we saw in § 10, the disturbed region which is generated behind the body cannot be reabsorbed because of the field. Therefore, the "trail" has the shape of a semi-infinite cylinder with its axis along $\mathbf{H}_0$, which also leads to an infinite value of N_q for $\mathbf{q} \perp \mathbf{H}_0$, i.e., for $\mathbf{q}_{\parallel} \rightarrow 0$. Of course, the fact that N_q tends to infinity in all the cases considered has no physical meaning, and is associated with the fact that we neglected the influence of the electric field on the motion of the ions and neglected collisions between particles. Upon account of these factors, as we shall see later, N_q assumes a finite, although rather large, value.

CHAPTER IV

ELECTRIC FIELD AROUND A BODY

§ 14. Approximate Calculation of the Electric Field

a. Introductory Remarks. The disturbances of the electron and ion concentrations around a moving body violate the quasi-neutrality of the plasma. As a result, an electric field appears here, which naturally has an effect itself on the distribution of charged particles. Therefore, the calculation of the electron and ion concentrations, and also of the electric field, requires the simultaneous solution of Eq. (1.30) for the potential together with (the solution of) the equations for the ion (1.23) and electron (1.24) distribution functions. If the diversity of the boundary conditions for the particles and field on the body's surface are also considered, then it becomes clear that the complete solution of such a problem is extremely difficult.

It is, however, possible to significantly simplify the system of Eqs. (1.23), (1.24), (1.30) if one takes into account a number of circumstances which are realized in the ionosphere.

In the first place, it is important to take into consideration that the body's velocity is much smaller than the thermal velocity of electrons, its surface is negatively charged, and the potential $|\varphi_0| \gg \varkappa T/e$. In this case, as indicated in §§3c and 4c, the electrons are distributed according to the Boltzmann law and the necessity of solving Eq. (1.24) is thereby eliminated.*

Furthermore, up to very great altitudes in the ionosphere, the Debye radius is small in comparison with the characteristic dimensions of a moving body, which is clear, for example, from Table IX where the corresponding values of the ratio R_0/D are presented for $R_0 = 1$ m. Therefore, we shall everywhere assume that $R_0 \gg D$.

Under these conditions Debye screening plays a very important role, as a result of which the concentrations of electrons and ions in a large part of the disturbed zone are close to each other. As we shall see below, this circumstance enables us to obtain an extremely simple expression, good in a large part of the disturbed region, for the potential of the electric field. Since the Debye radius is small, it is possible to separately investigate the double layer on the surface of a body. By the same token, the problem of the flux of particles (ions) being attracted to the body, and therefore also the question of the potential field on its surface, is significantly simplified.

Finally, since the body's velocity is considerably greater than the thermal velocity of the ions, disturbance of the ions occurs basically because of their interaction with the body itself. Therefore, in the first approximation, one can also neglect the influence of the electric field on the motion of ions. Strictly speaking, this statement is only valid for the zone ahead of the body, i.e., in the region of "compression." The electric

TABLE IX

Altitude	N_0	T	D, cm	R_0/D	$(R_0/D)^2$	$\ln (R_0/D)^2$
Layer E	10^5	$3 \cdot 10^2$	0.4	$2.5 \cdot 10^2$	$6 \cdot 10^4$	11
Layer F	10^6	10^3	0.2	$5 \cdot 10^2$	$2 \cdot 10^5$	13
1000 km	10^5	$3 \cdot 10^3$	1	10^2	10^4	9
3000 km	$7 \cdot 10^3$	$4 \cdot 10^3$	4	25	$6 \cdot 10^2$	6.5

*Here we shall not consider particular cases when a positive potential is specially given on the surface of the body (see Chapters IX and X).

field now acts for a rather long time on the particles occupying the rarefied region behind the body, so that its influence in general turns out to be important. Therefore, in the rarefied region, the results of calculations without taking the influence of the electric field into consideration are correct only to within a factor of the order of a few times unity. One can, to be sure, make them somewhat more correct if the influence of the field on the motion of the ions is taken into account in the hydrodynamic approximation. In this way, simple approximate expressions for the potential φ are obtained. A rigorous solution of the problem, with account of the influence of the electric field on the motion of the ions, is given in the following paragraphs for the region far distant from the body's surface.

b. Potential of the Field on the Body's Surface. Electrons may be absorbed and ions neutralized on the surface of a body. In this connection, the body itself becomes charged in the general case, since the number of electrons and the number of ions incident on it are not identical. The number of electrons and ions being absorbed essentially depends on the potential at the body's surface, its properties, and on the state of the plasma. It is clear that the body will be accumulating charges as long as the numbers of electrons and ions being absorbed on its surface per unit time are not equal. In other words, the total current incident on the body must be equal to zero; the potential of the field on the surface of the body is also determined by by this condition.*

In the case considered here of a rapidly moving body ($V_0 \gg v_i$), whose dimensions are much larger than the Debye radius ($R_0 \gg D$), both ions and electrons are captured for the most part by the front part of the body's surface: ions hit at the front of the body because of its large velocity; the region behind the body with a high negative potential (see below) hinders the incidence of electrons in this area.

The ion current incident on the front surface of a body under the conditions considered is almost no different from the current of neutral particles. The latter, as shown below in Chapter X, is determined by expression (10.8)

$$j_i = N_0 \left\{ (V_0 n) \left[\frac{1}{2} + \frac{1}{2} \Phi \left(\frac{V_0 n}{\sqrt{2\varkappa T / M_i}} \right) \right] + \sqrt{\frac{\varkappa T}{2\pi M_i}} \exp \left[-\frac{M_i (V_0 n)^2}{2\varkappa T} \right] \right\}, \tag{4.1}$$

where $\mathbf{n}$ is the outward normal to the body's surface (on the front surface, $\mathbf{V_0 n}$ is always positive). For a spherical body $\mathbf{V_0 n} = V_0 \cos \theta'$ where, for a given point on the surface, θ' is the polar angle measured from the z axis (i.e., from the direction $\mathbf{V}_0$). For angles θ' not too close to $\pi/2$, $(V_0 \cos \theta')/(2\varkappa T/M_i)^{1/2} \gg 1$ and

$$j_i \simeq N_0 V_0 \cos \theta'. \tag{4.2}$$

The electron current on the front part is simply determined by the expression

$$j_e = N_0 \sqrt{\frac{\varkappa T_e}{2\pi m}} \exp \left(\frac{e\varphi_S}{\varkappa T} \right), \tag{4.3}$$

which is valid for a body at rest [see Eq. (10.42)], since the body's velocity V_0 is small in comparison with the thermal velocity $v_e = (2\varkappa T/m)^{1/2}$ of the electrons. In Eq. (4.3), φ_S denotes the potential of the considered point $\mathbf{r}_S$ on the body's surface.

If the body's surface is a dielectric, then equality of the currents of absorbed electrons and ions must be satisfied at each point on the surface, i.e.,

$$(1 - R_i)\, j_i = (1 - R_e)\, j_e,$$

*Of course, it is possible to specially charge the body's surface up to an arbitrary potential, then the current will not be equal to zero. The determination of the dependence of the current strength on the potential of a body's surface also constitutes one of the problems of probe measurements and of the theory of a probe in a plasma (see Chapter X).

where R_e and R_i are the coefficients of reflection for electrons and ions. From here it follows that the potential on the surface is given by

$$\varphi_S = -\frac{\varkappa T}{e}\ln\left\{\frac{\sqrt{\frac{\varkappa T}{2\pi m}}(1-R_e)}{(1-R_i)(V_0 n)\left[\frac{1}{2}+\frac{A}{2}\right]+(1-R_i)\sqrt{\frac{\varkappa T}{2\pi M_i}}\exp\left[-\frac{M_i(V_0 n)^2}{2\varkappa T}\right]}\right\}, \tag{4.4}$$

where

$$A=\Phi\left(\frac{V_0 n}{\sqrt{2\varkappa T/M_i}}\right).$$

Here, as usual, it is assumed that $N_{i0} = N$ and the temperatures of ions and electrons are the same. In particular, for a spherical body with angles θ' not too close to $\pi/2$,

$$\varphi_S(\theta') \simeq -\frac{\varkappa T}{e}\ln\left[\sqrt{\frac{\varkappa T}{2\pi M V_0^2}}\frac{1-R_e}{(1-R_i)\cos\theta'}\right], \tag{4.5}$$

i.e., φ_S increases with increase of the angle θ'. The maximum value of φ_S is reached for $\theta' = \pi/2$; it equals

$$\varphi_{max} = -\frac{\varkappa T}{e}\ln\left(\sqrt{\frac{M_i}{m}}\frac{1-R_e}{1-R_i}\right).$$

If the body's surface is metallic, then the potential on it is constant and equal to φ_0. The value of φ_0 is, of course, determined from the condition that the total current onto the body's surface be equal to zero, i.e.,

$$J_i = \int j_i\,dS = J_e = \int j_e\,dS.$$

Using Eqs. (4.1) and (4.3), for a spherical body we obtain

$$N_0 V_0 \pi R_0^2 (1-R_i) = N_0\sqrt{\frac{\varkappa T}{2\pi m}}\,2\pi R_0^2(1-R_e).$$

provided that $V_0 \gg (2\varkappa T/M_i)^{1/2}$.

From here it follows that

$$\varphi_0 = -\frac{\varkappa T}{e}\ln\left[\sqrt{\frac{2\varkappa T}{\pi m V_0^2}}\frac{1-R_e}{1-R_i}\right]. \tag{4.7}$$

Naturally, the expressions obtained for the body's potential are of an approximate nature. For an exact calculation of φ_0 it is necessary to take a number of other processes into consideration, processes leading to a variation of the current onto the body's surface such as thermoelectric emission, the photoeffect, the "knocking out" of electrons or ions from the surface under the influence of corpuscular radiation, because of collisions with neutral atoms of the medium. The effectiveness of these processes strongly depends on the material of which the body's surface is composed.

In the regions of the ionosphere considered by us, if the coefficients of reflection for electrons and ions are not close to unity, the potential on the body's surface (given by Eqs. (4.6) and (4.7)) is of the order of 0.2 to 0.4 V. As $R_i \to 1$ the potential increases logarithmically.

There are indications in the literature, based on direct measurements, that the negative potential of an artificial satellite reaches a value of several volts. The reason for the appearance of such large negative potentials remains unclear up to now.*

c. Electric Field in the Region of Compression. As we saw above, the surface of a body may have a rather large potential $\varphi \sim -(3 \text{ to } 4)\varkappa T/e$. Therefore, the body absorbs only a small part of the electrons incident on it. In addition, the body's velocity is considerably smaller than the thermal velocity of the electrons, $V_0 \ll (2\varkappa T/m)^{1/2}$. All this leads to the fact that in the vicinity of the body, except for a small region immediately adjacent to its surface, as indicated above, the Boltzmann distribution

$$N = N_0 \exp\left(\frac{e\varphi}{\varkappa T}\right). \tag{4.8}$$

is valid for the electrons. Thus, the potential of the electric field satisfies Eq. (1.30):

$$\Delta\varphi = -4\pi e\left[N_i - N_0 \exp\left(\frac{e\varphi}{\varkappa T}\right)\right]. \tag{4.9}$$

Let us rewrite Eq. (4.9) in terms of the dimensionless variables $\varphi^* = e|\varphi|/\varkappa T$, $\xi = r/R_0$:

$$\Delta_\xi\varphi^* = \left(\frac{R_0}{D}\right)^2\left[\frac{N_i}{N_0} - \exp(-\varphi^*)\right], \tag{4.10}$$

where $D = (\varkappa T/4\pi e^2 N_0)^{1/2}$ is the Debye radius and values of $(R_0/D)^2$ for $R_0 = 1$ m in the ionosphere are given above in Table IX. It is clear from this same table that, for conditions of interest to us, a large quantity always stands in front of the right side of Eq. (4.10). It is therefore natural to seek a solution of the equation in the form of a series in powers of the small parameter $(D/R_0)^2$. In this connection $e^{-\varphi^*}$ plays a major role in Eq. (4.10), and it is convenient to rewrite it in the form

$$\varphi^* = -\ln\left[\frac{N_i(\xi)}{N_0} - \left(\frac{D}{R_0}\right)^2 \Delta_\xi\varphi^*\right]. \tag{4.11}$$

Expanding the solution in powers of $(D/R_0)^2$, i.e., assuming

$$\varphi^* = \varphi_0^* + \varphi_{(1)}^*\left(\frac{D}{R_0}\right)^2 + \ldots,$$

we have

$$\varphi_0^* = \ln\frac{N_0}{N_i(\xi)}. \tag{4.12}$$

---

*In fact, the basic processes which were not taken into consideration in the derivation of formulas (4.6) and (4.7) (thermoelectric emission, photoeffect) only lead to a decrease in the absolute value of the negative potential. In this case one can assume that the rocket or satellite retains a large negative potential, acquired for one or another reason in the lower layers of the atmosphere, only if the ions are extremely weakly neutralized by the surface of the body. In fact, because of the ionic current, the relaxation time of the surface potential is

$$\Delta t \sim \frac{C\varphi_0}{eJ_i} \sim \frac{\varphi_0}{e\pi R_0 V_0 N_0}\frac{1}{1-R_i},$$

where $C = R_0$ is the capacity of a sphere. We see that Δt becomes large only when R is very close to unity. For example, even for $\varphi_0 = 100$ V, $1 - R_i = 10^{-5}$, $R_0 \sim 1$ m, to ~ 1 sec, i.e., the relaxation time is still rather small.

One can then determine other terms of the series, namely $\varphi_{(1)}^{*}$, etc. This series converges rapidly if

$$\left(\frac{D}{R_0}\right)^2 \frac{N_0}{N_i} \frac{\Delta_\xi \ln \frac{N_0}{N_i}}{\ln \frac{N_0}{N_i}} \ll 1. \tag{4.13}$$

Because of reflection, the ion concentration in front of a body is somewhat higher than N_0, and condition (4.13) is satisfied everywhere, except for the region immediately adjacent to the body's surface of thickness on the order of the Debye radius (this region is usually called the double layer). Thus, the potential of the electric field is given by the expression

$$\varphi = \frac{\varkappa T}{e} \ln \frac{N_i(r)}{N_0}. \tag{4.14}$$

everywhere in front of the body, except for the region of the double layer. The physical meaning of expression (4.14) is extremely simple. In fact, for such a field, the electron concentration (4.8) identically coincides with the ion concentration. Consequently, the electric field (4.14) – is the field which ensures the quasi-neutrality of the plasma. So it must be under conditions when the dimensions of the disturbed region are much larger than the Debye radius (see §4).

The total concentration of ions ahead of a body is

$$N_i(r) = N_0 + N_{i2}(r) + \delta N_i(r),$$

where $N_{i2}(\mathbf{r})$ is the concentration of the ions reflected from the body, and $\delta N_i(\mathbf{r})$ is the change in the concentration of particles in the inflowing current, caused by the influence of the electric field. It is easy to convince oneself, however, that because of the large energy of the ions incident on the body, $MV_0^2/2 \gg \varkappa T \sim e|\varphi|$, the magnitude of δN_i is small in comparison with N_0, so that one can neglect it. Because of the same reason, the relative concentration of ions reflected from a body coincides with the relative concentration n_2/n_0 of neutral particles for the same law of reflection – it is only necessary to take the partial neutralization of the ions on the body's surface into consideration, i.e., the relation

$$\frac{N_{i2}(r)}{N_0} = R_i \frac{n_2(r)}{n_0},$$

where R_i is the coefficient for the reflection of ions from the surface of the body. Thus,

$$\varphi = \frac{\varkappa T}{e} \ln \left[1 + R_i \frac{n_2(r)}{n_0}\right]. \tag{4.15}$$

ahead of a body. As indicated above, the ratio $n_2(\mathbf{r})/n_0$ essentially depends on the nature of the scattering of particles by the body's surface. Expressions for $n_2(\mathbf{r})/n_0$ for specular reflection and diffuse scattering are presented above in §6.

Curves of different values of the potential (formula (4.15)) in the case of specular reflection of ions from the surface of a body with R_i close to unity are shown in Fig. 23.* It is clear from the figure that the electric field in front of a body is not large – of the order of $\varkappa T/e$. For small coefficients of reflection, and also at a

*As already indicated, for values of R_i close to unity, the potential on a body's surface increases logarithmically. However, even for $1 - R_i \sim 0.01$ the potential turns out to be not very high. In addition, this increase of φ manifests itself only in the double layer close to the surface.

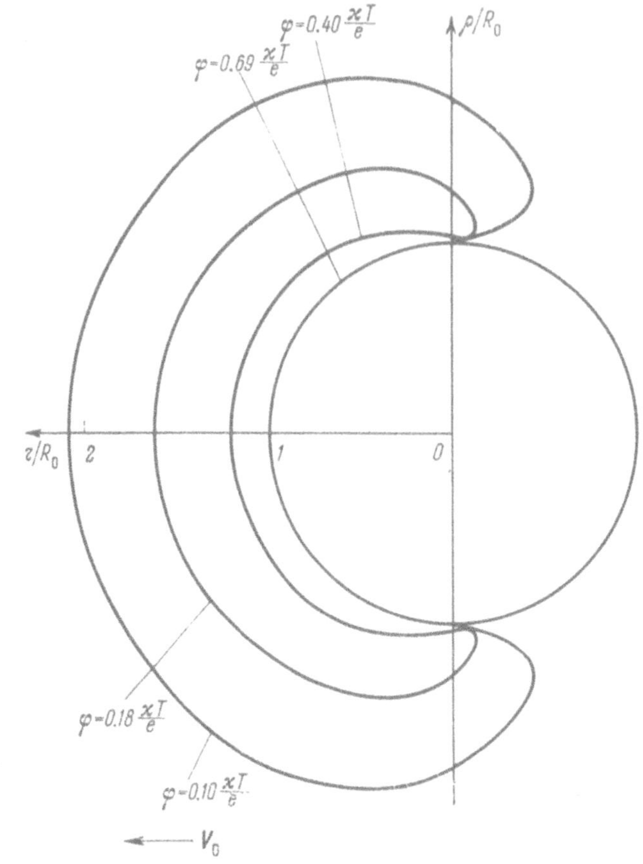

Fig. 23. Equipotential lines in front of a sphere for the case of specular reflection of ions from a body's surface.

considerable distance from the body's surface (provided that $R_i n_2/n_0 \ll 1$) the potential of the field is proportional to the excess concentration

$$\varphi = \frac{\varkappa T}{e} R_i \frac{n_2(r)}{n_0}. \tag{4.16}$$

Thus the field falls off with increasing distance from the body, in the same way as $n_2(\mathbf{r})$, i.e., it is proportional to $(R_0/r)^2$ (see §6).

d. Double Layer in Front of a Body. The quasi-neutrality of the plasma is violated near a body's surface – a double layer is created here. A decrease of the potential occurs, over the extent of this layer, from its value on the surface, determined by the formulas of §14b, to those values which the potential has in the quasi-neutral region close to the body (see §14c, e). The problem of calculating the potential distribution in such a double layer was considered by Jastrow and Pearse [55]. In this article, Poisson's equation for the potential was numerically integrated, where it was assumed that the distribution of particles and of the field is spherically symmetric and the electrons are distributed according to Boltzmann's law, but the concentration of ions is equal to its undisturbed value. These assumptions are quite unreasonable, since the body's motion strongly disturbs the concentration of ions, and the distribution is by no means spherically symmetric. Hence the results of [55] can only be qualitatively valid. The problem of the potential distribution in a flat double layer was investigated more correctly by Gintsburg [56]. In this article, however, the absorption of electrons on the surface of the body was not taken into consideration. As we shall see below, the nature of the double layer is significantly different in front of the body – in the region of compression, and behind the body – in the region of rarefaction. We consider the layer in front of the body in the present section. The double layer in the region of rarefaction will be considered in §14f.

The potential of the field in a double layer is described by Eq. (1.30). In this connection, the thickness of this layer is small for the conditions, when $R_0 \gg D$, of interest to us. In this case the term describing the variation of the potential in the direction normal to the surface plays the major role in the left side of Eq. (1.30), so that the equation turns into a one-dimensional equation. Denoting the distance along the normal to the surface by x, in place of Eq. (1.30) we than obtain*

$$\frac{d^2\varphi}{dx^2} = -4\pi e (N_i - N_e). \tag{4.17}$$

The ion concentration varies significantly only over distances on the order of the dimensions of the body, and therefore it is constant over the extent of a thin double layer, i.e.,†

$$N_i = N_0 \left(1 + R_i \frac{n_2(r_S)}{n_0}\right). \tag{4.18}$$

In the case of specular reflection $n_2(\mathbf{r}_S)/n_0 = 1$ and $N_i = N_0 (1 + R_i)$. An expression for the electron concentration N_e in the double layer close to a body which partially absorbs electrons will be derived in Chapter IX (see Eq. (9.16a)). We shall make use of this formula here:

$$N_e = N_0 \frac{1}{2} e^{\frac{e\varphi}{\varkappa T}} \left[1 + R_e + (1 - R_e)\Phi\left(\sqrt{\frac{e(\varphi - \varphi_0)}{\varkappa T}}\right)\right]. \tag{4.19}$$

Here, as usual, $\Phi(z) = \frac{2}{\sqrt{\pi}} \int\limits_0^z e^{-t^2} dt$ is the probability integral, R_e is the coefficient for the reflection of electrons. The electron distribution (4.19) differs from the Boltzmann distribution (4.8) because in it the absorption of part of the electrons by the body's surface is taken into account. Beyond the limits of the double layer, the distribution (4.19) coincides with (4.8) correct to exponentially small terms, if the negative potential of the body is sufficiently great ($-e(\varphi_0 - \varphi)/\varkappa T \gg 1$).

Substituting expressions (4.18) and (4.19) for N_i and N_e into Eq. (4.17) and changing to dimensionless variables $\varphi^* = e|\varphi|/\varkappa T$ and $\zeta = x/D$, we rewrite Eq. (4.17) in the form

$$\frac{d^2\varphi^*}{d\zeta^2} = 1 + R_i \frac{n_2(r_S)}{n_0} - \frac{1}{2} e^{-\varphi^*} \left[1 + R_e + (1 - R_e)\Phi\left(\sqrt{\varphi_0^* - \varphi^*}\right)\right]. \tag{4.20}$$

The boundary conditions have the following form: $\varphi^* = \varphi_0^* = e|\varphi_0|/\varkappa T$ for $\zeta = 0$ and $\varphi^* = \varphi_1^*$ for $\zeta \to \infty$, where value φ_1^* on the exterior boundary of the double layer is determined, of course, by the condition for quasi-neutrality, $N_i = N_e$, i.e.,

$$1 + R_i \frac{n_2(r_S)}{n_0} = \frac{1}{2} e^{-\varphi_1^*} \left[1 + R_e + (1 - R_e)\Phi\left(\sqrt{\varphi_0^* - \varphi_1^*}\right)\right]. \tag{4.21}$$

For sufficiently large negative values of φ_0 ($\varphi_0^* \gg 1$), from Eq. (4.21) we obtain

$$\varphi_1^* = \ln\left[1 + R_i \frac{n_2(r_S)}{n_0}\right]. \tag{4.22}$$

This value for φ_1^* coincides with Eq. (4.15).

*Of course, the surface is assumed to be sufficiently smooth so that its radius of curvature is everywhere $\gg D$.

†Here, as also earlier, the condition $e|\varphi| \ll MV_0^2$ is assumed to be satisfied, by virtue of which it is possible to neglect the influence of the field on the motion of the ions; it is possible, however, to also generalize the results of this section without particular difficulty to the case when the condition $e|\varphi| \ll MV_0^2$ is not satisfied.

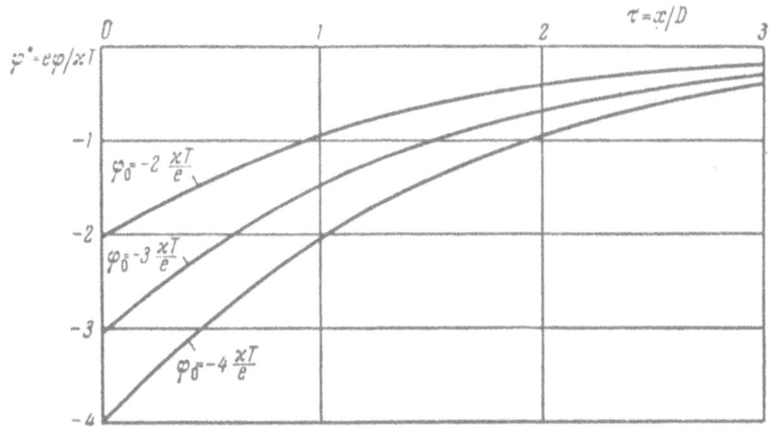

Fig. 24. Dependence of the field potential φ on the distance x/D inside the double layer around the front surface of a body for various values of the potential φ_0 on the surface. All quantities shown in the figure are dimensionless.

Since the right side of Eq. (4.20) does not depend on ζ, its solution is found by means of an elementary integration:

$$\zeta = \frac{1}{\sqrt{2}} \int_{\varphi^*}^{\varphi_0^*} \frac{d\varphi^*}{\left[\int_{\varphi_1^*}^{\varphi^*} f(\varphi^*)\, d\varphi^*\right]^{1/2}}, \tag{4.23}$$

where

$$f(\varphi^*) = 1 + R_i \frac{n_2(r_S)}{n_0} - \frac{1}{2}\left[1 + R_e + (1 - R_e)\,\Phi\left(\sqrt{\varphi_0^* - \varphi^*}\right)\right] e^{-\varphi^*}. \tag{4.24}$$

Also taking into consideration that

$$\int_{\varphi_1^*}^{\varphi^*} \Phi\left(\sqrt{\varphi_0^* - \varphi^*}\right) e^{-\varphi^*} d\varphi^* = \frac{2}{\sqrt{\pi}}\left(\sqrt{\varphi_0^* - \varphi^*} - \sqrt{\varphi_0^* - \varphi_1^*}\right)e^{-\varphi_0^*} - e^{-\varphi^*}\Phi\left(\sqrt{\varphi_0^* - \varphi^*}\right) +$$

$$+ e^{-\varphi_1^*}\Phi\left(\sqrt{\varphi_0^* - \varphi_1^*}\right) \approx e^{-\varphi_1^*} - e^{-\varphi^*}\Phi\left(\sqrt{\varphi_0^* - \varphi^*}\right) \tag{4.25}$$

(in the last expression it is assumed that $\varphi_0^* \gg 1$), we finally have

$$\zeta = \frac{1}{\sqrt{2}} \int_{\varphi^*}^{\varphi_0^*} \frac{d\varphi^*}{A}, \tag{4.26}$$

where

$$A = \left\{\left(1 + R_i \frac{n_2(r_S)}{n_0}\right)(\varphi^* - \varphi_1^*) - \frac{1 + R_e}{2}\left(e^{-\varphi_1^*} - e^{-\varphi^*}\right) - \frac{1 - R_e}{2}\left[e^{-\varphi_1^*} - e^{-\varphi^*}\Phi\left(\sqrt{\varphi_0^* - \varphi^*}\right)\right]\right\}^{1/2}.$$

In the case of a completely absorbing body, $R_i = 0$, $R_e = 0$, and consequently, $\varphi_1^* = 0$. Then instead of (4.26), we obtain

$$\zeta = \frac{1}{\sqrt{2}} \int_{\varphi^*}^{\varphi_0^*} \frac{d\varphi^*}{\left\{\varphi^* - 1 + \frac{1}{2} e^{-\varphi^*}\left[1 + \Phi\left(\sqrt{\varphi_0^* - \varphi^*}\right)\right]\right\}^{1/2}}. \tag{4.27}$$

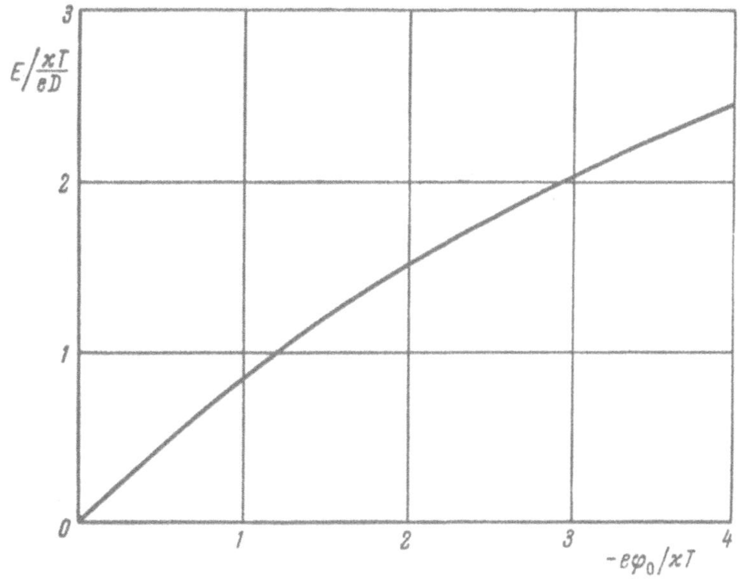

Fig. 25. Dependence of the electric field intensity E (on the front surface of a body) on its surface potential φ_0. All quantities shown in the figure are dimensionless.

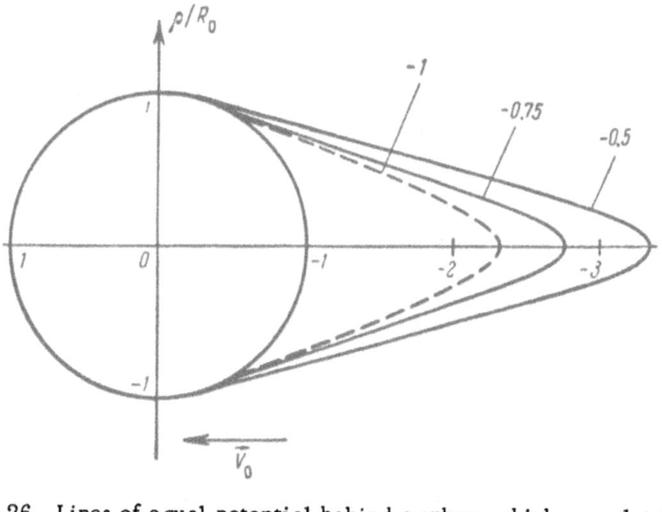

Fig. 26. Lines of equal potential behind a sphere which completely reflects the particles $(V_0/(2\varkappa T/M_i)^{1/2} = 8)$. Values of $(e\varphi/\varkappa T)$ ln $(R_0/D)^2$ are indicated in the figure.

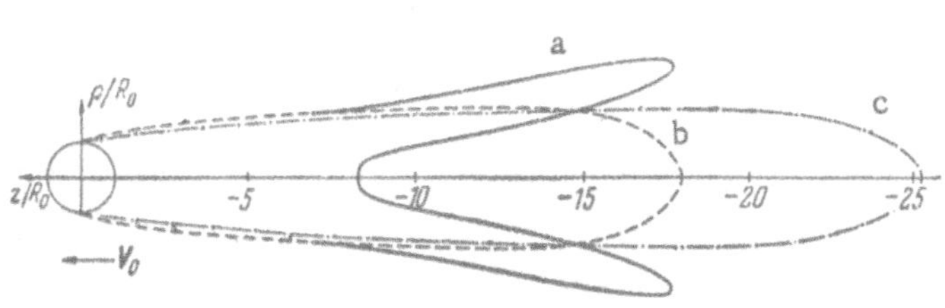

Fig. 27. Lines of equal potential $\varphi = -0.1(\varkappa T/e)$, calculated in various approximations.

This formula determines, in implicit form, the dependence of φ^* on ζ, i.e., the dependence of the potential on the distance to the body's surface. This dependence is shown in Fig. 24 for values of $e\varphi_0/\varkappa T = -\varphi_0^*$ equal to -2, -3, and -4. It is clear from the figure that in this case the thickness of the double layer is on the order of 1 to 2 Debye radii. It follows from Eq. (4.27) that, for large values of the potential of the field on the body, the thickness of the double layer increases in proportion to $(\varphi_0^*)^{1/2}$.

The electric field intensity is given by

$$E=-\frac{d\varphi}{dx}=\frac{\varkappa T}{eD}\frac{d\varphi^*}{d\zeta}=\frac{\varkappa T}{eD}\left(\frac{d\zeta}{d\varphi^*}\right)^{-1}.$$

As is evident from Eq. (4.26), for large values of the potential φ_0^* on the body's surface,

$$\frac{d\zeta}{d\varphi^*}=-\frac{1}{\sqrt{B}},$$

where

$$B=2\left(1+R_i\frac{n_2(r_S)}{n_0}\right)(\varphi^*-\varphi_1^*)-(1+R_e)(e^{-\varphi_1^*}-e^{-\varphi^*})-(1-R_i)\left[e^{-\varphi_1^*}-e^{-\varphi^*}\Phi\left(\sqrt{\varphi_0^*-\varphi^*}\right)\right].$$

Therefore,

$$E=-\sqrt{C},$$

where

$$C=4\pi N_0\varkappa T\left\{2\left(1+R_i\frac{n_2(r_S)}{n_0}\right)(\varphi^*-\varphi_1^*)-(1+R_e)\times\right.$$
$$\left.\times(e^{-\varphi_1^*}-e^{-\varphi^*})-(1-R_i)\left[e^{-\varphi_1^*}-e^{-\varphi^*}\Phi\left(\sqrt{\varphi_0^*-\varphi^*}\right)\right]\right\}. \tag{4.28}$$

In the ionosphere $D\sim 0.1$ cm, so that the electric field intensity near the surface of a body is of the order of a few volts per cm, if φ_0 is of the order of a few decavolts. The dependence of the electric field intensity (on the surface of the body) on φ_0 is shown in Fig. 25.

e. Electric Field in the Rarefied Region. Equation (4.10) is, of course, also valid in the region of rarefaction. By fulfilling the condition (4.13),

$$\left(\frac{D}{R_0}\right)^2\frac{N_0}{N_i}\frac{\Delta_\xi \ln\frac{N_0}{N_i}}{\ln\frac{N_0}{N_i}}\ll 1. \tag{4.13}$$

We find that the potential in the region of rarefaction is given by

$$\varphi=\frac{\varkappa T}{e}\ln\frac{N_i(r)}{N_0}. \tag{4.14}$$

At considerable distances behind the body, the ratio N_i/N_0 is not too small. Condition (4.13) is well satisfied in this case. In particular, as shown in Chapter II, at large distances behind the body (for $z > R_0V_0/v_i$)

$$-\delta N_i=N_0-N_i\ll N_0.$$

In this region the potential is proportional to the disturbance of the ion concentration:

$$\varphi=\frac{\varkappa T}{e}\frac{\delta N_i(r)}{N_0}. \tag{4.29}$$

At distances of order R_0 from a body, the rarefaction is very great, so that condition (4.13) is violated. Therefore, the double layer region behind a rapidly moving body is of the order of the dimensions of the body itself, i.e., it is much larger than the corresponding region in front. It is indicated by the dashed line in Fig. 26.

The concentration of ions behind a body was calculated in Chapters II and III without considering the influence of an electric field on the motion of the ions. Using the corresponding expressions for the concentration of ions in formula (4.14), one can find the potential of the electric field. For example, in the absence of a magnetic field, at large distances behind (when $z > R_0V_0/v_i$) a rapidly moving body of cross section S, with the aid of Eq. (2.29) we obtain

$$\varphi = -\frac{S_0 M_i V_0^2}{2\pi e z^2}\exp\left(-\frac{M_i V_0^2}{2\varkappa T}\cdot\frac{x^2+y^2}{z^2}\right). \tag{4.30}$$

The shape of the equipotential surfaces behind a body in the absence of a magnetic field are shown in Fig. 26. It is clear from the figure and from Eq. (4.14) that the potential of the field increases (in absolute value) as one approaches the body; the maximum value of φ is reached at the boundary of the double layer, where it is

$$\varphi_{\max} \approx -\frac{\varkappa T}{e}\ln\left(\frac{R_0}{D}\right)^2. \tag{4.31}$$

In the ionosphere $\ln (R_0/D)^2 \sim 10$ so that the potential φ close to the boundary of the double layer is an order of magnitude larger than $\varkappa T/e$, i.e., $\varphi \sim 1$ to 2 V. In front of the body, as indicated earlier, φ is only on the order of $\varkappa T/e$.

The influence of an electric field on the motion of the ions was not considered above. At the same time the potential of the field in the rarefied region is comparatively large, and the field acts on the ions, filling the rarefied region, for a sufficiently long time, in consequence of the large length of this region. Because of this, the rate of filling of the rarefied region by ions depends significantly on the field so that the results obtained without taking account of the influence of an electric field are correct only to within a factor of order unity.

The influence of an electric field on the motion of the ions was not considered above. At the same time, the potential of the field in the rarefied region is comparatively large, and the field acts on ions occupying the rarefied zone for a rather long time, as a consequence of the long extension of this zone. Because of this, the velocity (rate) of filling of the rarefied region by ions significantly depends on the field, so that the results obtained without account of the influence of an electric field are only valid to within a factor of the order of unity. To a rough approximation, it is possible to estimate the influence of an electric field if we note that the electric field should increase the pressure of the ions. At the same time, the electric field itself is caused by the electrons. Therefore, one can approximately take the field into consideration, replacing the pressure of the ions by the total pressure in the plasma which equals $\varkappa(N_eT_e + N_iT_i)$. Since the concentrations and temperatures of ions and electrons are equal, this leads to an effective doubling of the ion temperature. In this approximation the concentrations of ions and electrons in the rarefied region behind a body are, as before, determined by the expressions obtained in Chapters II and III; it is only necessary to replace T by 2T in them. This leads to contraction by a factor of $1/\sqrt{2}$ of the entire disturbed region. A comparison of the results of an approximate calculation of the ion concentration disturbance behind a body (curve c) with the solution obtained with account of the influence of an electric field on the motion of the ions in the region of a weakly disturbed plasma (curve a, also see § 15) is presented in Fig. 27. Curve b in this same figure is obtained from curve a by replacing T by 2T. It takes the influence of the field into account in the rough approximation indicated above.

f. Double Layer Behind a Body. The ion concentration in the double layer behind a body in the region of maximum rarefaction is negligibly small. On the boundary of the double layer, as we saw (see Eq. (4.31)), the potential of the field is rather large, and since it is negative, the electron concentration in this region is also small and does not exceed the value

$$N \leqslant N_0\exp\left(\frac{e\varphi_{\max}}{\varkappa T}\right) \approx N_0\left(\frac{D}{R_0}\right)^2. \tag{4.32}$$

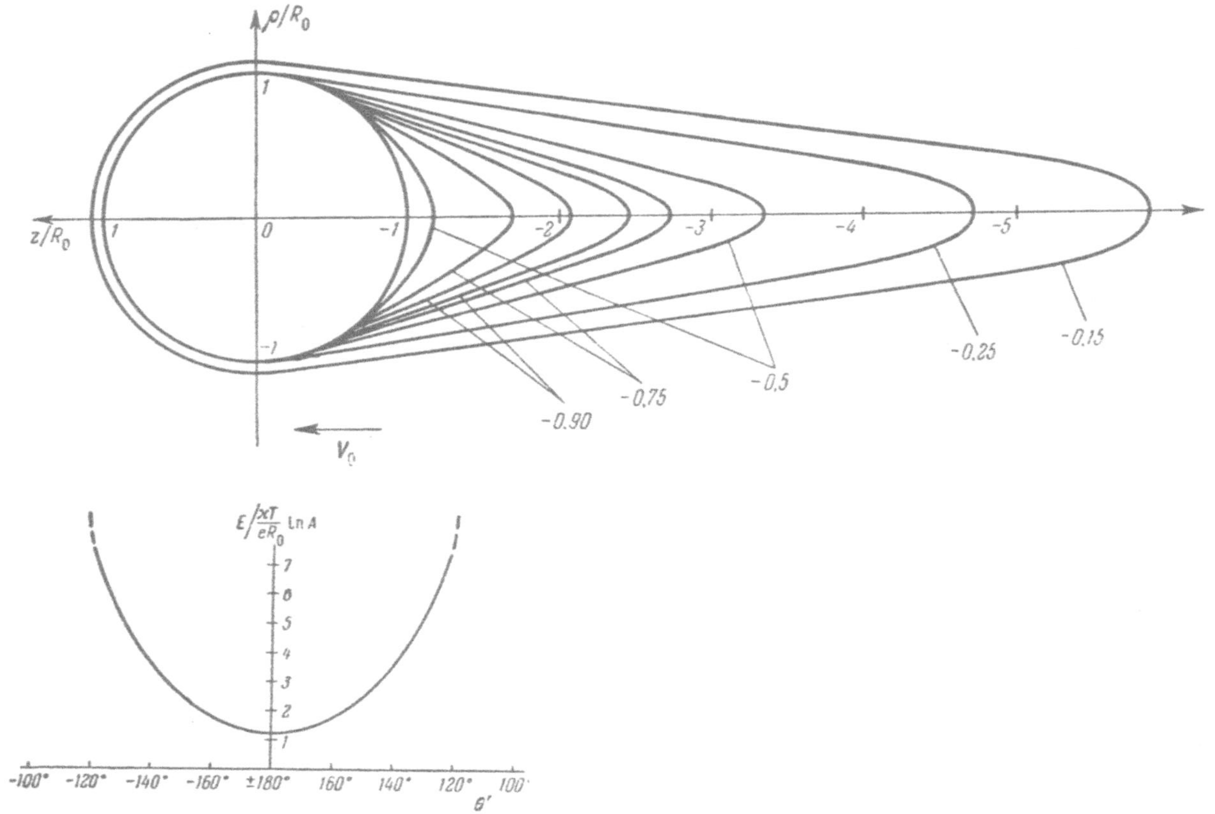

Fig. 28. a) Lines of equal potential in the vicinity of a metallic sphere which absorbs the particles incident on it ($\varphi_0 = 0.25\ (\varkappa T/e)\ \ln (R_0/D)^2$). Values of $e\varphi/[\varkappa T \ln (R_0/D)^2]$ are indicated on the figure. b) Dependence of the intensity of the field (on the surface of a metallic body) on direction.

Therefore, the influence of free charges on the electric field in the double layer behind a body is small, and correct to logarithmic terms of order $1/\ln (R_0/D)^2$, one can neglect this influence. Equation (4.9) is then written in the form

$$\Delta\varphi = 0. \tag{4.33}$$

Here the boundary conditions have the following form:

$$\varphi = -\frac{\varkappa T}{e} \ln\left(\frac{R_0}{D}\right)^2$$

on the boundary of the region of maximum rarefaction and

$$\varphi = \varphi(r_S)$$

on the surface of the body, where the value of $\varphi(\mathbf{r}_S)$ is determined as indicated in the paragraph of § 14b. In particular, the potential $\varphi(\mathbf{r}_S) = \varphi_0$ for a metallic body, and φ_0 is determined from the condition that the total current onto the body's surface is equal to zero (see Eq. (4.7)).

One can numerically obtain a solution of Eq. (4.33) for the given boundary conditions with the aid of the usual methods of electrostatics. Curves of different values of the potential φ for a metallic surface with $(MV_0^2/2\varkappa T)^{1/2} = 8$ and $\varphi_0 = 0.25\,(\varkappa T/e)\ln(R_0/D)^2$ are presented in Fig. 28a. It is clear from the figure that in the case under consideration, the potential of the field first increases (in absolute value) with increasing distance from the body up to the boundary of the double layer; the maximum value of φ on the boundary of the double layer is given by

$$\varphi_{\max} \approx -\frac{\varkappa T}{e} \ln\left(\frac{R_0}{D}\right)^2.$$

The dependence of the electric field intensity, near the surface of a spherical body in the rarefied zone, on the angle θ' is shown in Fig. 28b. The value of E is minimal in the direction opposite to the direction of the body's motion, i.e., for $\theta' = 180°$. With approach to $\theta' = 90°$, E increases considerably – up to a value of the order of $\varkappa T/eD$, where D is the Debye radius. In the opposite direction, as was shown in § 14d, the field intensity has the other sign; in absolute magnitude it is also of the order of $\varkappa T/eD$.

The potential in general must also preserve its form in the presence of other conditions on the surface of the body; for example, if only part of the ions or electrons are absorbed, or taking into account photoemission, thermal emission, and so forth. Only the form of the potential φ in the region of the double layer significantly depends on conditions at the surface. Of course, these remarks are valid only in the case when the dimensions of the body are much larger than the Debye radius. We also remark that Eq. (4.14) is applicable while also calculating the influence of a constant magnetic field (provided that the body is not moving strictly along the direction of the magnetic field), if the expressions obtained in Chapter III are substituted for $N_i(\mathbf{r})$. We recall that the magnetic field significantly affects the ion concentration only at rather large distances from the body, where the potential is already small.

We also note that if the presence of ions with different masses is taken into account, for example, the presence of a small number of light hydrogen ions, then this may strongly change the structure of the double layer behind the body.

Numerical integration of Poisson's equation for the electric field behind a body without separation into a double layer region and a quasi-neutral region is carried out in article [79].* The author considered an ellipsoid of revolution, moving at various angles with respect to its axis. The distribution of the ion concentration behind the body in [79] is assumed equal to the concentration of neutral particles, i.e., the calculation is carried out without account of the influence of an electric field on their motion, in the same approximation

*Article [79] became known to the authors during the period of preparation of the English edition of the book.

as above. The results of [79] indicate that in the region which we regarded as the region of quasi-neutrality, the concentration of electrons is actually very close to the concentration of ions. The electron concentration differs appreciably from the ion concentration only in a region of the order of the body's dimensions. Thus, the results of [79] confirm the correctness of the approximate calculation used above. They indicate that this calculation is correct to within a few percent. We also remark that the values of the relative difference in the concentrations of ions and electrons obtained in [79] are, naturally, in close agreement with the values of this quantity which follow from Eqs. (4.29) and (4.30).

§ 15. Influence of an Electric Field on the Motion of Ions

The electric field appearing in the vicinity of a body moving in a plasma was calculated in the preceding section. In this connection, the influence of the field on the motion of the ions was neglected. At the same time, an electric field has an important effect on the motion of ions in the region of rarefaction, as indicated above. Therefore, for a rigorous solution of the problem, it is here necessary to look for a simultaneous solution of the system of Eqs. (1.23), (1.24), and (1.30), which describe the distribution functions for electrons and ions and the electric field in the medium. In the general case, it has not been possible to find the solution of this system of equations, but one can investigate it by means of numerical calculations for specific cases.

However, at large distances from the body, the disturbances of the plasma are not large. Also the electric field is small here. Hence in a certain region, it is possible to linearize the system of Eqs. (1.23), (1.24), and (1.30). In addition, if the body is not moving along the magnetic field, then at large distances from the body the distribution of electrons is always the equilibrium distribution (1.29), and the potential of the electric field is determined by Eq. (4.9). In this case, the linearized system of Eqs. (1.23) and (4.9) has the form

$$v\frac{\partial f_i'}{\partial r}+\frac{e}{\varkappa T}f_{0i}\frac{\partial\varphi}{\partial r}\left(v+V_0\right)+\frac{e}{M_i c}\left[v+V_0,\ H_0\right]\frac{\partial f_i'}{\partial v}=\begin{cases}-\frac{rV_0}{r}f_{0i}\delta(r-R_0), & rV_0<0,\\ 0, & rV_0>0,\end{cases} \tag{4.34}$$

$$\Delta\varphi=-4\pi e\left(\delta N_i-N_0\frac{e\varphi}{\varkappa T}\right). \tag{4.35}$$

Here $f_i'=f_i-f_{0i}$ is the disturbance of the distribution function for the ions, $\delta N_i=N_i-N_0$ is the ion concentration disturbance:

$$\delta N_i=\int f_i' d^3v. \tag{4.36}$$

In Eqs. (4.34) and (4.35) it is assumed that f_{0i} is the Maxwellian distribution function, and only terms of zero or first order (of smallness) in f_i' and φ are kept. It is possible to obtain a rigorous analytic solution of the system of Eqs. (4.34) and (4.35), and it (this solution) is considered below.

a. Nonmagnetic Plasma. For simplicity, at the beginning we consider the case of a body moving in an unmagnetized plasma ($H_0=0$). In this connection, we remark that it is more convenient to determine not the functions $f_i'(\mathbf{r},\mathbf{v})$ and $\varphi(\mathbf{r})$ themselves, but their Fourier components $f_{\mathbf{q}}(\mathbf{v})$, $\varphi_{\mathbf{q}}$ with respect to the coordinates. After this, one can find f_i' and φ by means of the inverse Fourier transformation. However, one can already see the general nature of the influence of an electric field from the formulas for $N_{\mathbf{q}}$.

Changing (in the absence of a magnetic field) to Fourier components in Eq. (4.34), we obtain

$$iqvf_q+iq(v+V_0)\varphi_q f_{0i}\frac{e}{\varkappa T}=-f_{0i}V_0\pi R_0^2 G. \tag{4.37}$$

For small values of q when $qR_0 \ll 1$, $G(qR_0,\ \chi)=1$ so that the equation takes the form

$$iqvf_q+iq(v+V_0)\varphi_q f_{0i}\frac{e}{\varkappa T}=-f_{0i}V_0\pi R_0^2. \tag{4.38}$$

In the following section we shall see that it is possible to rigorously obtain an equation analogous to (4.38) for small values of q, without neglecting nonlinear (in the field) effects. In addition, for small values of q

$$N_q = \frac{N_0 e}{\varkappa T} \varphi_q. \tag{4.39}$$

as follows from Eq. (4.35). Substituting (4.39) into (4.38), we find

$$f_q + \frac{q(v+V_0)}{qv} \frac{f_{0i}}{N_0} N_q = \frac{i f_{0i} V_0 \pi R_0^2}{qv}. \tag{4.40}$$

Integrating both sides of (4.40) with respect to d^3v and taking (4.36) into consideration, we obtain

$$N_q = \frac{i\pi R_0^2 V_0 \int f_{0i} \frac{d^3v}{qv}}{2 + \frac{qV_0}{N_0} \int f_{0i} \frac{d^3v}{qv}} = \frac{i\pi R_0^2 V_0 \int f_{0i} \frac{d^3u}{qu - qV_0 - i\delta}}{2 + \frac{qV_0}{N_0} \underset{(\delta \to +0)}{\int} f_{0i} \frac{d^3u}{qu - qV_0 - i\delta}}.$$

In order to assign a meaning to the improper integrals, in accordance with the discussion on p. 45 we replaced $\mathbf{q}\mathbf{V}_0$ by $\mathbf{q}\mathbf{V}_0 + i\delta$ $(\delta \to +0)$. With the aid of formulas (2.90) and (2.91), we finally have

$$N_q = \frac{\pi R_0^2 N_0 \left(\frac{M_i V_0^2}{2\varkappa T}\right) Q(a)}{q\,[2 + iaQ(a)]}, \tag{4.41}$$

where the function Q(a) is, as before, defined by Eq. (2.92). For finite values of q the quantity N_q is approximately equal to expression (4.41) multiplied by the function $G(qR_0, \chi)$. Formula (4.41) differs from formula (2.93), obtained without consideration of the influence of an electric field, by the presence of the denominator which depends on the angle χ between $\mathbf{q}$ and $\mathbf{V}_0$. In this connection, if

$$a = \sqrt{\frac{M_i V_0^2}{2\varkappa T}} \cos\chi \gg 1 \tag{4.42}$$

(which is satisfied in any case when the angle χ is not close to $\pi/2$), then

$$Q(a) \approx \frac{i}{a} \tag{4.43}$$

and the denominator is equal to unity, i.e., under such conditions the electric field is unimportant. But if

$$a \ll 1, \quad \text{i.e.,} \quad \cos\chi \ll \sqrt{\frac{\varkappa T}{M_i V_0^2}}, \tag{4.44}$$

then the denominator is equal to two, i.e., in this case the electric field decreases N_q by a factor of two. It is of interest to note that the last result corresponds exactly to the hydrodynamical approximation considered in the preceding section.

The results obtained have a simple physical meaning. In fact, the electric field may have an important effect on the motion of ions only in that case when the field does not vary too rapidly, i.e., during the time required for significant variation of the field an ion is able to travel a distance on the order of a wavelength. The frequency of the component of the field in a fixed (stationary) coordinate system is equal to $\mathbf{q}\mathbf{V}_0$. Therefore, the field may have a significant effect on the motion of an ion only if $v_i/\mathbf{q}\mathbf{V}_0 \sim (\varkappa T/M_i)^{1/2}/\mathbf{q}\mathbf{V}_0 \gtrsim 1/q$, i.e., only when $a \ll 1$. We see that an electric field, although it does not qualitatively change the results obtained without its consideration, nevertheless may appreciably change the value of N_q. In this connection, one should keep in mind that the linear approximation probably somewhat overestimates the influence of an electric field, since in it one actually assumes that, in the region where the field is sufficiently large, the ion concentration is close to N_0. In reality, the ion concentration in the region of maximum value of the field is smaller than N_0, which must lead to a decrease in the influence of the field.

We note that if, in the denominator of (4.41), we replace a by $(\omega/q)(M_i/2\varkappa T)^{1/2}$ and we equate this denominator to zero, then we obtain a dispersion equation describing the propagation of ionic plasma waves, i.e., an equation describing the relation between the frequency ω and the wave vector $\mathbf{q}$ in such a wave. This circumstance is not of great importance, however, in the present case since the ionic plasma waves are strongly attenuated, and consequently the zeros of the denominator lie in the complex plane far away from the real axis and do not manifest themselves for real values of ω and $\mathbf{q}$. If, however, the temperature of electrons in the plasma were much higher than the temperature of the ions (in the ionosphere, there is apparently no basis for such an assumption), then these waves would be much more weakly damped out, which would lead to additional maxima in the expressions for f_q, N_q, and φ_q for conditions when $\mathbf{q}\mathbf{V}_0/q$ would be equal to the velocity of an ionic plasma wave.

It is possible to obtain an expression for $\delta N(\mathbf{r})$ from the formula for N_q by taking the inverse Fourier transformation [70]. It is impossible to carry out this transformation analytically to the end. In order to estimate the influence which an electric field turns out to have on $\delta N(\mathbf{r})$, we shall carry out the corresponding calculation at large distances from the body, where the disturbances of the density decrease $\sim 1/r^2$. Here we may directly use formula (4.41) for small values of q. It is clear from (4.41) that N_q has the form

$$N_q = \frac{B(a_0 \cos\chi)}{q} a_0^2 \pi R_0^2 N_0, \tag{4.45}$$

where $a_0 = (M_i V_0^2/2\varkappa T)^{1/2}$, and χ is the angle between $\mathbf{q}$ and $\mathbf{V}_0$. Therefore

$$\delta N(r) = \frac{a_0^2 \pi R_0^2 N_0}{(2\pi)^3} \int \frac{B}{q} e^{i\mathbf{q}\mathbf{r}} d^3 q. \tag{4.46}$$

Introducing a spherical coordinate system with axis along $\mathbf{V}_0$ and denoting, as before, the angle between $\mathbf{r}$ and $\mathbf{V}_0$ by ϑ, we have:

$$d^3 q = \cos\chi \, d\chi \, d\varphi q^2 \, dq, \quad \mathbf{q}\mathbf{r} = qr(\cos\vartheta\cos\chi + \sin\vartheta\sin\chi\cos\varphi). \tag{4.47}$$

Substituting (4.47) into (4.46) and integrating with respect to $d\varphi$, we obtain

$$\delta N(r) = \frac{a_0^2 \pi R_0^2 N_0}{(2\pi)^2} \int B(a_0 \cos\chi) \exp(iqr\cos\vartheta\cos\chi) \, J_0(qr\sin\vartheta\sin\chi) \, q \, dq \sin\chi \, d\chi. \tag{4.48}$$

The integral with respect to dq in Eq. (4.48) diverges for large values of q. This is associated with the fact that expression (4.45) for N_q is valid only for small values of q, so that in fact the integrand decreases for large values of q and the integral converges. Therefore, in order to assign the correct meaning to the integral with respect to dq, we must make it convergent. For this purpose, we carry out the following substitution in the exponent of the exponential:

$$\cos\vartheta\cos\chi \rightarrow \cos\vartheta\cos\chi + i\delta \quad (\delta \rightarrow +0).$$

Now one can evaluate the integral with respect to dq with the aid of the formula

$$\int_0^\infty e^{-\gamma t} J_0(\beta t) \, t \, dt = \frac{\gamma}{(\gamma^2 + \beta^2)^{3/2}} \quad (\mathrm{Re}\,\gamma > 0).$$

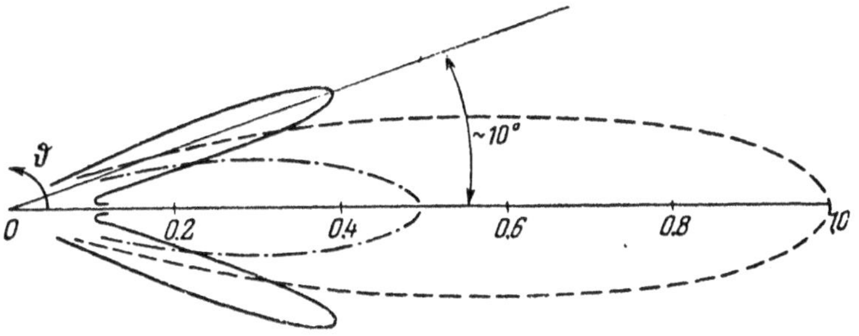

Fig. 29. Plot of the function $B_1(\vartheta)$ for $a_0 = V_0/(2\varkappa T/M_i)^{1/2} = 8$.

After simple calculations, for $\cos\vartheta < 0$ (i.e., behind the body) we obtain

$$\delta N(r) = \frac{a_0^2 \pi R_0^2 N_0}{r^2 (2\pi)^2}\, i \int_0^{\pi} \frac{\cos\vartheta B\,(a_0 \cos\chi)\cos\chi \sin\chi\, d\chi}{(\sin\vartheta - \cos\chi + i\delta)^{3/2}(\sin\vartheta + \cos\chi - i\delta)^{3/2}},$$

where $\delta \to +0$. Denoting the integral by $-i4\pi^2 B_1(\vartheta, a_0)$, we have

$$\delta N(r) = -\frac{R_0^2 N_0 M_i V_0^2}{2r^2 \varkappa T} B_1(\vartheta, a_0) \quad . \tag{4.49}$$

In the absence of electric and magnetic fields $B_1(\vartheta, a_0)$ according to (2.29) is given by the formula

$$B_1 = \exp(-a_0^2\vartheta^2) = \exp\left(-\frac{M_i V_0^2}{2\varkappa T}\vartheta^2\right). \tag{4.50}$$

If the influence of the field is taken into consideration in the hydrodynamic approximation, then

$$B_1 = \frac{1}{2}\exp(-a_0^2\vartheta^2/2). \tag{4.50a}$$

It is possible to evaluate the integral in (4.49) numerically; a graph of the function $B_1(\vartheta, a_0)$ in polar coordinates for the value $a_0 = 8$ is given in Fig. 29. For comparison, the corresponding curve (4.50) without account of the influence of the electric field is shown by the dashed line, and the dot-dashed line indicates the function B_1 from Eq. (4.50a). It is clear from the figure that an electric field significantly alters the concentration of particles. In particular, the greatest rarefaction is now no longer achieved on the z axis. One can say that an electric field leads to some focussing of the particles towards the axis. The order of magnitude of the disturbance behind a body, of course, is not strongly changed. It is of most interest that the maximum disturbance, with account of the influence of a field, behind a body is reached not on the axis of motion, but on the surface of a cone with an opening angle $2\theta_m$, where $\theta_m \sim 10°$.

This cone is similar to the Mach cone in ordinary gas dynamics. The growth of disturbances on the cone reflects the Cherenkov excitation of longitudinal ionic-sound waves by a body moving through a plasma with supersonic velocity. In fact, for $T_e = T_i$, the dispersion law for such waves [74] in the absence of collisions has the form

$$\omega + i\gamma = \sqrt{\frac{2\varkappa T}{M}}\,(1.3)\,(1 + 0.23\,i)\,K, \tag{4.51}$$

where γ characterizes the attenuation of the wave and K is the wave number (real). Therefore, the opening angle ϑ_M of the cone corresponds closely to the velocity of ionic-sound waves, i.e.,

$$\vartheta_M = \arcsin(\omega/V_0 K) = 1.3/a_0 \approx 9.2°$$

In a single-temperature plasma the ionic-sound waves are rather strongly damped, as is clear from Eq. (4.51). Therefore, the Mach cone is "washed out."

Up to now, we have assumed that the temperatures of ions and electrons are the same. Now let us assume that $T_e \gg T_i$. An investigation of this case is of some practical interest. The point is that when $T_e >> T_i$ (see p. 109 of [17]), then for a description of the motion in a quasi-neutral plasma one can neglect the thermal motion of the ions and use the hydrodynamic equations which have the form:

$$(\partial N/\partial t) + \mathrm{div}(N\,\mathbf{v}) = 0, \tag{4.52a}$$

$$N\left[(\partial \mathbf{v}/\partial t) + (\mathbf{v}\cdot\nabla)\mathbf{v}\right] = v_T^2\,\nabla N, \tag{4.52b}$$

where $v_T = (\varkappa T_e/M_i)^{1/2}$ is the velocity of sound in the plasma. Equations (4.52a) and (4.52b) coincide with the usual equations of gas dynamics for the case of isothermal motion of the gas. At large distances behind a body ($z \gg a_0R_0$), the disturbances are small and one can linearize Eqs. (4.52). In this connection, it is possible to take the presence of a moving body into account, if terms describing the absorption (neutralization) of ions on the body's surface and the transfer of momentum from the ions to the body are introduced into the equation. The equation of continuity in the system of coordinates attached to the body then takes the form:

$$N_0\,\mathrm{div}\,\mathbf{u} + \mathbf{V}_0\nabla N = \tilde{I}\,\delta(\mathbf{r}), \tag{4.53}$$

where I is the number of ions absorbed per unit time and $\mathbf{u} = \mathbf{v} + \mathbf{V}_0$. In order to take the transfer of momentum from the ions to the body into account, we rewrite Eq. (4.52b) in the form of an equation for the conservation of momentum:

$$\frac{\partial}{\partial t}(M_i N v_i) + \sum_K \frac{\partial}{\partial x_K} N(M_i\, v_i\, v_K + M_i\, v_T^2\delta_{iK}) = 0.$$

Now it is clear that friction will be taken into account, if the term $-\overline{\mathbf{F}}_i\delta(\mathbf{r})$ is added to the right side of this equation, where $\overline{\mathbf{F}}_i$ is the momentum transferred by the ions to the body per unit time. In the stationary case, after linearization the equation is written in the form

$$-M_i N(\mathbf{V}_0\nabla)\,\mathbf{u} + M_i\, v_T^2\nabla\,\delta N = (-\,\mathbf{F} + \tilde{I}\,\mathbf{V}_0 M_i)\,\delta(r) \tag{4.54}$$

Here Eq. (4.53) is also utilized. If it is assumed that $\tilde{I} = 0$, then Eqs. (4.53) and (4.54) coincide with those used by Dokuchaev [16]. Therefore, the following calculations are similiar to the calculations in [16].

Taking the Fourier transform with respect to coordinates in Eqs. (4.53) and (4.54), and elminating from them the Fourier component of $\mathbf{u}$, we find

$$N_{\mathbf{q}} = \mathrm{i}\,\frac{\mathbf{q}\cdot\mathbf{F}_i - 2(\mathbf{q}\,\mathbf{V}_0)M_i\,\tilde{I}}{M\{(\mathbf{q}\,\mathbf{V}_0)^2 - q^2\,v_T^2\}}.$$

If we assume that all ions are absorbed by the body, then

$$I = -\pi R_0^2 N_0 V_0, \qquad \mathbf{F}_i = -\pi R_0^2 N_0 V_0\,(M_i\mathbf{V}_0)$$

so that

$$N_{\mathbf{q}} = i\,\frac{\pi R_0^2 N_0\, V_0\,(\mathbf{q}\mathbf{V}_0)}{\{(\mathbf{q}\mathbf{V}_0)^2 - q^2\,v_T^2\}}.$$

Taking the inverse Fourier transform, we obtain

$$\delta N(\vartheta) = -\frac{R_0^2\,a_0^2\,N_0}{z^2(1-\beta^2\tan^2\vartheta)^{3/2}}, \tag{4.55}$$

where $\beta^2 = a_0^2 - 1$ and

$$a_0^2 = V_0^2/v_T^2 = M V_0^2/2\varkappa T_e.$$

It is clear from Eq. (4.55) that the angular distributions of the distrubances at large distances behind a body are significantly different in the gas-dynamic approximation and in the approximations considered above. The characteristic feature of the gas-dynamic approximation is the "conversion" of the disturbance to infinity on the Mach cone. The physical reason for this notation is: shock waves, whose intensity falls off with distance in proportion to $r^{-4/3}$ (see [21], p. 557), are created on the Mach cone in gas-dynamics. This also leads to a divergence in the asymptotic terms $\sim 1/r^2$. Thus, in the gas-dynamic approximation, the disturbances of the particle concentration behind a body decrease much more slowly than $1/r^2$.

We note in conclusion that the influence of an electric field on the motion of ions was investigated in article [57] by Davis and Harris. A simultaneous numerical integration of the equations of motion of the ions and of Poisson's equation was carried out in this article. However, the authors disregarded the thermal motion of the ions which, in our opinion, is not permissible since it is just the thermal velocities which play a decisive role in filling the rarefied zone. Therefore, one cannot expect that the results of this article, represented in the form of rather intricate graphs, give a quantitatively valid description of the disturbed region. Nevertheless, this article is of interest since it is clear that a complete quantitative solution of the problem of the interaction of a body with a plasma can only be obtained with the aid of numerical calculations. From what has been said above, it is clear that Davis and Harris in their calculations could have used the equations of gas-dynamics. This, undoubtedly, would simplify the calculations. In this connection, at large distances from a body, the corresponding results must coincide with Eq. (4.55). Therefore, any attempts to develop effective methods for such calculations are of great value. It is of interest that the results of [57] also indicate some focusing of ions by the field. We also mention the work of Sengupta [58]. In this article, the disturbances around a cylinder moving in a plasma are calculated under the assumption that the distribution function f may be represented in the form of a product of two factors, one of which only depends on the coordinates, and the other depends only on velocities. Such a form of the function f holds only in the case of an equilibrium Boltzmann distribution. Therefore, the corrections to this distribution obtained by the author hardly have any real meaning.

b. Plasma in a Magnetic Field. Now let us investigate the influence of an electric field on the motion of ions in the case when the plasma is located in an external magnetic field. The calculation will be carried out, as above, in the linear approximation. Equation (4.34) in Fourier components is written in the form

$$\frac{e}{M_i c}[\mathbf{u}\mathbf{H}_0]\frac{\partial f_q}{\partial \mathbf{u}} + i\mathbf{q}(\mathbf{u}-\mathbf{V}_0) f_q + i\mathbf{q}\mathbf{u}\varphi_q \frac{f_{0i}}{N_0}\frac{e}{\varkappa T} = -\pi R_0^2 f_{0i} G, \tag{4.56}$$

or for $qR_0 \to 0$

$$\frac{e}{M_i c}[\mathbf{u}\mathbf{H}_0]\frac{\partial f_q}{\partial \mathbf{u}} + i\mathbf{q}(\mathbf{u}-\mathbf{V}_0) f_q + i\mathbf{q}\mathbf{u}\varphi_q \frac{e}{\varkappa T}\frac{f_{0i}}{N_0} = -\pi R_0 V_0 f_{0i}. \tag{4.56a}$$

Solving this equation in exactly the same manner as Eq. (3.32) (see Eq. (4.39)), we find the following expression for N_q:

$$N_q = -\frac{\pi R_0^2 V_0}{\Omega}\,\frac{\displaystyle\int\int_0^\infty f_{0i}(\mathbf{u}) \exp A \, d^3u \, dx}{1 + \dfrac{i}{\Omega N_0}\displaystyle\int\int_0^\infty f_{0i}(\mathbf{u}) B \exp A \, d^3u \, dx}. \tag{4.57}$$

Here

$$A=\left[-i\left(\frac{q_{\|}u_{\|}-qV_0}{\Omega}x+\frac{2q_{\perp}u_{\perp}}{\Omega}\sin\frac{x}{2}\right)\right],$$

$$B=\left[q_{\|}u_{\|}+q_{\perp}u_{\perp}\cos\left(\beta+\frac{x}{2}\right)\right].$$

It is possible to integrate the integral with respect to dx in the denominator by parts. Here it is necessary to substitute

$$q_{\perp}u_{\perp}\cos\left(\beta+\frac{x}{2}\right)=q_{\perp}u_{\perp}\cos\beta\cos\frac{x}{2}-q_{\perp}u_{\perp}\sin\beta\sin\frac{x}{2}$$

and to take into consideration that the integral of the second term is equal to zero because of the fact that the integrand is an odd function of β. As a result we obtain

$$N_q=-\frac{\pi R_0^2 V_0}{\Omega}\frac{\int\int_0^{\infty} f_{0i}(u)\exp A\, d^3u\, dx}{2+\frac{iqV_0}{\Omega N_0}\int\int_0^{\infty} f_{0i}(u)\exp A\, d^3u\, dx}, \tag{4.58}$$

where, as also above,

$$A=\left[-i\left(\frac{q_{\|}u_{\|}-qV_0}{\Omega}x+\frac{2q_{\perp}u_{\perp}}{\Omega}\sin\frac{x}{2}\right)\right].$$

Now calculating the integral over d^3u with the aid of formulas (3.38)-(3.39), we finally have

$$N_q=-\frac{\pi R_0 N_0 V_0}{\Omega}\frac{\int_0^{\infty}\exp\left[-\frac{iqV_0}{\Omega}x-\frac{\varkappa T}{2M_i\Omega}C\right]dx}{2+\frac{iqV_0}{\Omega}\int_0^{\infty}\exp\left[\frac{iqV_0}{\Omega}x-\frac{\varkappa T}{2M_i\Omega}C\right]dx}, \tag{5.59}$$

where

$$C=\left(q_{\|}^2x^2+4q_{\perp}^2\sin^2\frac{x}{2}\right).$$

Expression (4.59) differs from (3.40), obtained without taking account of an electric field, by the presence of the denominator. As we shall see later on, this denominator may play an important role.

As in Chapter III, let us carry out a more detailed investigation of formula (4.59) under the conditions $q_{\|}\to 0$, $\mathbf{q}\mathbf{V}_0/\Omega\to p$, $p=0, \pm1, \pm2, \ldots$ In this case, transforming the integral with the aid of formulas (3.47) and (3.43), we obtain

$$\left.\begin{aligned}N_q&=-\frac{\pi R_0^2 N_0 V_0}{q_{\|}}\frac{\sqrt{\frac{M_i}{2\varkappa T}}Q(a_1)\exp\left(-\frac{q_{\perp}^2\varkappa T}{M_i\Omega^2}\right)I_p\left(\frac{q_{\perp}^2\varkappa T}{M_i\Omega^2}\right)}{2+i\frac{qV_0}{q_{\|}}\sqrt{\frac{M_i}{2\varkappa T}}Q(a_1)\exp\left(-\frac{q_{\perp}^2\varkappa T}{M_i\Omega^2}\right)I_p\left(\frac{q_{\perp}^2\varkappa T}{M_i\Omega^2}\right)},\\ a_1&=\frac{qV_0-p\Omega}{q_{\|}}\sqrt{\frac{M_i}{2\varkappa T}}.\end{aligned}\right\} \tag{4.60}$$

The behavior of expression (4.60), however, is different in the two cases, when p = 0 and when p ≠ 0. Let p = 0. Then

$$N_q = -\frac{\pi R_0^2 N_0 V_0}{q_{\|}} \frac{\sqrt{\frac{M_i}{2\varkappa T}}\, Q(a_2) \exp\left(-\frac{q_\perp^2 \varkappa T}{M_i \Omega^2}\right) I_0\left(\frac{q_\perp^2 \varkappa T}{M_i \Omega^2}\right)}{2 + i a_2 Q(a_2) \exp\left(-\frac{q_\perp^2 \varkappa T}{M_i \Omega^2}\right) I_0\left(\frac{q_\perp^2 \varkappa T}{M_i \Omega^2}\right)}, \tag{4.61}$$

where $a_2 = (\mathbf{q}\mathbf{V}_0/q_{\|})(M_i/2\varkappa T)^{1/2}$ and the function $Q(a_2)$ is determined by formula (2.92). If $q_{\|} \to 0$, but a_2 remains finite, then $N_\mathbf{q}$ tends to infinity like $1/q_{\|}$, as in the case of the absence of an electric field. If a body is moving along the field, then instead of (3.51) we obtain

$$N_q = -\frac{\pi R_0^2 N_0}{q_{\|}} \frac{i \exp\left(-\frac{q_\perp^2 \varkappa T}{M_i \Omega^2}\right) I_0\left(\frac{q_\perp^2 \varkappa T}{M_i \Omega^2}\right)}{2 - \exp\left(-\frac{q_\perp^2 \varkappa T}{M_i \Omega^2}\right) I_0\left(\frac{q_\perp^2 \varkappa T}{M_i \Omega^2}\right)}. \tag{4.62}$$

If now p ≠ 0, then, when $q_{\|} \to 0$ but $a_1 \to$ const, the integrals in the numerator and denominator which are tending to infinity are reduced and,

$$N_q = \frac{i\pi R_0^2 N_0 V_0}{qV_0} = \frac{i\pi R_0^2 N_0 V_0}{p\Omega}, \tag{4.63}$$

$$q_{\|} \to 0, \quad a_1 \to \text{const}. \tag{4.64}$$

Thus, in this case the influence of an electric field is extremely important. Due to its effect, $N_\mathbf{q}$ remains finite under condition (4.64) whereas, without account of the field, $N_\mathbf{q}$ would tend to infinity. Hence it follows that it is necessary to take the electric field into consideration for a calculation of the scattering of electromagnetic waves by the trail of a body. This, of course, does not mean that an electric field exerts such a strong influence on the quantity of $\delta N(\mathbf{r})$. The point is that the infinite value of $N_\mathbf{q}$ under condition (4.64) is associated with the strict periodicity of $\delta N(\mathbf{r})$ in space. An electric field somewhat disturbs this periodicity, which immediately leads to the vanishing of the infinity, except for the case p = 0. Meanwhile for any kind of measurements associated with the spatial distribution of the ions, these small changes of the periodicity are not very important; hence it is reasonable to use the formulas of Chapter III for $\delta N(\mathbf{r})$. Unfortunately, very tedious numerical calculations are required in order to carry out the inverse Fourier transformation on $N_\mathbf{q}$ given by Eq. (4.59) and to obtain $\delta N(\mathbf{r})$ taking into account electric and magnetic fields. We note, however, that for motion of the body at an angle to the field, the disturbances with account of an electric field decrease in proportion to 1/r, the same as before.

§ 16. General Properties of the Equations Which Take Account of an Electric Field

a. Influence of Nonlinear Effects. In the present section we shall investigate the question of how close the properties, obtained in the preceding section, of the solutions of the linearized equations are to the properties of the solutions of the exact equations. In this connection, it is more convenient to use the equations for the Fourier components. We shall investigate the properties of these equations for $q \to 0$, i.e., at large distances from the body.

We first consider the equation without a magnetic field. For this, in the exact equation (1.23) we set $f_i = f_{0i} + f'_i$, $\mathbf{H}_0 = 0$ and we go over to Fourier components, i.e., we express f'_i in terms of f_q, multiply the equation by $e^{-i\mathbf{qr}}$ and integrate with respect to d^3r. We obtain

$$iqv\, f_q - \frac{e}{M_i}\frac{\partial f_{0i}}{\partial v}\, iq\varphi_q = \frac{e}{(2\pi)^3 M_i}\int iq_1 \varphi_{q_1} \frac{\partial f'_{q-q_1}}{\partial v}\, d^3q + \oint \frac{A_i}{|\nabla F|}\, e^{-i\mathbf{qr}}\, dS. \tag{4.65}$$

Now in Eq. (4.65) we go over to the limit $q \to 0$. In this connection, on the right side we may simply set $q=0$. In order to do this, it is necessary that the integral with respect to d^3q_1 is convergent for $q=0$. But we shall see later on that the functions f_q and φ_q are proportional to $1/q$ for small values of q so that the integral converges. Thus, as $q \to 0$

$$iqvf_q + \frac{e}{\varkappa T} iuqf_{0i}\varphi_q = I(v), \tag{4.65}$$

where

$$I(v) = \oint \frac{A_i}{|\nabla F|} dS + \frac{e}{(2\pi)^3 M_i} \int iq\varphi_q \frac{\partial f_{-q}}{\partial v} d^3q = \oint \frac{A_i}{|\nabla F|} dS + \frac{e}{M_i} \int \frac{\partial \varphi}{\partial r} \frac{\partial f_i'}{\partial v} d^3r. \tag{4.67}$$

Equation (4.66) has the same form as Eq. (4.56a), with only this difference: $-\pi R_0 V_0 f_{0i}$ on the right side of Eq. (4.56a) has been replaced by the function $I(\mathbf{v})$ which depends on the properties of the electric field in the near zone, at comparatively small distances from the body.

Now we write Eq. (4.9) for the potential and expand $\exp(e\varphi/\varkappa T)$ in powers of φ; then

$$\Delta\varphi - 4\pi e N_0 \left[\frac{e\varphi}{\varkappa T} + \frac{1}{2} \left(\frac{e\varphi}{\varkappa T} \right)^2 + \ldots \right] = -4\pi e \int f_i' \, d^3v.$$

Going over to Fourier components, we find

$$\left(q^2 + 4\pi e N_0 \frac{e}{\varkappa T} \right) \varphi_q + \frac{1}{2} \frac{e^2}{(\varkappa T)^2} \int \varphi_{q_1} \varphi_{q-q_1} \frac{d^3q_1}{(2\pi)^3} + \ldots = 4\pi e \int f_q \, d^3v.$$

For small values of q, the term which is linear in φ_q is of order $1/q$, and all higher terms tend to constants as $q \to 0$, and one can neglect them in the same way as terms containing q^2. Therefore, for small values of q, Eq. (4.39)

$$N_q = N_0 \frac{e\varphi_q}{\varkappa T}$$

is rigorously true. Simultaneously solving (4.66) and (4.39), we obtain

$$N_q = \frac{-i \int \frac{I(v)}{qv - i\delta} d^3v}{2 + iaQ(a)} \qquad (\delta \to +0), \tag{4.68}$$

where

$$a = \frac{qV_0}{q} \sqrt{\frac{M_i}{2\varkappa T}}.$$

It is immediately obvious from Eq. (4.68) that, upon consideration of nonlinear (in the electric field) effects, N_q also behaves like $1/q$ for small values of q. By the same token, it can be rigorously demonstrated that the disturbances of concentration and of potential fall off like $1/r^2$ at large distances from the body.

It is, of course, impossible to calculate the function $I(\mathbf{v})$ in general form. In the case when the scattering of ions by the body's surface is elastic, one can obtain a useful formula for $I(\mathbf{v})$ in terms of the law for the scattering of particles by the body and by the electric field around it.

In fact, it is clear from the form of Eq. (4.66) that $I(\mathbf{v})d^3v$ is the number of ions acquiring a velocity $\mathbf{v}$ per unit time as a result of scattering by the body and by the field surrounding it. We recall in this connection that the left side of the kinetic equation is the total time derivative of the distribution function. This enables one to obtain a formula for $I(\mathbf{v})$ by repeating the arguments usually developed for the determination of the collision integral.

Let an ion, moving near a body with "impact" parameter ρ and azimuthal angle φ, acquire a velocity $\mathbf{v}$ after scattering. Then the initial velocity of the ion is $\mathbf{v}_1(\mathbf{v}, \rho, \varphi)$, where the function $\mathbf{v}_1(\mathbf{v}, \rho, \varphi)$ is determined by the scattering law, and the number of ions acquiring a velocity $\mathbf{v}$ per unit time is simply the number of ions incident with velocity $\mathbf{v}_1(\mathbf{v}, \rho, \varphi)$, i.e.,

$$\varrho \, d\varrho \, d\varphi v N_0 \left(\frac{M_i}{2\pi\varkappa T}\right)^{3/2} \exp\left\{-\frac{M_i \, [v_1(v, \varrho, \varphi) + V_0]^2}{2\varkappa T}\right\}. \tag{4.69}$$

In writing down Eq. (4.69), it is assumed that the incident-ion distribution is a Maxwellian out to infinity in the fixed coordinate system and that the scattering is elastic, i.e., $v_1 = v$. In order to determine $I(\mathbf{v})$, it is still necessary to subtract from (4.69) the number of ions colliding with the body which have the velocity $\mathbf{v}$ before collision. This number equals

$$\varrho \, d\varrho \, d\varphi v N_0 \left(\frac{M_i}{2\pi\varkappa T}\right)^{3/2} \exp\left[-\frac{M_i (v + V_0)^2}{2\varkappa T}\right].$$

Thus,

$$I(v) = N_0 \left(\frac{M_i}{2\pi\varkappa T}\right)^{3/2} v \int \varrho \, d\varrho \, d\varphi \left\{\exp\left\{-\frac{M_i \, [v_1(v, \varrho, \varphi) + V_0]^2}{2\varkappa T}\right\} - \exp\left\{-\frac{M_i \, [v + V_0]}{2\varkappa T}\right\}\right\} =$$

$$= f_{0i}(u) \, v \int \varrho \, d\varrho \, d\varphi \left\{\exp\left[-\frac{V_0 \delta v(v, \varrho, \varphi)}{\varkappa T}\right] - 1\right\}, \tag{4.70}$$

where $\delta\mathbf{v} = \mathbf{v}_1 - \mathbf{v}$ is the change in an ion's velocity upon scattering. It is clear from Eq. (4.70) that $I(\mathbf{v})$ naturally has the form

$$I(v) = -f_{0i} v S_0(v), \tag{4.71}$$

where $S_0(\mathbf{v})$ is the effective cross section of the body for the scattering of particles with velocity $\mathbf{v}$.

We see that one can calculate $I(\mathbf{v})$ simply if the law for the scattering of particles by a body is known while taking into account an electric field. Of course, formula (4.70) does not enable us to calculate $I(\mathbf{v})$ in the general case because the electric field itself around the body is unknown. Nevertheless, this formula is convenient for approximate calculations.

Up to now we have considered the case of a nonmagnetic plasma. If, however, the body's motion takes place in the presence of a magnetic field, then similiar calculations (although requiring a more complicated basis) lead to an equation for f_q, differing from (4.66) only by the addition of a term containing the magnetic field:

$$\frac{\partial f_q}{\partial u} \frac{e}{M_i c} [u H_0] + iqv f_q + \frac{e}{\varkappa T} iqu f_{0i} \varphi_q = I(v). \tag{4.72}$$

Here the function $I(\mathbf{v})$ is given by the same formula (4.67); however, the specific dependence of I on $\mathbf{v}$ in the presence of a magnetic field may be completely different than without the field. If the magnetic field is not too strong, i.e.,

$$\varrho_H = \frac{c \sqrt{M_i 2\varkappa T}}{e H_0} \gg R_0, \tag{4.73}$$

then it generally has no effect on the motion of ions at distances $\sim R_0$, which give the largest contribution to the integral (4.67). Therefore, it is possible to calculate the function $I(\mathbf{v})$ under condition (4.73) by solving the nonlinear problem without a magnetic field. The solution of Eq. (4.72) has the form

$$N_q = \frac{\frac{1}{\Omega}\int\int\limits_0^\infty I(v)\exp A\,dx\,d^3u}{2+\frac{iqV_0}{\Omega}\int\limits_0^\infty \exp\left[i\frac{qV_0}{\Omega}x - \frac{\varkappa T}{2M_i\Omega^2}C\right]dx}, \tag{4.74}$$

where

$$A = \left\{-i\left[\frac{q_{\parallel}u_{\parallel} - qV_0}{\Omega} + \frac{2q_{\perp}u_{\perp}}{\Omega}\cos\left(\beta + \frac{x}{2}\right)\sin\frac{x}{2}\right]\right\},$$

$$C = \left(q_{\parallel}^2x^2 + 4q_{\perp}^2\sin^2\frac{x}{2}\right).$$

It is not difficult to show that formula (4.74) leads, independently of the form of $I(\mathbf{v})$, to the same qualitative features of the behavior of N_q as (4.59) does. In particular, as before the ion concentration disturbance $\delta N(\mathbf{r})$ decreases according to a $1/r$ law at large distances from the body in the ($\mathbf{v}_0\,\mathbf{H}_0$) plane.

In concluding this section, we note that the function $I(\mathbf{v})$ is related in a simple way to the retarding force which acts on a body. Actually, returning to coordinate space, in place of Eqs. (4.66) and (4.72) we have

$$\frac{df_i}{dt} = I(v)\,\delta(r), \tag{4.75}$$

where df_i/dt is the total time derivative of the distribution function. Multiplying (4.75) by $M_i\mathbf{v}$ and integrating with respect to d^3v, we obtain to the left the time derivative of the total momentum of the ions, i.e., taken with the opposite sign, the force acting on the body as a result of the interaction with the ions. Thus,

$$F_i = -M_i\int vI(v)\,d^3v. \tag{4.76}$$

In order to obtain the total force, it is necessary to add a similar expression for the neutral particles to Eq. (4.76). In this connection, it is necessary to consider the neutral particles even in the case when the unperturbed plasma is almost completely ionized, since neutral particles may be created upon neutralization of ions on the body's surface. The momenta of such particles, escaping (flying away) from the body, give a contribution to the force $\mathbf{F}$.

b. Relation of Concentration Disturbances to the Dielectric Permittivity of the Plasma. We already mentioned that because of the presence in the equations for f_q of terms which are linear in the electric field, a denominator, whose zeros are related to the dispersion law for longitudinal plasma oscillations, appears in the expressions for $N_{\mathbf{q}}$. We can immediately show that the appearance of this denominator is not accidental. More precisely, the terms which are linear in the potential lead to the multiplication of N_q for small values of $\mathbf{q}$ by a factor which is simply related to the dielectric permittivity of the plasma. The proof will be given in a form which is also suitable to the case when an external magnetic field and collisions between particles are present.

As shown in the preceding section, for small values of q the Fourier components of the distribution function f_q satisfy an equation which one can symbolically write in the form

$$L[f_q] = -iquf_{0i}\frac{e}{\varkappa T}\varphi_q + I(v), \tag{4.77}$$

where $L[f]$ is a linear operator acting on functions of the velocity $\mathbf{v}$. An equation of the form (4.77) is valid in the presence of a magnetic field while taking collisions into account. The point is that collisions, under the condition (1.2), are important only at very large distances from the body, where the disturbances are small. Therefore, the collision integral may be linearized in any case.

It follows from Eq. (4.77) that

$$f_q = -\frac{ie}{\varkappa T}\varphi_q L^{-1}[quf_{0i}] + L^{-1}[I(v)], \tag{4.78}$$

where L^{-1} is the operator whose inverse is L. Integrating (4.78) with respect to d^3v and taking (4.39) into consideration, we obtain

$$N_q = \frac{\int L^{-1}[I(v)]\,d^3v}{1+\int L^{-1}\left[qu\frac{f_{0i}}{N_0}\right]d^3v} = \frac{\int f_1(v)\,d^3v}{1+\int f_2(v)\,d^3v}, \tag{4.79}$$

where the functions f_1 and f_2 satisfy the equations

$$\left.\begin{aligned} L[f_1] &= I(v), \\ L[f_2] &= iqu\frac{f_{0i}}{N_0}. \end{aligned}\right\} \tag{4.80}$$

In order to connect the function f_2 with the dielectric permittivity, we impose an external longitudinal electric

$$E = E_0 e^{iqr} = -iq\varphi.$$

on the plasma. Under the influence of this field, the electron concentration is changed by an amount

$$\delta N = N_0\frac{e\varphi}{\varkappa T}.$$

The change in the ion concentration, $\delta N_i = \int f\,d^3v$, is determined with the aid of the equation $L[f] = iquf_{0i}\,(e/\varkappa T)\varphi$. Using (4.80), we obtain

$$\delta N_i = -\frac{N_0 e\varphi}{\varkappa T}\int f_2\,d^3v. \tag{4.81}$$

As is well known, the divergence of the dielectric polarization of a medium is equal to minus the density of the charges appearing under the action of the field, i.e.,

$$\operatorname{div}\mathbf{P} = -e(\delta N_i - \delta N) = \frac{N_0 e^2\varphi}{\varkappa T}\left\{\int f_2\,d^3v + 1\right\}.$$

On the other hand, the components of **P** are determined from the relations

$$P_l = \sum_k \frac{\varepsilon_{lk}-\delta_{lk}}{4\pi}E_k = -i\sum_k \frac{\varepsilon_{lk}q_k - q_l}{4\pi}\varphi,$$

where $\varepsilon_{lk}(\mathbf{q})$ is the dielectric permittivity tensor of the medium, taking spatial dispersion into account. Thus,

$$\int f_2\,d^3v + 1 = -iD^2\sum_{l,k}(\varepsilon_{lk}q_l q_k - q^2) \approx -iD^2\sum_{l,k}\varepsilon_{lk}q_l q_k, \tag{4.82}$$

and finally*

$$N_q = -\frac{\int f_1(v)\, d^3v}{\sum\limits_{l,k} \varepsilon_{lk} q_l q_k D^2}. \tag{4.83}$$

We now remember that we are carrying out the calculations in a coordinate system attached to the body. Upon transformation to a fixed coordinate system, a field with wave vector $\mathbf{q}$ which was static in the moving coordinate system acquires a frequency $\omega = \mathbf{q}\mathbf{V}_0$. Therefore

$$\varepsilon_{lk} = \varepsilon_{lk}(\mathbf{q}\mathbf{V}_0,\ \mathbf{q}), \tag{4.84}$$

where $\varepsilon_{lk}(\omega, \mathbf{q})$ is the dielectric permittivity tensor of a stationary plasma. As is well known, the relation between the frequency and wave vector of longitudinal waves propagating in a plasma is determined by the equation†

$$\sum_{l,k} q_l q_k \varepsilon_{lk}(\omega,\ \mathbf{q}) = 0.$$

If the dispersion law determined from this equation is $\omega = \omega(\mathbf{q})$, then the expression for N_q will have poles for those values of $\mathbf{q}$ which satisfy the relation

$$\mathbf{q}\mathbf{V}_0 = \omega(\mathbf{q}), \tag{4.85}$$

Since we are only interested in real values of $\mathbf{q}$, these poles exist only for real values of $\mathbf{q}$ or for complex values with a small imaginary part.

For this it is necessary, first of all, that $\omega(\mathbf{q})$ be almost real, i.e., that the waves are weakly damped. In the second place, it is necessary that the phase velocity of the waves, in the region of weak attenuation, be less than the velocity of the body's motion. In fact, it follows from (4.85) that

$$\mathbf{q}\mathbf{V}_0 = qV_0 \cos\chi = \omega(q), \quad V_0 > \frac{\omega(q)}{q}. \tag{4.86}$$

If Eq. (4.85) has an almost real root (i.e., very small imaginary part), then this leads to a strong scattering of radiowaves for values of the wave vector $\mathbf{q}$ close to this root. In the spatial distribution $\delta N(\mathbf{r})$ in this case a comparatively sharp Mach cone appears with an opening corresponding to the phase velocity of the longitudinal wave (see § 15a). As it must, condition (4.86) coincides with the condition for Cherenkov radiation, in which a body excites waves in a plasma.

---

*One can also carry out similar calculations for the case when the condition $qD \ll 1$ is not satisfied. In this case formula (4.83) is valid just for the disturbance of electrons, N_q; but N_{qi} is related to N_q by the formula

$$N_{qi} = \frac{N_q}{(1 + q^2D^2)}.$$

†In fact, for longitudinal waves it follows from the equation div $\mathbf{D} = 0$ that

$$i \sum_{l,k} q_l \varepsilon_{lk} E_k = i \sum_{l,k} q_l q_k \varepsilon_{lk} \varphi = 0.$$

Thus, φ is not equal to zero only if

$$\sum_{l,k} q_l q_k \varepsilon_{lk} = 0.$$

c. Law of Similarity. It is, of course, impossible to calculate the ion concentration disturbance at arbitrary distances from a body in general form, with account of an electric field. Provided

$$a_0 = \sqrt{\frac{M V_0^2}{2\varkappa T}} \gg 1, \; R_0 \gg D,$$

one can, however, obtain a similarity law which is valid in the absence of a magnetic field. For this it is necessary to set $\mathbf{H} = 0$, $Y_i = 0$ in Eq. (1.23). If the region external to the body's surface is considered, then $\delta(F) = 0$. Then, using expression (4.14) for φ, we obtain:

$$(V_0 + u_z)\frac{\partial f}{\partial z} + \mathbf{u}_1 \frac{\partial f}{\partial \mathbf{z}_1} - \frac{\varkappa T}{M_i}\frac{\partial}{\partial z}\ln\left(\int f d^3 r\right)\frac{\partial f}{\partial u_z} - \frac{\varkappa T}{M_i}\frac{\partial}{\partial \mathbf{r}_1}\ln\left(\int f d^3 r\right)\frac{\partial f}{\partial \mathbf{u}_1} = 0. \tag{4.87}$$

Here, as usual, the z axis is chosen in the direction $\mathbf{V}_0$ of the body's motion; $\mathbf{u}_1$ and $\mathbf{r}_1$ are the projections of $\mathbf{u}$ and $\mathbf{r}$ on the plane perpendicular to $\mathbf{V}_0$.

We go over to dimensionless variables

$$t = -\frac{Z}{R_0}, \quad \rho = \frac{\mathbf{r}_1}{R_0}, \quad \mathbf{w} = \frac{\mathbf{u}_1}{\sqrt{\frac{2\varkappa T}{M}}}$$

neglecting terms containing the large quantity a_0 in their denominators. Then, taking into consideration that in the undisturbed region of the plasma the velocity distribution of ions in the beam is Maxwellian, we rewrite Eq. (4.87) for the function g, where

$$g(t, \rho, \mathbf{w}) = \frac{f}{N_0}\exp\left(\frac{M_i u_z^2}{2\varkappa T}\right)\left(\frac{2\pi\varkappa T}{M_i}\right)^{3/2}$$

in the form

$$\frac{\partial g}{\partial t} - \mathbf{w}\frac{\partial g}{\partial \rho} + \frac{1}{2}\frac{\partial}{\partial \rho}\ln\left(\int g\, d^2 w\right)\frac{\partial g}{\partial \mathbf{w}} = 0 \tag{4.88}$$

As before, one can write the boundary condition for Eq. (4.88) in the form (2.10). Let the maximum cross section S_0 of the body be a circle of radius R_0. Then the boundary condition takes the form

$$g(0, \rho, \mathbf{w}) = \begin{cases} e^{-w^2} & \text{for } \rho > 1, \\ 0 & \text{for } \rho < 0. \end{cases}$$

We see that Eq. (4.88) and its boundary conditions do not contain, in the case under investigation, any parameters characterizing the velocity or dimensions of the body. Therefore the function determined from (4.88) and its boundary conditions is a "universal" function. Returning to the initial variables and taking the cylindrical symmetry of the problem into consideration, we arrive at the following formula

$$f = N_0\left(\frac{M}{2\pi\varkappa T}\right)^{3/2} e^{\frac{-M_i u_z^2}{2\varkappa T}}\, g\left(u_1\sqrt{\frac{M_i}{2\varkappa T}}, \frac{r_1}{R_0}, \frac{z}{R_0}\sqrt{\frac{2\varkappa T}{M_i V_0^2}}\right),$$

which also expresses the similarity principle. The concentration of particles behind a body may then be represented in the form

$$N(\vartheta, z) = N_0\, h\left(\frac{z}{R_0}\tan\vartheta, \frac{z}{R_0}\sqrt{\frac{2\varkappa T}{MV_0^2}}\right)$$

where

$$\tan\vartheta = r_1/z,\ h = \int g\, d^2 w.$$

The similarity law established here may, in particular, be used for the simulation of the considered phenomena under laboratory conditions. We also mention that this similarity law is analogous to a law which exists in ordinary gas dynamics for motion with large Mach number [75].

One can additionally simplify Eq. (4.88) under the condition

$$t \ll 1,\ |\xi| \ll 1,\ \xi = 1 - \rho.$$

Under such a condition, one can regard the problem as a planar problem and, in addition, there is no characteristic spatial dimension in the problem, so it is self-reproducing. In this connection, all quantities depend only on the ratio $\xi/t = \tau$. Setting

$$g = g(\tau, w_\xi)$$

(the ξ axis is perpendicular to the boundary of the cross section S_0), we obtain the following equation for the function g:

$$\frac{\partial g}{\partial t}(w_\xi + \tau) - \frac{\partial g}{\partial w_\xi}\frac{\partial}{\partial t}\left(\ln \int g\, dw_\xi\right) = 0.$$

The boundary conditions for this equation are written in the form:

$$g \to e^{-w_\xi^2} \text{ for } \tau \to -\infty,$$

$$g \to 0 \text{ for } \tau \to \infty.$$

§ 17. Disturbances in the Vicinity of a Point Body

In this section we shall consider the disturbances created by a rapidly moving charged point body, i.e., a body with dimensions much smaller than the Debye radius. In this connection we shall only be interested in large (in comparison with the Debye radius) distances from the body. The calculations will be carried out for the Fourier components. Naturally, the function f_q satisfies the general equation (4.66) without a magnetic field and Eq. (4.72) in the presence of a magnetic field. In the equation for the potential of the field it is necessary, of course, to add to the right side of (4.9) the term $-4\pi eZ\,\delta(\mathbf{r})$ containing the body's charge, where we shall regard the body's charge eZ as sufficiently small [see condition (4.93) below]. Linearizing Eq. (4.9), we obtain

$$\Delta\varphi = 4\pi\frac{e^2\varphi}{\varkappa T}N_0 - 4\pi e\int f_i'\, d^3v - 4\pi eZ\delta(r),\quad -q^2\varphi_q = 4\pi N_0\frac{e\varphi_q}{\varkappa T} - 4\pi eN_q - 4\pi eZ.$$

Neglecting the term containing q^2, instead of Eq. (4.39) we have

$$\frac{e\varphi_q}{\varkappa T} = \frac{N_q}{N_0} + \frac{Z}{N_0}. \tag{4.89}$$

The expressions for φ_q and N_q, as we know, contain terms $\sim 1/q$ which are large for small vaues of q. However, when the charge eZ is small, these terms contain the small factor e^2Z^2; therefore, there is a large range of values of q where both terms on the right side of (4.89) are of the same order. We note that if the charge eZ is not small, then it is impossible to now abandon the term eZ, since this exceeds the accuracy of the approximation used. Simultaneously solving Eqs. (4.66) and (4.89), we obtain

$$N_q = \frac{-i \int \frac{I(v) - iquf_{0i}Z}{qvN_0} d^3v}{2 + iaQ(a)}, \tag{4.89a}$$

where

$$a = \frac{qV_0}{q} \sqrt{\frac{M_i}{2\varkappa T}}.$$

According to Eq. (4.71), with regard to order of magnitude we have

$$I(v) \sim f_{i0} V_0 S_{\text{Coul.}} \tag{4.90}$$

In formula (4.90), $S_{\text{Coul.}}$ is the effective cross section for the scattering of ions with velocity v by the Coulomb field of the charge. Therefore, the second term under the integral in (4.89a) is the main term at distances

$$r = \frac{1}{q} \ll \sqrt{\frac{2\varkappa T}{M_i V_0^2}} \frac{Z}{N_0 S_{\text{Coul.}}}. \tag{4.91}$$

In the opposite limiting case, the first term is indeed the principal one.

Let us mention here that the problem under investigation was solved earlier by Kraus and Watson [44]. The system of equations used by them differs from ours by the absence of the term with I(v), so that it is possible to obtain their answer from Eq. (4.89a) by setting $I(v) = 0$. From what has been said above, it is clear therefore that the results of [44] are correct only for values of r satisfying the inequality (4.91) and are not valid in the opposite limiting case, i.e., at large distances from a body. This stems from the fact that the authors of [44] carried out their calculation according to perturbation theory, i.e., they expanded the solution in powers of the charge Z, limiting the expansion to the first term. Meanwhile the quantity I(v), as we shall see below, formula (4.96), is proportional to Z^2. If calculations analogous to those of Kraus and Watson are carried out to second order in perturbation theory, i.e., retaining terms of order Z^2, then a result agreeing with ours is, of course, obtained.

For the evaluation of $I(\mathbf{v})$ we shall use formula (4.66). Here it is necessary to keep in mind that values of ρ much smaller than the Debye radius introduce the principal contribution to the integral with respect to $d\rho$ in (4.70). The field is a pure Coulomb field at such distances. The scattering angle for an ion in a Coulomb field is given by the Rutherford formula (see p. 70 in [27]). If the body's charge is sufficiently small, the ion's scattering angle ϑ, i.e., the angle between $\mathbf{v}$ and $\mathbf{v}_1$, is also small. In this case it is possible to approximately represent the Rutherford formula in the form

$$\vartheta = \frac{2Ze^2}{M_i v^2} \frac{1}{\varrho} \approx \frac{2Ze^2}{M_i V_0^2 \varrho}.$$

Let us expand the right side of (4.70) in powers of ϑ up to terms of order ϑ^2. Such an expansion is possible if the terms containing ϑ in the exponent of the exponential in (4.70) are small. This leads to the condition

$$\sqrt{\frac{M_i V_0^2}{2\varkappa T}} \vartheta \sim \frac{Ze^2}{\sqrt{M_i V_0^2 2\varkappa T}} \frac{1}{\varrho} \ll 1, \tag{4.92}$$

which must be satisfied in any case for $\rho \lessdot D$. As a result, the following condition is imposed on the charge:

$$\frac{Ze^2}{\sqrt{M_i V_0^2 2\varkappa T}} \frac{1}{D} \ll 1. \tag{4.93}$$

Expanding and integrating (4.70) with respect to $d\varphi$, we obtain

$$I(v) = 2\pi f_{0i} V_0 \left\{ -\frac{M_i V_0^2}{2\varkappa T} + \frac{M_i}{(2\varkappa T)^2} [V_0^2 v^2 - (V_0 v)^2] \right\} \times$$

$$\times \int \vartheta^2(\varrho)\, \varrho\, d\varrho = -2\pi f_{0i} V_0 \left\{ \frac{M_i V_0^2}{2\varkappa T} - \frac{M_i}{(2\varkappa T)^2} [V_0^2 u^2 - (V_0 u)^2] \right\} \int \vartheta^2(\varrho)\, \varrho\, d\varrho. \tag{4.94}$$

The integral on the right side of (4.94) diverges logarithmically. At large ρ the integral must be cut off at $\rho \sim D$, and at small distances it must be cut off at those values of ρ for which condition (4.92) ceases to be satisfied, i.e., when

$$\varrho \sim \varrho_1 = \frac{Ze^2}{\sqrt{M_i V_0^2 2\varkappa T}}. \tag{4.95}$$

Finally we have*

$$I(v) = -\frac{4\pi Z^2 e^4}{M_i \varkappa T V_0^3} f_{0i} \cdot \ln \frac{D}{\varrho_1} \left\{ V_0^2 - \frac{M_i}{2\varkappa T} [V_0^2 u^2 - (V_0 u)^2] \right\}. \tag{4.96}$$

For the evaluation of (4.94) it was assumed that the field is a pure Coulomb field up to distances $\rho \sim \rho_1$, for which the dimension of the body should satisfy the condition

$$R_0 \lessdot \varrho_1 = \frac{Ze^2}{\sqrt{M_i V_0^2 2\varkappa T}}. \tag{4.97}$$

If the size of the body, R_0, is larger than the value of ρ_1 given by formula (4.95), then as the lower limit it is obviously necessary to take not ρ_1 but R_0.

Substituting (4.96) into (4.89a) and expressing, as usual, the integral in the numerator in terms of the probability function of imaginary argument, $Q(a)$, we finally obtain

$$N_q = -\frac{K}{2 + iaQ(a)}, \tag{4.98}$$

where

$$K = \{(2\pi Z^2 e^4 N_0)/[qV_0^3 (2\varkappa^3 T^3 M_i)^{1/2}]\} \ln \frac{D}{\varrho_1} [V_0^2 - (V_0 n)^2]\, [(1 - 2a^2) Q(a) + 2ia] + Z,$$

$$n = q/q.$$

In the presence of an external magnetic field, the simultaneous solution of Eqs. (4.72) and (4.89) gives

*It is clear from (4.86) that the cross section $S_{\text{Coul.}}$, entering into the estimate (4.90) and the inequality (4.91), equals in order of magnitude

$$S_{\text{Coul.}} \sim \frac{4\pi Z^2 e^4}{M_i \varkappa T V_0^2} \ln \frac{D}{\varrho_1}.$$

$$N_q = \frac{\frac{1}{\Omega}\int\int\limits_0^{\infty} [I(v) - iquf_{0i}Z/N_0] \exp\{A\}\, dx\, d^3u}{2 + \frac{iqV_0}{\Omega}\int\limits_0^{\infty} \exp\{B\}\, dx}, \tag{4.99}$$

where

$$A = -i\left[\mu x + 2\gamma\cos\left(\beta + \frac{x}{2}\right)\sin\frac{x}{2}\right],$$

$$B = i\frac{qV_0}{\Omega}x - \frac{\varkappa T}{2M_i\Omega^2}\left(q_{\parallel}^2 x^2 + 4q_{\perp}^2 \sin^2\frac{x}{2}\right)$$

[the notation in Eq. (4.99) is the same as in Eq. (3.33)]. If the Larmor radius of the ions is larger than D, the magnetic field does not change the form of I(**v**). Carrying out the integration with respect to d^3u in the numerator, where I(**v**) is given by formula (4.96), we finally obtain

$$N_q = \frac{\frac{2\pi Z^2 e^4 N_0}{V_0^3 (2\varkappa^3 T^3 M_i)^{1/2}} \ln\frac{D}{\varrho_1}\left[\frac{2\varkappa T}{M_i\Omega^2}\int\limits_0^{\infty} C(x)\, dx + Z\right]}{2 + i\frac{qV_0}{\Omega}\int\limits_0^{\infty} \exp B\, dx}, \tag{4.100}$$

where the expression for B is given above, and

$$C(x) = \left[V_0^2\left(q_{\parallel}^2 x^2 + q_{\perp}^2 4\sin^2\frac{x}{2}\right) - \left(q_{\parallel}V_{0\parallel}x + q_{\perp}V_{1\perp}2\sin\frac{x}{2}\right)^2 \exp B\right].$$

Here $\mathbf{V}_{1\perp}$ is the vector obtained by rotating $\mathbf{V}_{0\perp}$ through an angle $-x/2$.

The problem of the motion of a point charge in a plasma in the presence of a magnetic field was also investigated by Greifinger [51]. The author carried out his calculation in the first approximation with respect to Z, i.e., as in [44], neglecting the term containing I(**v**). In addition, he neglected the thermal motion of the ions; therefore, his results have meaning only for the case when the temperature of the ions is much less than the temperature of the electrons ($T_i \ll T_e$). In order to obtain expressions for N(**r**) from Eqs. (4.98) and (4.100), it is necessary to take the inverse Fourier transform. Here the first term in Eq. (4.98) leads to a decrease of the disturbance with distance like $1/r^2$, and the second term leads to a $1/r^3$ decrease. Numerical calculations, similar to those made for the calculation of N(**r**) according to formula (4.51), are necessary for the determination of the angular dependence of N(**r**). An analytic expression for that part of N(**r**) which is proportional to $1/r^3$ was obtained in [44] for the case when $T_i \ll T_e$.

Two terms also appear in the expression δN(**r**) for motion in a magnetic field. One falls off with increasing distance in the plane $\mathbf{V}_0\mathbf{H}_0$ like 1/r, and the other falls off like $1/r^2$. As you go away from this plane, the disturbances decrease exponentially.

We remark in conclusion that, besides the disturbances near a moving point charge, the problem of disturbances caused by the motion of a thin cylinder, for example, an antenna, may also be of interest. Such a problem was investigated by Rand [52]. The author used an approximation which is essentially similar to the approximation of Kraus and Watson. In particular, his result is also not valid at large distances from the cylinder.

§ 18. Disturbance of a Plasma by a High-Frequency Electric Field

At small distances from a radiating antenna, placed in a plasma, the intensity of the high-frequency electric field may become extremely high. This leads to field stresses on charged particles and, consequently, to a disturbance of the homogeneous plasma [28]. The present section is devoted to an investigation of these disturbances.*

In order to solve such a problem, it is necessary, as usual, to investigate the system of Eqs. (1.23), (1.24), and (1.25) which determine the distribution functions of electrons and ions and the electric field in the plasma. Here a time-dependent variable electric field of frequency ω, created by external sources, is given on the surface of the antenna.

Our problem reduces to the determination of the time-averaged disturbances of the distribution functions for electrons and ions, the constant electric field, the amplitude of the variable electric field of frequency ω, and the amplitudes of the corresponding variable disturbances of the distribution functions. First we shall determine the dependence of the time-averaged disturbance of the distribution functions and the constant electric field on the amplitude of the variable electric field in a plasma. Then we shall also find the amplitude of the variable field for a number of specific cases.

a. Force Acting on an Electron in an Inhomogeneous Variable Electric Field. As already indicated above, first it is necessary to calculate the average (more precisely, the average over a period of the field) disturbance of the plasma, caused by a variable electric field, i.e., the constant component of this disturbance. For this, we determine the average value of the force acting on charged particles in an inhomogeneous, variable electric field. Naturally, the average force acting on ions is m/M_i times smaller than the force acting on electrons, so that in a rapidly oscillating field it is inversely proportional to the mass of the particle (see formula (4.96) above, and also p. 120 of [27]). Therefore, it is only necessary to consider the direct influence of a variable field on the motion of electrons. The motion of ions changes only because of the Coulomb interaction with electrons.†

The expression for the force per unit volume acting on an electron in an inhomogeneous, alternating electric field has the form (see [30])

$$f=\sum_{i,k}\frac{1}{16\pi}\operatorname{grad}\left[E_i^*E_k\frac{\partial\varepsilon_{ik}}{\partial N}N\right]-\frac{1}{16\pi}\sum_{i,k}\frac{E_i^*E_k}{16\pi}\operatorname{grad}\varepsilon_{ik}, \tag{4.101}$$

where E_k is the complex amplitude of the electric field, E_i^* is the complex conjugate of E_i, ε_{ik} is the dielectric permittivity tensor of the medium, and N is the concentration of electrons.

It is possible to transform the derivative $\partial\varepsilon_{ik}/\partial N$ if we use the fact that the tensor $(\varepsilon_{ik}-\delta_{ik})/4\pi$ is the polarization tensor of the plasma, which is proportional to the concentration of electrons. Thus, for example, in the absence of a magnetic field

$$\varepsilon_{ik}{}^{\prime}=\left(1-\frac{4\pi Ne^2}{m\omega^2}\right)\delta_{ik}$$

or

$$\frac{\varepsilon_{ik}-\delta_{ik}}{4\pi}=\frac{e^2}{m\omega^2}N. \tag{4.102}$$

Therefore

$$\frac{\partial}{\partial N}(\varepsilon_{ik}-\delta_{ik})=\frac{\varepsilon_{ik}-\delta_{ik}}{N},\quad N\frac{\partial\varepsilon_{ik}}{\partial N}=\varepsilon_{ik}-\delta_{ik}, \tag{4.103}$$

---

*This effect was first considered by Getmantsev and Denisov [29]. Disturbance of the plasma by a high-frequency field was also investigated by Volkov [76].

†Here only an alternating field whose frequency is much larger than the Larmor frequency of the ions ($\omega\gg\Omega$) and the plasma frequency of the ions ($\omega\gg\Omega_0=m\omega_0/M_i$) is considered. Of course, in the case of low frequencies, when $\omega\lessgtr\Omega$ or $\omega\lessgtr\Omega_0$, an alternating field turns out to have a much more important influence on the motion of the ions.

so that the average force acting per unit volume is

$$f = \frac{1}{16\pi} \sum_{i,k} (\varepsilon_{ik} - \delta_{ik}) \operatorname{grad} E_i^* E_k. \tag{4.104}$$

In a nonmagnetic plasma ($\mathbf{H}_0 = 0$)

$$f = -\frac{e^2 N}{4m\omega^2} \operatorname{grad} E^* E, \tag{4.105}$$

and the formula of Miller [28] is obtained for the force f/N related to one particle:

$$F = \frac{f}{N} = -\operatorname{grad}\left(\frac{e^2 |E|^2}{4m\omega^2}\right). \tag{4.106}$$

Let us demonstrate that expression (4.106) has a simple physical meaning. For this, we consider the motion of an electron in an alternating field $\mathbf{E} \cos \omega t$. Under the influence of the field, an electron oscillates with frequency ω; the variable component of its velocity is given by

$$v_\infty = \frac{eE}{m\omega} \sin \omega t.$$

The energy of the electron's oscillatory motion, averaged over the period of the field, is obviously given by

$$\overline{\mathscr{E}}_\infty = \frac{m}{2} \overline{v_\infty^2} = \frac{e^2 |E|^2}{2\omega^2 m} \overline{\sin^2 \omega t} = \frac{e^2 |E|^2}{4m\omega^2}.$$

Therefore, if the electric field is inhomogeneous, i.e., if its amplitude E varies in space, $E = E(r)$, then an average force

$$F = -\operatorname{grad} \overline{\mathscr{E}}_\infty = -\operatorname{grad} \frac{e^2 |E|^2}{4m\omega^2}.$$

acts on the electron. It is also understood that expression (4.106) for the average force acting on an electron is valid only under the assumption that the field amplitude changes little over the distance which the electron travels during the period of oscillation. This leads to the condition

$$R \gg \frac{v_e}{\omega}, \quad R \gg \left(\frac{e|E|}{m\omega}\right)\frac{1}{\omega}, \tag{4.107}$$

where $v_e = (2\varkappa T/m)^{1/2}$ is the thermal velocity of electrons, $e|E|/m\omega$ is the velocity of electrons corresponding to the ordered motion, and R is a characteristic distance over which the electric field amplitude varies significantly. Condition (4.107) leads, in particular, to a limitation on the intensity of the field

$$|E| \ll \frac{m\omega^2 R}{e}. \tag{4.108}$$

Quite similarly, one can find the average force acting on an electron in an alternating electric field when a constant magnetic field $\mathbf{H}_0$ is present in the plasma. In this case, the dielectric permittivity tensor has the following form (see [31]):

$$\left.\begin{aligned}\varepsilon_{11} &= 1 - \frac{4\pi e^2 N}{m\omega^2}, \quad \varepsilon_{22} = \varepsilon_{33} = 1 - \frac{4\pi e^2 N}{m(\omega^2 - \omega_H^2)}, \\ \varepsilon_{23} &= -\varepsilon_{32} = -\frac{i4\pi e^2 N}{m}\frac{\omega_H}{\omega(\omega^2 - \omega_H^2)}, \quad \varepsilon_{12} = \varepsilon_{21} = \varepsilon_{13} = \varepsilon_{31} = 0,\end{aligned}\right\} \tag{4.109}$$

where the 1 axis is directed along the magnetic field and $\omega_H = eH_0/mc$ is the Larmor frequency of the electrons. Substituting (4.109) into Eq. (4.101) and taking relationship (4.103) into consideration, we find that

$$\boldsymbol{F} = \frac{f}{N} = -\operatorname{grad} \sum_{i,k} \frac{e^2 E_i^* E_k \alpha_{ik}}{4m\omega^2}, \tag{4.110}$$

where

$$\alpha_{ik} = \frac{m(\delta_{ik} - \varepsilon_{ik})\omega^2}{4\pi N e^2} = \begin{pmatrix} 1 & 0 & 0 \\ 0 & \dfrac{\omega^2}{\omega^2 - \omega_H^2} & i\dfrac{\omega_H \omega}{\omega^2 - \omega_H^2} \\ 0 & -i\dfrac{\omega_H \omega}{\omega^2 - \omega_H^2} & \dfrac{\omega^2}{\omega^2 - \omega_H^2} \end{pmatrix}. \tag{4.111}$$

In the absence of a magnetic field, $\alpha_{ik} = \delta_{ik}$ and expression (4.110) coincides with (4.106).

b. Influence of an Inhomogeneous, Alternating Electric Field on the Plasma. We saw above that if conditions (4.106) and (4.107) are satisfied, then an additional force acts on electrons when an alternating electric field is present in the plasma. Therefore, in the kinetic equation for the time-averaged part of the electron distribution function, instead of the force $-e\mathbf{E} = \operatorname{grad} e\varphi$ it is necessary to consider the total force acting on an electron, i.e.,

$$\boldsymbol{F} = \operatorname{grad}\left[e\varphi - \sum_{i,k} \frac{\alpha_{ik} e^2 E_i^* E_k}{4m\omega^2}\right]. \tag{4.112}$$

Consequently, in the formula for the Boltzmann distribution which usually determines the concentration of electrons, it is necessary to replace $e\varphi$ by the complete expression inside the square brackets of (4.112). Thus,

$$N = N_0 \exp\left(\frac{e\varphi - \sum\limits_{i,k} \dfrac{\alpha_{ik} e^2 E_i^* E_k}{4m\omega^2}}{\varkappa T}\right), \tag{4.113}$$

and Poisson's equation (4.9) takes the form

$$\Delta\varphi = -4\pi e\left[\int f_i d^3 v - N_0 \exp\left(\frac{e\varphi - \sum\limits_{i,k} \dfrac{\alpha_{ik} e^2 E_i^* E_k}{4m\omega^2}}{\varkappa T}\right)\right], \tag{4.114}$$

where f_i as before is the distribution function of the ions. Therefore, the problem reduces to the simultaneous solution of Eq. (4.114) and Eq. (1.23) for the distribution function of the ions. In addition, of course, it is necessary to determine the complex amplitude of the alternating electric field from Maxwell's equations.

Now let us suppose that the characteristic size of the inhomogeneity (for example, the radius of the antenna) is much larger than the Debye radius. In this case, at a sufficiently large distance from the surface of the antenna (outside of the transition layer which is formed on this surface), one can neglect the term $\Delta\varphi$ in Poisson's equation, as was already done in § 14c. From Eq. (4.114) it then follows that

$$\varphi = \frac{\varkappa T}{e} \ln \frac{N_i}{N_0} + \sum_{i,\,k} \alpha_{ik} \frac{eE_i^* E_k}{4m\omega^2} . \tag{4.115}$$

This relation replaces formula (4.14), which was obtained in the absence of an alternating electric field.

The determination of the distribution of ions requires the solution of the kinetic equation for f_i with the value of φ determined by Eq. (4.155). We consider, for simplicity, only the case of an antenna at rest in the plasma. We also suppose that one can neglect the absorption of ions on the surface (neutralization). Then one can regard the distribution of ions, as well as the distribution of electrons, as an equilibrium distribution in the vicinity of the antenna. This means that the concentration of ions is given by the Boltzmann formula

$$N_i = N_0 \exp\left(-\frac{e\varphi}{\varkappa T} \right) . \tag{4.116}$$

In this case, as is clear from Eq. (4.115), the potential of the constant electric field is

$$\varphi = \sum_{i,\,k} \alpha_{ik} \frac{eE_i^* E_k}{8m\omega^2} . \tag{4.117}$$

Correspondingly, the concentration of electrons and ions is given by

$$N_e = N_i = N_0 \exp\left(-\sum_{i,\,k} \alpha_{ik} \frac{e^2 E_i^* E_k}{8\varkappa T m\omega^2} \right) . \tag{4.118}$$

In particular, in the absence of a magnetic field $\alpha_{ik} = \delta_{ik}$ and (see [29]):

$$N_i = N_e = N_0 \exp\left(-\frac{e^2 |E|^2}{8\varkappa T m\omega^2} \right) . \tag{4.119}$$

From here it is clear that at sufficiently large values of the amplitude of the alternating field, i.e., when

$$|E| \gtrsim \sqrt{8\varkappa T m}\, \frac{\omega}{e} \approx 2.0\, \frac{\omega}{10^8} \sqrt{\frac{T}{1000^\circ}} \ \text{V/cm}, \tag{4.20}$$

the disturbance produced in the plasma by the field is extremely large.

Associated with the presence of a magnetic field $\mathbf{H}_0$, if the antenna is directed parallel to $\mathbf{H}_0$, then the intensity of the variable electric field is perpendicular to $\mathbf{H}_0$, and

$$N_i = N_e = N_0 \exp\left\{ -\frac{e^2}{8m\varkappa T (\omega^2 - \omega_H^2)} \left(|\mathbf{E}|^2 + \frac{i\omega_H}{\omega} \frac{[\mathbf{E}^* \mathbf{E}]\, \mathbf{H}_0}{H_0} \right) \right\} . \tag{4.121}$$

If the antenna is directed perpendicular to $\mathbf{H}_0$, then

$$N_i = N_e = N_0 \exp\left\{ -\frac{e^2}{8m\varkappa T} \left(\frac{|E_\parallel|^2}{\omega^2} + \frac{|E_\perp|^2}{\omega^2 - \omega_H^2} + \frac{i\omega_H}{\omega(\omega^2 - \omega_H^2)} \frac{[\mathbf{E}^* \mathbf{E}]\, \mathbf{H}_0}{H_0} \right) \right\} .$$

c. Amplitude of the Alternating Electric Field in a Plasma. In the preceding section, the disturbances in the concentration of electrons and ions were expressed in terms of the complex amplitude of the alternating electric field. Now it is necessary to calculate this amplitude. As we shall see below, a transition layer in which the field distribution has a complicated nature is formed near the surface of the antenna. In this section, we shall assume that the point of observation is located outside of this layer.

It is possible to calculate the amplitude of the field independently of the determination of N(**r**) only in the case when the frequency of the field is much larger than the Larmor frequency ω_0 of the plasma:

$$\omega \gg \omega_0. \tag{4.122}$$

In this case, in order to determine **E**(**r**) one can regard the dielectric permittivity of the plasma as equal to unity, i.e., calculate **E**(**r**) according to the same formula as for a vacuum. If, however, condition (4.122) is not satisfied, then a disturbance of N(**r**) changes the dielectric permittivity of the plasma. Since N(**r**) itself depends on **E**, the equation for **E** becomes nonlinear. As is well known, the appearance of harmonics of the fundamental frequency is possible in every nonlinear problem, i.e., oscillations with frequencies 2ω, 3ω, and so forth appear. Consideration of these harmonics would strongly complicate the solution of the problem. We shall show, however, that under condition (4.97) the amplitude of the harmonics is small and it is possible to disregard them. In order to prove this, let us write down the equation for the variable (alternating) component of the electron distribution function. It is precisely this equation which turns out to be nonlinear in a strong field. We consider that, under condition (4.97), one can neglect the term containing derivatives with respect to coordinates in this equation. Afterwards, the equation takes the form*

$$\frac{\partial f_e}{\partial t} + \frac{\partial f_e}{\partial v}\frac{eE}{m} = 0.$$

Multiplying this equation by $-e\mathbf{v}$ and integrating with respect to d^3v, we bring it to the form

$$\frac{\partial \hat{j}}{\partial t} - \frac{e^2}{m} N_e(r) E = 0, \tag{4.123}$$

where $\hat{j} = -e\int v f_e \, d^3v$ is the current density.

As already stated, $N_e = N_i$ under condition (4.97). But because of the large mass of the ions, the variable field has little influence on their motion. Therefore, $N_i(\mathbf{r})$ is equal to its time-averaged value, and does not contain a high-frequency component. Thus, the equation relating $\hat{\mathbf{j}}$ with **E** turns out to be an equation with time-independent coefficients (although these coefficients also depend on the amplitude **E**), so that higher harmonics do not appear. It also follows from Eq. (4.123) that under condition (4.97) $\hat{\mathbf{j}}$ is related to $\hat{\mathbf{E}}$ by the usual formula in which N should be understood as the average value of N(**r**). This means that we may use for **E** the usual macroscopic equations of a field in a medium, in which the dielectric permittivity is expressed in terms of the average concentration N(**r**), determined by formulas (4.113) or (4.118). In the latter case

$$\varepsilon_{ik}(r) = \delta_{ik} - \frac{\omega_0^2}{\omega^2} \exp\left(-\frac{e^2}{8m\omega^2 \varkappa T} \alpha_{lm} E^* E_m\right) \alpha_{ik}.$$

If no magnetic field is present,

$$\varepsilon(r) = 1 - \frac{\omega_0^2}{\omega^2} \exp\left(-\frac{e^2}{8m\omega^2 \varkappa T} |E|^2\right). \tag{4.124}$$

At distances from the antenna which are small in comparison with its length and in comparison with the wavelength of the radiated waves, one can regard the electric field as directed perpendicular to the antenna. Its amplitude only depends on the distance ρ to the antenna. In this case, in order to determine the field, only one of Maxwell's equations is sufficient (for us), namely the equation

$$\operatorname{div} \mathscr{D} = 0$$

*For simplicity, we present the proof without a magnetic field.

($\mathscr{D}$ is the electric displacement vector), which in this case reduces to the equation

$$\frac{1}{\varrho}\frac{d}{d\varrho}\varrho\mathscr{D}=0. \tag{4.125}$$

Integrating Eq. (4.125), we obtain

$$\varrho\mathscr{D}=C, \tag{4.126}$$

or considering that $\mathscr{D}=\varepsilon E$,

$$\varrho\varepsilon E=C, \tag{4.126a}$$

where C is an integration constant.

In order to determine the constant C, let us consider some point near the body's surface, but outside the transition layer.

At this point one can set

$$\varrho=\varrho_0, \quad \varepsilon(\varrho)=\varepsilon(\varrho_0),$$

where ρ_0 is the radius of the antenna.

The electric field at this point can be expressed in terms of the surface charge density σ_0 on the antenna with the aid of the formula

$$E(\varrho_0)=\frac{4\pi\sigma_0}{\varepsilon(\varrho_0)}.$$

which is well known from electrostatics. Substituting this equation into Eq. (4.126), we find that

$$C=4\pi\sigma_0\varrho_0.$$

On the other hand, on the surface of the antenna itself, at the very beginning of the transition layer where the plasma is still not polarized and the charge is not screened, the surface density σ_0 is related to the electric field intensity by the same equation as in vacuum, i.e.,

$$\sigma_0=\frac{E_0}{4\pi},$$

where E_0 is the intensity of the field on the surface itself, determined by the external source of the antenna's excitation. Substituting the last two relations into (4.126) and (4.126a), we finally obtain

$$\mathscr{D}(\varrho)=\frac{\varrho_0}{\varrho}E_0 \tag{4.127}$$

and

$$E(\varrho)=\frac{\varrho_0}{\varrho}\frac{E_0}{\varepsilon(\varrho)}. \tag{4.128}$$

The dependence of ε on ρ is determined by $N(\rho)$, i.e., in the final calculation of formula (4.119). Hence formulas (4.128) and (4.119) represent a system of equations whose simultaneous solution enables one to determine $N(\rho)$ and $E(\rho)$. Below we shall carry out this solution for certain special cases. We recall that formulas (4.119) and (4.128) are only applicable outside the transition layer on the surface. The structure of this layer requires special investigation and is described in § 18f.

d. Disturbances Produced in a Plasma by an Alternating Electric Field of High Frequency. Let us consider a high-frequency electric field, $\omega \gg \omega_0$, when

$$\varepsilon = 1 - \frac{\omega_0^2}{\omega^2} \approx 1,$$

so that the field in the plasma is

$$E = \frac{\varrho_0}{\varrho} E_0. \tag{4.129}$$

Here, as usual, E_0 is the field at the surface of the antenna, ρ_0 is the radius of the antenna, and ρ is the distance from the axis of the antenna. Using expression (4.129) in formulas (4.112), (4.117), and others, it is possible to determine the concentration of electrons and ions. In particular, in the neighborhood of an antenna at rest

$$N_i = N_e = N_0 \exp\left\{-\frac{e^2E^2}{8\varkappa T m\omega^2}\frac{\varrho_0^2}{\varrho^2}\right\}. \tag{4.130}$$

From here it is clear that if the amplitude of the field on the surface of the antenna

$$E_0 \gtrsim \sqrt{8\varkappa Tm}\,\frac{\omega}{e} \approx 2{,}0\,\frac{\omega}{10^8}\sqrt{\frac{T}{1000^\circ}} \quad \text{V/cm}, \tag{4.131}$$

then the alternating field strongly disturbs the plasma over distances on the order of ρ_0. For large values of ρ the disturbance falls off in proportion to $1/\rho^2$.

If the antenna is moving, then according to (4.114) the potential of a constant electric field near the antenna is determined by the expression

$$\varphi = \frac{eE_0^2\varrho_0^2}{4m\omega^2\varrho^2} + \frac{\varkappa T}{e}\ln\frac{N_i}{N_0}. \tag{4.132}$$

The calculation of the ion and electron concentration disturbances now requires only consideration of the kinetic equation for the distribution function (1.23) of the ions containing the potential of the electric field, expressed by formula (4.132). Up to now, such a problem has not been solved.

We note, however, that the potential φ turns out to have an influence on the disturbance of the ions around a rapidly moving body only in the case when

$$e\varphi \gtrsim \frac{M_iV_0^2}{2}.$$

Therefore, a high-frequency field strongly disturbs the ions around a rapidly moving body, provided

$$E_0 \gtrsim \sqrt{M_i m}\,\frac{\omega V_0}{e}. \tag{4.133}$$

e. Disturbances Produced in a Plasma by an Electric Field of Arbitrary Frequency. Resonance Amplification of the Effect for $\omega \approx \omega_0$. Let us consider an alternating electric field of arbitrary frequency ω, which we shall assume to be rather weak. The condition limiting the amplitude of the field in this connection will be indicated below. In this case, in order to determine the amplitude of an alternating field in the plasma, its influence on the plasma can be neglected to the first approximation, and according to (4.128) we obtain

$$E = \frac{E_0\varrho_0}{\varrho\varepsilon} = \frac{E_0\varrho_0}{\varrho\left(1 - \omega_0^2/\omega^2\right)}, \tag{4.134}$$

where $\omega_0^2 = 4\pi e^2 N_0/m$. Substituting this expression into Eqs. (4.112), (4.114), and (4.116), one can determine the disturbances of the electron and ion concentrations. In particular, in the neighborhood of an antenna at rest

$$\delta N_i = \delta N_e = -N_0 \frac{c^2 E_0^2 \varrho_0^2 \omega^2}{8\varkappa T m \varrho^2 (\omega_0^2 - \omega^2)^2}. \tag{4.135}$$

At high frequencies $\omega \gg \omega_0^2$ this expression for δN naturally coincides with Eq. (4.130).

In the vicinity of the Langmuir frequenciy (as $\omega \to \omega_0$) both the electric field's amplitude as well as the disturbances produced by it increase resonantly. In this region, one cannot completely neglect the influence of an alternating electric field on the concentration of particles in the plasma. The amplitude of the field is now simultaneously determined by Eqs. (4.128) and (4.124). If the field is weak, then one can expand the exponent in (4.124), and we obtain (as $\omega \to \omega_0$)

$$\varepsilon \approx \varepsilon_0 + \frac{e^2}{8m\omega_0^2 \varkappa T} |E|^2 = \varepsilon_0 + \beta |E|^2, \tag{4.136}$$

$$\left.\begin{aligned} \varepsilon_0 &\approx \frac{2(\omega - \omega_0)}{\omega_0}, \\ \beta &= \frac{e^2}{(8m\omega_0^2 \varkappa T)}. \end{aligned}\right\} \tag{4.137}$$

Substituting (4.136) into (4.128), we obtain an equation for the amplitude of the field

$$\varepsilon_0 E + \beta |E|^2 E = E_0 \frac{\varrho_0}{\varrho}. \tag{4.138}$$

Multiplying this equation by its complex conjugate, we obtain

$$(\varepsilon_0 + \beta |E|^2)^2 = \frac{\varrho_0^2}{\varrho^2} |E_0|^2. \tag{4.139}$$

It is of interest to note that this equation has the same form as the equation for the amplitude of forced vibrations of an anharmonic oscillator without damping near resonance (see [27], p. 113). Equation (4.139) is a cubic equation with respect to $|E|^2$ and in general has three roots – $|E_1|^2$, $|E_2|^2$, $|E_3|^2$. The dependence of these roots on ε_0 is schematically indicated in Fig. 30a. We see that if ε_0 is larger than a certain value of ε_K, then each value of $|E_0|^2$ corresponds to one value of $|E|^2$. If, however, $\varepsilon < \varepsilon_K$, then to each value of the field amplitude outside the plasma there corresponds three values of the field inside the plasma, i.e., hysteresis occurs. It is possible, however, to show that states corresponding to the middle root $|E_3|^2$ are unstable and cannot be realized. In order to determine the critical value ε_K we note that for

$$\varepsilon = \varepsilon_K, \quad \frac{d\varepsilon_0}{d|E|^2} = 0.$$

Differentiating (4.139) with respect to $|E|^2$, we obtain

$$(\varepsilon_0 + \beta |E|^2)^2 + 2|E|^2 \left(\frac{d\varepsilon_0}{dE^2} + \beta\right) = 0. \tag{4.140}$$

Setting $d\varepsilon_0/d|E|^2 = 0$ and simultaneously solving (4.139) and (4.140), we obtain

$$\varepsilon_K = -\frac{3}{2^{2/3}} (\beta |E|^2)^{2/3} (\rho_0/\rho)^{4/3} \tag{4.141}$$

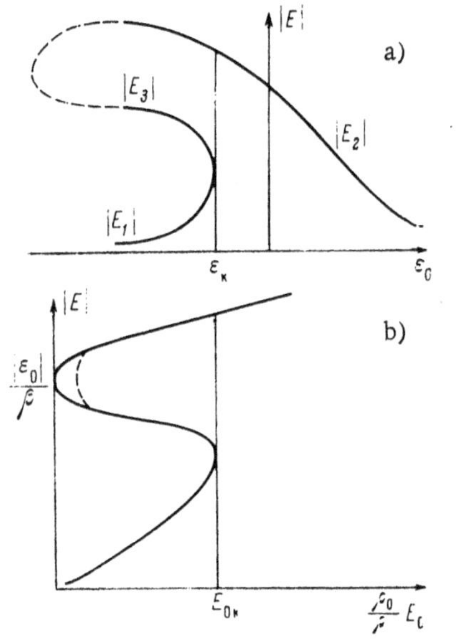

Fig. 30. Curve a shows the dependence of the intensity of the field in a plasma on ε_0 (or the frequency), and curve b shows the dependence on the amplitude of the external field.

For this value of ε_0, the quantity $|E|^2$ may take two values, $|E_{1K}|^2$ and $|E_{2K}|^2$:

$$|E_{1K}|^2 = \frac{|E_0|^{2/3}}{2^{2/3}\varkappa^{2/3}} (\rho_0/\rho)^{4/3}, \quad |E_{2K}|^2 = 2|E_{1K}|^2.$$

It is not difficult to show that the root $|E_2|^2$ increases without limit as ε_0 decreases. This, however, is associated with the neglect of damping, i.e., dissipation of energy. Upon taking dissipation, the second and third branches merge at some negative value ε_0, as indicated by the dashed line in the figure.

The dependence of the amplitude of the field on $(\rho_0/\rho)|E_0|$ has a significantly different character depending on the value of ε_0. For $\varepsilon_0 > 0$ this dependence has a monotonic and single-valued character. For $\varepsilon_0 < 0$ hysteresis occurs. Since $\rho_0/\rho \leq 1$ always, a hysteresis region exists only for a sufficiently strong field on the surface of the antenna, i.e., when

$$|E_0| > E_{0K} = \frac{2^{1/2}|\varepsilon_0|^{3/4}}{3^{3/4}\beta^{1/2}} \quad (\varepsilon_0 < 0). \tag{4.142}$$

The dependence of $|E|$ on $(\rho_0/\rho)|E_0|$ for $\varepsilon_0 < 0$ is schematically shown in Fig. 30b (the dashed line represents the path of the curve with account of damping). At exact resonance, i.e., when $\varepsilon_0 = 0$, $|E|$ is given by the formula

$$|E| = \left(\frac{\varrho_0 |E_0|}{\beta}\right)^{1/3} \frac{1}{\varrho^{1/3}}. \tag{4.143}$$

We see that at resonance the electric field falls off very slowly with increasing distance from the antenna – according to a $\rho^{-1/3}$ law (but not according to a ρ^{-1} law as it does far from resonance).

f. Transition Layer for an Alternating Electric Field. As already stated, the formulas for E and δN obtained above on the basis of the macroscopic theory are not applicable in the transition layer immediately adjacent to the surface of the antenna. As we shall see below, the thickness of this layer increases strongly as $\omega \to \omega_0$. We therefore limit the discussion to just this case, which is the most interesting case. In order to calculate the structure of the transition layer, it is necessary to also consider, in addition to ε which we have considered up to now, corrections of another kind, associated with the dependence of E on coordinates, i.e., associated with spatial dispersion. Spatial dispersion appears in the dependence of ε not only on frequency ω but also on wave vector **q**. For small values of **q**, when

$$qD \ll 1, \tag{4.144}$$

one can expand ε in powers of **q**. This expansion has the form (see [17], p. 88)

$$\varepsilon(\omega, q) \approx \varepsilon(\omega) - 3(qD)^2 \left(\frac{\omega_0}{\omega}\right)^4. \tag{4.145}$$

As long as the nonlinear corrections and the corrections with respect to q are small, one can consider them independently. Therefore, for $\omega \approx \omega_0$ the final expression for ε has the form

$$\varepsilon \approx \varepsilon_0 + \beta |E|^2 - 3(qD)^2. \tag{4.146}$$

Formula (4.146) means that the electric displacement vector $\mathcal{D}$ ($\mathbf{r}$) is related to $\mathbf{E}(\mathbf{r})$ by the equation

$$\mathcal{D} = \varepsilon_0 E + \beta |E|^2 E + 3D^2 \Delta E, \tag{4.147}$$

since, in coordinate space $\mathbf{q}^2 E$ is equivalent to $-\Delta \mathbf{E}$. In the case of axial symmetry

$$\mathcal{D} = \varepsilon_0 E + \beta |E|^2 E + 3D^2 \frac{1}{\varrho} \frac{d}{d\varrho} \varrho \frac{dE}{d\varrho} \cdot$$

Substituting the last expression for $\mathcal{D}$ into Eq. (4.127), we obtain an equation determining the distribution of the field in the plasma:

$$3D^2 \frac{1}{\varrho} \frac{d}{d\varrho} \varrho \frac{dE}{d\varrho} + \varepsilon_0 E + \beta |E|^2 E = \frac{\varrho_0}{\varrho} E_0. \tag{4.148}$$

It is necessary to solve Eq. (4.148) with the boundary condition $E = E_0$ for $\rho = \rho_0$ (i.e., on the boundary of the plasma).

Since, however, for $\varepsilon_0 \to 0$ the value of E inside the plasma is much larger than E_0, one can with the required accuracy regard the boundary value of E to be equal to zero:

$$E = 0, \quad \varrho = \varrho_0. \tag{4.149}$$

E must tend to zero as $\rho \to \infty$. Let us rewrite Eq. (4.148) in terms of dimensionless variables. For this, we set

$$E = \frac{E_0}{(\beta E_0^2)^{1/3}} y, \quad \varrho = \frac{D\sqrt{3}}{(\beta E_0^2)^{1/6}} t, \quad \varrho_0 = \frac{D\sqrt{3}}{(\beta E_0^2)^{1/6}} t_0, \tag{4.150}$$

after which Eq. (4.148) takes the form

$$\left.\begin{aligned} &\frac{1}{t} \frac{d}{dt} t \frac{dy}{dt} + \mu y + |y^2| y = \frac{t_0}{t}, \\ &\mu = \frac{\varepsilon_0}{(\beta E_0^2)^{1/3}}. \end{aligned}\right\} \tag{4.151}$$

It is clear from (4.151) that if $\mu \gg 1$, in other words, if

$$\varepsilon_0 \gg (\beta E_0^2)^{1/3} = \left(\frac{e^2 E_0^2}{8m\omega_0^2 \varkappa T} \right)^{1/3}, \tag{4.152}$$

then one can neglect the nonlinear term in the equation. In this case the distribution of the field is determined by the linear equation

$$3D^2 \frac{1}{\varrho} \frac{d}{d\varrho} \varrho \frac{dE}{d\varrho} + \varepsilon_0 E = \frac{\varrho_0 E_0}{\varrho}. \tag{4.153}$$

We shall not write the solution of this equation, which can be expressed in terms of Bessel functions.

If we are interested in distances from the surface of the antenna which are small in comparison with its radius, so that

$$x = \varrho - \varrho_0 \ll \varrho_0, \tag{4.154}$$

then (4.151) can be approximately written in the form

$$\frac{d^2y}{d\xi^2}+\mu y+|y^2|y=1,$$

where

$$\xi=\frac{(\beta E_0^2)^{1/6}}{D\sqrt{3}}x. \tag{4.155}$$

In this case, instead of Eq. (4.153) we have

$$3D^2\frac{d^2E}{dx^2}+\varepsilon_0 E=E_0. \tag{4.156}$$

This equation with the boundary condition E = 0, x = 0 has the solution

$$\left.\begin{aligned} E&=\left(\frac{E_0}{\varepsilon_0}\right)\left\{1-\exp\left[-\left(\frac{x}{D}\right)\sqrt{\frac{|\varepsilon_0|}{3}}\right]\right\}, \quad \varepsilon_0<0,\\ E&=\left(\frac{E_0}{\varepsilon_0}\right)\left\{1-\exp\left[\left(\frac{ix}{D}\right)\sqrt{\frac{\varepsilon_0}{3}}\right]\right\}, \quad \varepsilon_0>0. \end{aligned}\right\} \tag{4.157}$$

Expressions (4.157) coincide, as they must, with those obtained by Landau directly with the aid of the kinetic equation [32]. We note, however, that the small Landau damping term is absent in the second formula of (4.157). Equations (4.148) and (4.156) do not describe this damping. With account of damping

$$E=\left(\frac{E_0}{\varepsilon_0}\right)\left\{1-\exp\left[\left(\frac{ix}{D}\right)\sqrt{\frac{\varepsilon_0}{3}}-\left(\frac{3x}{D}\right)\sqrt{\frac{\pi}{2\varepsilon_0}}\exp\left(-\frac{3}{2\varepsilon_0}\right)\right]\right\}. \tag{4.157a}$$

From formulas (4.157) and (4.157a) it is clear that the amplitude of the field and the thickness of the surface layer increase as ε_0 decreases. The characteristic distance X, over which the field varies, turns out to be on the order of

$$X\sim D\sqrt{\frac{3}{|\varepsilon_0|}}\gg D.$$

This inequality justifies the admissibility of the expansion (4.145), and consequently (justifies) all subsequent calculations. For very small values of ε_0, however, the linear equation (4.156) is inapplicable. One can obtain an analytic solution of the more general equation (4.155) only for the case of real y. This solution has a complicated form, and we shall not write it down here. If ε_0 is so small that the inequality

$$\mu\ll 1,$$

is satisfied, the reverse of inequality (4.152), then one can neglect the term containing μ in Eq. (4.155). In this case, the field is determined by the equation

$$\frac{d^2y}{d\xi^2}+y^3=1. \tag{4.158}$$

which does not contain the parameters. It is clear from Eqs. (4.150) and (4.158) that the amplitude of the field in a plasma and the characteristic distance over which it significantly varies are respectively given by

$$\left.\begin{aligned} E&\sim\frac{E_0}{(\beta E_0^2)^{3/2}},\\ X&\sim\frac{D\sqrt{3}}{(\beta E_0^2)^{1/6}}. \end{aligned}\right| \tag{4.159}$$

From (4.159) it follows that, in order to be able to regard the field as weak and expand the exponent in (4.124), it is necessary that the condition

$$\beta \mid E_0 \mid^2 \ll 1 \tag{4.160}$$

be satisfied, or

$$\frac{\mid E_0 \mid^2}{16\pi} \ll 2\varkappa T N_0. \tag{4.161}$$

We note that under condition (4.160) the necessary inequality $X \gg D$ is automatically satisfied.*

In concluding this section, we present without derivation the formula obtained in a linear approximation in [32] for the distribution of the field near a planar boundary of the plasma, when the frequency is far from ω_0. This formula has the form

$$E(x) = \frac{E_0}{\varepsilon} + \frac{2E_0}{\sqrt{3}\,\varepsilon^2}\left(\frac{\omega}{\omega_0}\right)^{2/3}\left(\frac{x}{D}\right)^{2/3}\exp\left\{-\frac{3}{4}\left(\frac{\omega x}{\omega_0 D}\right)^{2/3} + i\left[\frac{3\sqrt{3}}{4}\left(\frac{\omega x}{\omega_0 D}\right)^{2/3} + \frac{2\pi}{3}\right]\right\}. \tag{4.162}$$

Here, as usual, $\varepsilon = 1 - \omega_0^2/\omega^2$, $\omega_0 = (4\pi e^2 N_0/m)^{1/2}$. From here it is clear that the amplitude of an alternating electric field rapidly approaches the constant value E_0/ε_0 as one goes away from the surface of the body. Here the thickness of the transition layer is of the order of the Debye radius D, and only in the case of a low-frequency field, $\omega \ll \omega_0$, does it increase somewhat in proportion to ω_0/ω.

---

*The effect considered here of resonance excitation of a plasma by an alternating field for $\omega \approx \omega_0$ must, of course, appear during probe investigations of a plasma with the aid of a high-frequency field. In particular, one can show that, for $\omega \to \omega_0$ and a constant field E_0 the constant component of the probe current increases resonantly. It is possible that this effect has been observed in experiments (see, for example, [67] and [76]).

CHAPTER V

CONSIDERATION OF THE INFLUENCE OF COLLISIONS BETWEEN PARTICLES ON THE DISTURBANCE OF THE CONCENTRATIONS OF ELECTRONS AND IONS

§ 19. Effective Number of Collisions

In the preceding chapters we saw that the Fourier component $N_{\mathbf{q}}$ of the electron concentration becomes very large when the product $\mathbf{q}\mathbf{V}_0$ and the longitudinal component $q_{\parallel}$ tend to zero, if the plasma is located in a magnetic field. In the absence of a magnetic field, $N_{\mathbf{q}}$ increases rapidly when $q \to 0$. However, for these values of $\mathbf{q}$ the formulas obtained earlier are inapplicable, because the abrupt increase of $N_{\mathbf{q}}$ in these cases is associated with the fact that very large regions, whose dimensions may be comparable with the mean free path of the ions, begin to play a role in the integral for $N_{\mathbf{q}}$. Therefore, for small values of $\mathbf{q}\mathbf{V}_0$ and $q_{\parallel}$ (or small values of q if no magnetic field is present) it is necessary to take the collisions of ions with themselves and with other particles into consideration. Because of the very complicated form of the collision integrals (1.31), (1.32), however, a rigorous solution of the problem with collisions is possible only in certain limiting cases (see §21). Therefore, in order to obtain a comparatively simple general formula, we shall use the method of the effective number of collisions, a method widely used in physics. Of course, in connection with such a method of calculation, the formulas obtained have only an interpolational character.

Usually the effective number of collisions ν is introduced by adding the term $Y = -\nu(f - f_{0i})$ to the right side of the kinetic equation. However, such a form of the collision integral is unsatisfactory in our case, since it violates the law of conservation of the number of particles $\left(\int Y d^3v \equiv 0\right)$, and this may lead to a significant error in the calculation of the spatial distribution of the ions. Therefore, we shall write the collision integral in the form

$$Y_{\text{eff}} = -\nu\left(f_i - \frac{f_{0i}}{N_0}\int f_i\, d^3v\right) = -\nu\left(f_i' - \frac{f_{0i}}{N_0}\int f_i'\, d^3v\right). \tag{5.1}$$

From Eq. (5.1) it is immediately clear that

$$\int Y_{\text{eff}}\; d^3v = 0. \tag{5.2}$$

In order of magnitude, the effective number of collisions ν is equal to the sum of the collisions of ions with neutral particles and with other ions, i.e., (see [34] and [35]),

$$\nu = 2.2 n_0 \sigma_{in} \sqrt{\frac{\varkappa T}{M_i}} + \frac{10/3\, e^4 N_0 \Lambda}{(\varkappa T)^{3/2} M_i^{1/2}}. \tag{5.3}$$

In this formula σ_{in} is the effective cross section for the collisions of an ion with neutral particles, and Λ is the Coulomb logarithm, i.e.,

$$\Lambda = \frac{3}{2}\ln\left(0.37\frac{\varkappa T}{e^2 N_0^{1/3}}\right).$$

Adding expression (5.1) to the right side of Eq. (4.72), we obtain

$$\frac{e}{M_i c}[uH_0]\frac{\partial f_q}{\partial u}+iqvf_q+iqu\frac{e}{\varkappa T}f_{0i}\varphi_q=I(v)-\nu\left(f_q-\frac{f_{0i}}{N_0}\int f_q\,d^3v\right). \quad (5.4)$$

Since

$$\int f_q\,d^3v=N_q=N_0\frac{e\varphi_q}{\varkappa T},$$

then one can reduce (5.4) to the form

$$\frac{e}{M_i c}[uH_0]\frac{\partial f_q}{\partial u}+i(qv-i\nu)f_q+i(qu+i\nu)\frac{e}{\varkappa T}f_{0i}\varphi_q=I(v). \quad (5.5)$$

This equation is similar to Eq. (4.56) and one can solve it by the same method. If I(**v**) is evaluated in accordance to (4.56), then the final answer has the form

$$N_q=-\frac{\dfrac{\pi R_0^2N_0V_0}{\Omega}\displaystyle\int_0^\infty \exp A\,dx}{2+\dfrac{iqV_0-2\nu}{\Omega}\displaystyle\int_0^\infty \exp A\,dx}, \quad (5.6)$$

where

$$A=\left[\frac{iqV_0-\nu}{\Omega}x-\frac{\varkappa T}{2M_i\Omega^2}\left(q_\parallel^2x^2+4q_\perp^2\sin^2\frac{x}{2}\right)\right].$$

It is immediately clear from Eq. (5.6) that one can neglect the influence of collisions on the distribution of ions and electrons in the case when

$$qV_0\gg\nu. \quad (5.7)$$

The influence of collisions is most important for small values of $\mathbf{q}\mathbf{V}_0$ and $q_\parallel$, when

$$qV_0\lesssim\nu\ll\Omega,\quad q_\parallel\sqrt{\frac{\varkappa T}{M_i}}\lesssim\nu\ll\Omega. \quad (5.8)$$

We note here that the condition $\nu\ll\Omega$ is satisfied in the ionosphere at all altitudes of interest to us.

Upon fulfillment of conditions (5.8), the periodic factor $\exp[-(2q_\perp^2\varkappa T/M_i\Omega^2)\sin^2 x/2]$ inside the integral oscillates rapidly (see p. 62), and one can replace it by the average value over a period. In this case, the same calculations which were also used in the derivation of formula (4.60) lead to the following result:

$$N_q=-\frac{\dfrac{\pi R_0^2N_0V_0}{q_\parallel}\sqrt{\dfrac{M_i}{2\varkappa T}}Q(a_2)B}{2+\dfrac{iqV_0-2\nu}{iqV_0-\nu}ia_2Q(a_2)B}, \quad (5.9)$$

where

$$B=\exp\left(-\frac{q_\perp^2\varkappa T}{M_i\Omega^2}\right)I_0\left(\frac{q_\perp^2\varkappa T}{M_i\Omega^2}\right)$$

and

$$a_2=\frac{qV_0+i\nu}{q_\parallel}\sqrt{\frac{M_i}{2\varkappa T}}. \quad (5.10)$$

Now if $\mathbf{q}\mathbf{V}_0$ and $q_{\|}$ tend to zero, then $N_{\mathbf{q}}$ still does not tend to infinity, but is determined with the aid of the following formula:

$$N_q = -\frac{\frac{\pi R_0^2 N_0 V_0}{2\nu} \exp\left(-\frac{q_\perp^2 \varkappa T}{M_i \Omega^2}\right) I_0\left(\frac{q_\perp^2 \varkappa T}{M_i \Omega^2}\right)}{1-\exp\left(-\frac{q_\perp^2 \varkappa T}{M_i \Omega^2}\right) I_0\left(\frac{q_\perp^2 \varkappa T}{M_i \Omega^2}\right)}. \tag{5.11}$$

Here let us briefly also consider the formulas for $N_{\mathbf{q}}$, taking collisions into account, but without the presence of a magnetic field. These formulas are only of theoretical value, since in the upper ionosphere condition (3.41) is more stringent than (5.7) so that a magnetic field always manifests itself more strongly than collisions. Adding Y_{eff} to the right side of Eq. (4.37) and solving it, we find

$$N_q = -\frac{i \int \frac{I(v)}{qv - i\nu} d^3v}{2 + \frac{iqV_0 - 2\nu}{N_0} \int \frac{f_{0i}}{iqv + \nu} d^3v}. \tag{5.12}$$

If

$$qV_0 \ll \nu, \quad q\sqrt{\frac{\varkappa T}{M_i}} \ll \nu, \tag{5.13}$$

then

$$N_q = \frac{\int I(v)\, d^3v - \frac{iq}{\nu} \int uI(v)\, d^3v}{\left(-iqV_0 + \frac{2q_\perp^2 \varkappa T}{\nu M_i}\right)}. \tag{5.14}$$

Substituting $I(\mathbf{v})$ in the form $-\pi R_0^2 V_0 f_{0i}$ into Eq. (5.12), we obtain

$$N_q = -\frac{\pi R_0^2 N_0 V_0}{q} \frac{\left(\frac{M_i}{2\varkappa T}\right)^{1/2} Q(a_3)}{2 + \left(a_3 + i\nu\sqrt{M_i/2\varkappa T}\right) Q(a_3)}, \tag{5.15}$$

where

$$a_3 = \frac{qV_0 + i\nu}{q} \left(\frac{M_i}{2\varkappa T}\right)^{1/2}.$$

We note here that expression (5.14) coincides with the formulas for N_q which we shall obtain in §22 from the equation for the diffusion of ions. Such a result is quite natural since (5.13) is just exactly the condition for the applicability of the diffusion approximation.

§20. Simplification of the Equation Containing a Collision Integral of General Form

We saw in the preceding section that upon fulfillment of conditions (5.8) it is possible to considerably simplify the solution of the kinetic equation with an effective number of collisions. This is not accidental. We shall now show that conditions (5.8) enable us to carry out simplifications directly in the kinetic equation of general form. These simplifications enable us, in particular, to obtain the solutions of the equation with the exact collision integral in certain limiting cases, and to compare them (these exact solutions) with the solution of the equation containing the approximate collision integral (5.1).

First we note that, since the disturbances of the distribution functions are still small at those distances from the body at which collisions are important, we may linearize the collision integral, i.e., in it all distribution functions can be set equal to

$$f_i = f_{0i} + f'_i$$

and terms quadratic in f'_i are neglected. After this, the collision integral becomes a linear operator, acting on the function f'_i.

In order to simplify the following calculations, it is convenient to use the theorem proved in § 16b, according to which one can omit the term containing the potential φ from the kinetic equation, taking the influence of φ into consideration by the introduction of the dielectric permittivity. It follows from formulas (4.79)-(4.83) that one can represent N_q in the form

$$N_q = \frac{\int f_1 \, d^3v}{1 + \int f_2 \, d^3v} = \frac{\int f_1 \, d^3v}{\sum\limits_{l,k} \varepsilon_{lk} q_l q_k D^2} . \tag{5.16}$$

Here, if the nature of the operator L is introduced in (4.80), then in the case considered by us, f_1 and f_2 satisfy the equations

$$i(qu - qV_0) f_1 + \frac{e}{M_i c} [uH_0] \frac{\partial f_1}{\partial u} = I(u), \tag{5.17}$$

$$i(qu - qV_0) f_2 + \frac{e}{M_i c} [uH_0] \frac{\partial f_2}{\partial u} = iqu \frac{f_{0i}}{N_0} . \tag{5.18}$$

We shall see that in order to determine N_q it is necessary to separately determine the two functions f_1 and f_2. Let us go on to the solution of this problem.

In cylindrical coordinates, Eq. (5.17) takes the form

$$\frac{\partial f_1}{\partial \beta} - i(\mu + \gamma \cos \beta) f_1 + \frac{1}{\Omega} Y [f_1(u')] = \frac{I(u)}{\Omega} , \tag{5.19}$$

where

$$\mu = \frac{q_{\parallel} u_{\parallel} - qV_0}{\Omega} , \quad \gamma = \frac{q_{\perp} u_{\perp}}{\Omega} .$$

Now we make the following change of variables

$$f_1(u) = \exp(i\gamma \sin \beta) g(u). \tag{5.20}$$

Then $g(\mathbf{u})$ satisfies the equation

$$\frac{\partial g}{\partial \beta} - i\mu g = -\exp(-i\gamma \sin \beta) \frac{1}{\Omega} Y [\exp(i\gamma' \sin \beta') g(u')] - \exp(-i\gamma \sin \beta) \frac{I(u)}{\Omega} , \tag{5.21}$$

where

$$\gamma' = \frac{q_{\perp} u'_{\perp}}{\Omega} .$$

In order to arrive at the desired (sought for) simplification, we expand $g(\mathbf{u})$ in a Fourier series with respect to β:

$$\begin{aligned} g(u) = g_0(u_{\parallel}, u_{\perp}) + g_1(u_{\parallel}, u_{\perp}) e^{i\beta} + g_{-1}(u_{\parallel}, u_{\perp}) e^{-i\beta} + \ldots \\ \ldots + g_m(u_{\parallel}, u_{\perp}) e^{im\beta} + \ldots \quad (m = 0, \pm 1, \pm 2, \ldots). \end{aligned} \tag{5.22}$$

Now let us estimate the order of the individual terms in Eq. (5.22) under condition (5.8). One can determine the order of all terms of the series, except the zero term, by comparing the largest first term on the left side of (5.21) with the right side. Thus

$$g_m \sim \frac{I(u)}{m\Omega} \qquad (m \neq 0).$$

For the zero term, the derivative $\partial g_0/\partial\beta$ vanishes and it is necessary to estimate g_0 by comparing the second or third terms with the right side. In this connection, if condition (5.8) is satisfied,

$$g_0 \sim \frac{\Omega}{qV_0}\frac{I(u)}{\Omega} \gg g_m.$$

As a result we may limit ourselves in the expansion for g to only the term g_0, having taken the dc Fourier component of Eq. (5.21), i.e., having averaged it with respect to the angle β. Finally, $g = g_0(u_\parallel, u_\perp)$ satisfies the equation*

$$i(q_\parallel u_\parallel - qV_0)\,g(u_\parallel, u_\perp) - \frac{1}{2\pi}\int_0^{2\pi} d\beta \exp(-i\gamma\sin\beta)\,Y\,[\exp(i\gamma'\sin\beta')\,g(u'_\parallel, u'_\perp)] =$$

$$= \frac{1}{2\pi}\int_0^{2\pi} \exp(-i\gamma\sin\beta)\,I(u)\,d\beta. \tag{5.23}$$

If, as we shall assume later on, I(u) does not depend on the angle β, for example, if it has the form $-\pi R_0^2 V_0 f_{0i}$, then

$$\frac{1}{2\pi}\int_0^{2\pi} \exp(-i\gamma\sin\beta)\,I(u)\,d\beta = \frac{1}{2\pi}\int_0^{2\pi} \exp(-i\gamma\sin\beta)\,d\beta I(u) = J_0(\gamma)\,I(u). \tag{5.24}$$

One can subject Eq. (5.18) to a similar simplification.† For this, it is necessary to introduce a new unknown function $h(u_\parallel, u_\perp)$, related to f_2 by the equation

$$f_2 = \frac{f_{i0}}{N_0} + h(u_\parallel, u_\perp)\exp\left(i\,\frac{q_\parallel u_\perp}{\Omega}\sin\beta\right). \tag{5.25}$$

The equation for the function h introduced in this manner differs from the equation for g only by the replacement

$$I \to \frac{iqV_0}{N_0} f_{0i}. \tag{5.26}$$

Therefore, one can obtain all formulas for h from the formulas for g, without making any additional calculations. However, we do not need these formulas since in the two limiting cases which we shall analyze in detail in this section, one can immediately write down the results for the expression $1 + \int f_2\, d^3v$, which enters into the desired formula (5.16) for N_q. In fact, if

---

*The idea of the approximation made here is similar to the idea of the approximation developed by Budker and Belyaev [36, 37]. However, we do not assume that f'_i and φ are slowly varying in the direction perpendicular to the field. Therefore, our formulas go over into the formulas obtained in [36, 37] only in the limiting case $\gamma \ll 1$.

†If we wanted to directly simplify the equation for f_q without using the relations of § 16b, then we should take

$$f_q = -\frac{e}{\varkappa T}\varphi_q f_{0i} + \exp(i\gamma\sin\beta)\,g_q(u_\parallel, u_\perp)$$

and average the equation for g_q with respect to β.

$$q_\perp \sqrt{\frac{\varkappa T}{M_i}} \ll \Omega, \qquad q_\parallel \sqrt{\frac{\varkappa T}{M_i}} \ll q V_0, \tag{5.27}$$

then the right side of Eq. (5.18), and together with it also the function f_2, turns out to be small so that one can neglect the integral of f_2. Hence

$$\sum_{l,\,k} \varepsilon_{lk} q_l q_k D^2 = 1 + \int f_2\, d^3v \approx 1. \tag{5.28}$$

In the other limiting case, when

$$q_\perp \sqrt{\frac{\varkappa T}{M_i}} \gg \Omega, \tag{5.29}$$

because of rapid oscillations of the exponent in $\exp(i\gamma \sin \beta)$ the right side of the equation for h becomes small. Therefore, one can neglect the second term containing h in Eq. (5.25). Therefore, $f_2 \approx f_{0i}/N_0$ and

$$\sum_{l,\,k} \varepsilon_{lk} q_l q_k D^2 = 1 + \int f_2\, d^3v \approx 2. \tag{5.30}$$

In order to illustrate the approximation developed here, let us apply Eq. (5.23) to the collision integral (5.2), for which an exact solution was already obtained in the preceding section. Substituting (5.2) into (5.23) and evaluating the integral with respect to $d\beta$, we have

$$[i(q_\parallel u_\parallel - qV_0) + \nu]\, g - \nu \frac{f_{0i}}{N_0} J_0(\gamma) \int J_0\left(\frac{q_\perp u_\perp}{\Omega}\right) g\, d^3u = J_0(\gamma)\, I(u). \tag{5.31}$$

From here we find that

$$\int f_1\, d^3v = \int J_0\left(\frac{q_\perp u_\perp}{\Omega}\right) g\, d^3u = \frac{\dfrac{1}{i} \displaystyle\int \frac{J_0^2\left(\frac{q_\perp u_\perp}{\Omega}\right) I(u)}{i(q_\parallel u_\parallel - qV_0) + \nu}\, d^3u}{1 - \dfrac{\nu}{N_0} \displaystyle\int \frac{J_0^2\left(\frac{q_\perp u_\perp}{\Omega}\right) f_{0i}}{i(q_\parallel u_\parallel - qV_0) + \nu}\, d^3u}.$$

Writing down the analogous formula for the function f_2 and evaluating the integral with respect to d^3u, we immediately obtain, as we should, formula (5.9).*

In this case, when $q_\perp u_\perp/\Omega \sim (q_\perp/\Omega)(\varkappa T/M_i)^{1/2} \gg 1$, one can use asymptotic expressions for J_0^2:

$$J_0^2(x) \approx \frac{2\cos^2\left(x - \frac{\pi}{4}\right)}{\pi x} = \frac{1}{\pi x}\left[1 + \cos\left(2x - \frac{\pi}{2}\right)\right]. \tag{5.32}$$

*Here, in order to evaluate the integrals with respect to $du_\perp$, it is necessary to use the formula

$$\int_0^\infty e^{-t^2} J_0^2(\alpha t)\, t\, dt = \frac{1}{2} e^{-\frac{\alpha^2}{2}} I_0\left(\frac{\alpha^2}{2}\right).$$

The second term inside the square brackets, $\cos(2x - \pi/2)$, gives a small contribution to the integral and, therefore, one can omit it. Taking (5.30) into consideration, we finally obtain

$$N_q = \frac{\Omega}{2q_{\perp}\pi} \int \frac{I(u)}{i(q_{\parallel}u_{\parallel} - qV_0) + \nu} \frac{d^3u}{u_{\perp}}. \tag{5.33}$$

In the other limiting case, when $q_{\parallel}(\varkappa T/M_i)^{1/2} \ll qV_0$ and $q_{\perp}(\varkappa T/M_i)^{1/2} \ll \Omega$, after simple calculations we obtain

$$N_q = \frac{\int I(u)\, d^3u}{-iqV_0 + \dfrac{\nu q_{\perp}^2 \varkappa T}{\Omega^2 M_i}}. \tag{5.34}$$

In the following section we shall compare formulas (5.33) and (5.34) with the formulas obtained from the equations containing the exact collision integrals.

§ 21. Solution of the Equations Containing the Exact Collision Integrals in Various Limiting Cases

In the present section we shall derive formulas for N_q with the aid of equations containing the exact collision integrals (1.31) and (1.32). In this connection, it is possible to obtain an answer in explicit form for the limiting cases given by inequalities (5.27) and (5.29).

Let us initially assume that inequality (5.29) is satisfied, and let us consider the case of collisions of ions with neutral particles, i.e., the collision integral (1.31). Linearizing this expression and substituting it into Eq. (5.23), we obtain

$$i(q_{\parallel}u_{\parallel} - qV_0)g + g\int Wf_{0n}(u')\, d^3u'\, d^3u_1\, d^3u_1' - \frac{1}{2\pi}\int_0^{2\pi} \exp(-i\gamma\sin\beta)\, d\beta \int Wg(u_1) f_{0n}(u') \exp(i\gamma_1 \sin\beta_1) \times d^3u'\, d^3u_1\, d^3u_1' = J_0(\gamma) I(u), \qquad \gamma_1 = \frac{q_{\perp}u_{1\perp}}{\Omega}. \tag{5.35}$$

Here $W = W(\mathbf{v}, \mathbf{v}', \mathbf{v}_1, \mathbf{v}_1')$ is the probability for a collision of particles in which the particles' velocities are changed from $\mathbf{v}_1$ to $\mathbf{v}$, and from $\mathbf{v}_1'$ to $\mathbf{v}'$. Now let us take into consideration that, under condition (5.29) $\exp(i\gamma\sin\beta)$ is a rapidly oscillating function. Therefore, integrals with respect to β of expressions containing this function turn out to be small. The third term on the left side of Eq. (5.35) is such a small term, so that it is possible to neglect it. After this, (5.35) reduces to the form of an equation containing an effective number of collisions, namely:

$$[i(q_{\parallel}u_{\parallel} - qV_0) + \nu_{in}^{(1)}(u)]\, g = J_0(\gamma) I(u). \tag{5.36}$$

The effective number of collisions, now depending on velocity, is given by the formula

$$\nu_{in}^{1} = \int Wf_{0n}(u')\, d^3u'\, d^3u_1\, d^3u_1'. \tag{5.37}$$

Solving Eq. (5.36) and replacing J_0 by the asymptotic expression (5.32), we obtain

$$N_q = \frac{1}{2\pi}\frac{\Omega}{q_{\perp}} \int \frac{I(u)}{i(q_{\parallel}u_{\parallel} - qV_0) + \nu_{in}^{(1)}(u)}\, d^3u. \tag{5.38}$$

Formula (5.38) is similar to formula (5.33) with only this difference: now ν depends on velocity.

Now let us go to the case of collisions between ions, i.e., to collision integral (1.32). It is clear in advance that those terms in the collision integral Y in which the unknown function is integrated with respect to **v'** again turn out to be small becuase of the oscillating factor. Therefore, one can immediately write Y in the form

$$Y = -\sum_k \frac{1}{M_i} \frac{\partial j_k}{\partial u_k},$$

$$j_k = \sum_l \frac{\pi e^4 \Lambda}{M_i} \int \left[f_q \frac{\partial f_{0i}(u_1)}{\partial u_{1l}} - f_{0i}(u_1) \frac{\partial f_q}{\partial u_l} \right] \frac{w^2 \delta_{lk} - w_l w_k}{w^3} d^3 u_1$$

$$= -\frac{\pi e^4 \Lambda}{M_i} \sum_l \left[\frac{\partial f_q}{\partial u_l} + \frac{M_i u_l}{\varkappa T} f_q \right] A_{lk}, \tag{5.39}$$

where

$$A_{lk} = \int f_{0i}(u_1) \frac{w^2 \delta_{lk} - w_l w_k}{w^3} d^3 u_1, \quad w = u - u_1.$$

One can express the integral for A_{lk} in terms of the probability integral. Omitting these transformations, we immediately write down the result:

$$A_{lk} = N_0 \sqrt{\frac{\varkappa T}{2\pi M_i}} \left[a(u^2) \frac{u_l u_k}{u^2} + b \delta_{lk} \right], \tag{5.40}$$

where

$$a = \left[\left(\frac{M_i u^2}{\varkappa T} - 3 \right) \sqrt{\frac{\pi \varkappa T}{2 M_i u^2}} \Phi\left(\sqrt{\frac{M_i u^2}{2\varkappa T}} \right) + 3 \exp\left(-\frac{M_i u^2}{2\varkappa T} \right) \right] u^{-2},$$

$$b = \left[\left(\frac{M_i u^2}{\varkappa T} - 1 \right) \sqrt{\frac{\pi \varkappa T}{2 M_i u^2}} \Phi\left(\sqrt{\frac{M_i u^2}{2\varkappa T}} \right) + \exp\left(-\frac{M_i u^2}{2\varkappa T} \right) \right] u^{-2},$$

$$\Phi(x) = \frac{2}{\sqrt{\pi}} \int_0^x e^{-t^2} dt.$$

If we now return to Eq. (5.23), we note that

$$\gamma \sin \beta = \frac{q_\perp u_\perp}{\Omega} \sin \beta = \frac{q_1 u_\perp}{\Omega}, \tag{5.41}$$

where $\mathbf{q}_1$ is a vector perpendicular to $\mathbf{q}_\perp$ and $\mathbf{H}_0$, and $|\mathbf{q}_1| = |\mathbf{q}_\perp|$. Substituting (5.39) and (5.41) into (5.23), we take into consideration that in the case when $(q_\perp/\Omega)(\varkappa T/M_i)^{1/2}$ is large we should only differentiate the exponent with respect to u. Integration with respect to dβ reduces to an average over all directions of the vector $\mathbf{q}_1$ in a plane perpendicular to $\mathbf{H}_0$. Therefore, the second term on the left side of Eq. (5.23) can be brought to the form

$$\frac{1}{2\pi} \int_0^{2\pi} d\beta \exp(-i\gamma \sin \beta) Y [\exp(i\gamma' \sin \beta') g]$$

$$= \frac{\pi e^4 \Lambda}{M_i^2 \Omega^2} g \sum_{l,k} \overline{q_{1l} q_{1k}} A_{lk} = \frac{\pi e^4 \Lambda}{M_i^2 \Omega^2} q_\perp^2 \left(b + \frac{1}{2} a \frac{u_\perp^2}{u^2} \right) N_0 \left(\frac{\varkappa T}{2\pi M_i} \right)^{1/2} g = \frac{q_\perp^2 \varkappa T}{M_i \Omega^2} \nu_{ii}^{(1)}(u) g \tag{5.42}$$

where

$$\nu_{ii}^{(1)} = \frac{e^4 \Lambda N_0}{M_i} \left(\frac{\pi \varkappa T}{2 M_i} \right)^{1/2} \left(b + \frac{1}{2} a \frac{u_\perp^2}{u^2} \right).$$

Substituting (5.42) into (5.23) and solving the resulting equation, we have

$$N_q = \frac{1}{2\pi}\frac{\Omega}{q_\perp}\int \frac{I(u)}{i(q_{\parallel}u_{\parallel} - qV_0) + \frac{q_\perp^2 \varkappa T}{M_i \Omega^2}\nu_{ii}^{(1)}(u)} \frac{d^3u}{u_\perp}. \tag{5.43}$$

Formula (5.43) differs from (5.33) not only in regard to the dependence of ν on **u**, but also with regard to the large coefficient $q_\perp^2 \varkappa T/M_i\Omega^2$ in front of ν. This means that in that region where the actual collision integral has the form (1.32), the interpolated formula (5.6) gives an overestimated value of N_q for $q_\perp^2 \varkappa T/M_i\Omega^2 \gg 1$.

Now let us go on to the other limiting case, when

$$q_\perp \sqrt{\frac{\varkappa T}{M_i}} \ll 1, \quad q_\parallel \sqrt{\frac{\varkappa T}{M_i}} \ll qV_0. \tag{5.27}$$

In this case, as we shall see below, collisions are important only when

$$qV_0 \sim \frac{\nu q_\perp^2 \varkappa T}{M_i \Omega^2} \ll \nu. \tag{5.44}$$

Now we again introduce the vector $\mathbf{q}_1$ (see Eq. (5.41), we expand the left side of Eq. (5.23) in powers of $\mathbf{q}_1\mathbf{u}$ and we neglect terms containing $q_\parallel u_\parallel$. Then

$$-iqV_0 g - \frac{1}{2\pi}\int_0^{2\pi} d\beta \left(1 - \frac{iq_1 u}{\Omega}\right) Y\left[\left(1 + \frac{iq_1 u'}{\Omega}\right) g\right] = I(u). \tag{5.45}$$

In order to estimate the function g, we integrate (5.45) with respect to d^3u. Here the main term containing Y drops out in virture of the identity (5.2). Therefore

$$\int g\, d^3u \sim \frac{\int I\, d^3u}{qV_0}, \quad g \sim \frac{I}{qV_0}. \tag{5.46}$$

If we substitute a quantity of such order into (5.45), then a large term $\sim(\nu/\mathbf{q}\mathbf{V}_0)\mathbf{I}$ appears on the left side, so that the left side of the equation turns out to be much larger than the right side. Therefore, in order for the equation to have a solution, the function Y[g] must identically reduce to zero. Therefore, it is necessary to look for a solution of the form

$$g = \frac{N_q}{N_0} f_{0i}. \tag{5.47}$$

After this, Y[g] vanishes so that $Y[f_{0i}] \equiv 0$. The coefficient for f_{0i} in (5.47) can thus be picked out at once in order to fulfill the condition $N_q = \int g\, d^3u$, which in the case considered follows from (5.16) to (5.28). Having, substituted (5.47) into (5.45) and having integrated with respect to d^3u, we obtain

$$N_q = \frac{\int I(u)\, d^3u}{-iqV_0 + \nu_i^{(2)} \frac{q_\perp^2 \varkappa T}{M_i \Omega^2}}, \tag{5.48}$$

where

$$\nu_i^{(2)} = -\frac{M_i}{2\varkappa T N_0}\int d^3u\, u_\perp Y[u'_\perp f_{0i}(u')]. \tag{5.49}$$

Formula (5.49) has the same form as formula (5.33), which was obtained with the aid of the effective number of collisions. We note that it is clear from Eq. (5.48) that collisions are actually important upon fulfillment of condition (5.44).

Now let us evaluate $\nu_i^{(2)}$ separately for collisions of ions with neutral ($\nu_{in}^{(2)}$) and charged ($\nu_{ie}^{(2)}$) particles. Substituting the collision integral Y defined by formula (1.31) into (5.49), we find that

$$\nu_i^{(2)} = \nu_{in}^{(2)} = -\frac{M_i}{2\varkappa T N_0} \int W \left[u_\perp^2 f_{0i}(u) f_{0n}(u') - (u_\perp u_{1\perp}) f_{0i}(u_1) f_{0n}(u_1')\right] d^3u\, d^3u'\, d^3u_1\, d^3u_1'. \tag{5.50}$$

The situation is somewhat more complicated in that region where collisions of charged particles play the principal role. The integral for the collisions of ions with ions satisfies the momentum conservation law, by virtue of which

$$\int p Y_{ii}\, d^3u = \int \frac{u}{M_i} Y_{ii}\, d^3u = 0. \tag{5.51}$$

Therefore, Y_{ii} drops out of (5.50), and by Y in this formula, one should understand only the collision integral Y_{ie} referring to collisions of ions with electrons. After linearization, Y_{ie} has the form

$$Y = -\frac{1}{M_i} \sum_k \frac{\partial j_k}{\partial u_k}, \tag{5.52}$$

where

$$j_k = \sum_l \frac{\pi e^4 \Lambda}{M_i} \left[\frac{\partial f_q}{\partial u_l} + \frac{M_i u_l f_q}{\varkappa T}\right] B_{lk}.$$

Here

$$B_{lk} = \int f_{0e}(u') \frac{w^2 \delta_{lk} - w_l w_k}{w^3} d^3u' \approx \int f_{0e}(u') \frac{u'^2 \delta_{lk} - u_l' u_k'}{u'^3} d^3u' = \frac{2}{3} \delta_{lk} \int f_{0e} \frac{d^3u}{u} = \frac{4}{3} N_0 \sqrt{\frac{m}{2\pi \varkappa T}}\, \delta_{lk},$$

if it is taken into consideration that $\mathbf{w} = \mathbf{u} - \mathbf{u}' \approx -\mathbf{u}'$, since the velocity of the electrons is much larger than the velocity of the ions. Substituting (5.52) into (5.49), we obtain

$$\nu^2 = \nu_{ie}^{(2)} = \frac{2}{3} \frac{\sqrt{2\pi} e^4 \Lambda}{\sqrt{M_i} (\varkappa T)^{3/2}} \sqrt{\frac{m}{M_i}} N_0. \tag{5.53}$$

In order of magnitude, this expression is $(m/M_i)^{1/2}$ times smaller than the effective number of collisions ν. Thus, whereas for large values of $(q_\perp/\Omega)(\varkappa T/M_i)^{1/2}$ formula (5.6) gives an exaggerated value of N_q for the collisions of charged particles, in the region $(q_\perp/\Omega)(\varkappa T/M_i)^{1/2} \ll 1$ formula (5.6) gives an underestimated value of N_q.

§ 22. Disturbances of the Concentration of Electrons and Ions in the Diffusion Approximation (Lower Ionosphere)

The motion of a body in the lower layers of the ionosphere takes place under conditions when the mean free path is still not so large. Therefore, a calculation of the disturbances at distances much larger than the mean free path from the body may be of interest. This question is of special interest in connection with the problem of the scattering of radiowaves by the "trail" of a body. It was already seen in § 19 that if [see Eq. (5.7)] $\mathbf{qV}_0 \gg \nu$, where ν is the number of ion collisions, then one can neglect the influence of collisions on the Fourier component N_q of the electron density, in terms of which the effective scattering cross section is also calculated. Meanwhile the scattering of radiowaves increases with decreasing values of $\mathbf{q}$. In the lower ionosphere, the case when the opposite inequalities are satisfied is

$$q\sqrt{\frac{2\varkappa T}{M_i}} \ll \nu_{in}, \quad \text{i.e.,} \quad \lambda \gg 2\pi l_i, \quad qV_0 \ll \nu_{in}, \tag{5.54}$$

which will also be considered in the present section.*

In order to calculate the disturbances associated with the fulfillment of conditions (5.54), we might, of course, still use the solution obtained earlier of the kinetic equation containing an effective collision integral (5.1). In this case, however, it is not necessary. In fact, the inequality (5.54) expresses the fact that the characteristic distance $1/q$ over which the particle density varies (significantly) is large in comparison with the mean free path. But this is just exactly the condition for the applicability of the macroscopic theory. Therefore, it is possible to solve this problem with the aid of the macroscopic equations for the diffusion of electrons and ions.† In this connection, since the disturbances of the electron and ion concentrations are small at large distances from a body, one can, of course, linearize these equations.

The equation for the steady-state diffusion of ions has the form

$$\operatorname{div} j^i = \tilde{I}\delta(r), \tag{5.55}$$

where $\mathbf{j}^i$ is the current density of the ions, $\tilde{I}$ taken with the opposite sign is the total number of ions absorbed by the body per unit time. The delta-function $\delta(\mathbf{r})$ on the right side of Eq. (5.55) expresses the fact that ions are absorbed only at the point where the body is located. The quantity $\tilde{I}$ is related to the collision integral $I(\mathbf{v})$, describing collisions of ions with the body, by the obvious equation

$$\tilde{I} = \int I(v)\, d^3v. \tag{5.56}$$

Therefore, if as usual we assume that $I(\mathbf{v}) = -\pi R_0^2 V_0 f_{0i}$, then

$$\tilde{I} = -\pi R_0^2 N_0 V_0. \tag{5.57}$$

The current density $\mathbf{j}^i$ of the ions in the diffusion approximation may only depend on the first derivatives of the ion concentration with respect to coordinates and on the electric field intensity. In addition, as already stated, we must linearize the equation. Therefore, all of these quantities should enter into $\mathbf{j}^i$ linearly, and the general expression for $\mathbf{j}^i$ which possesses such properties has the form

$$j_m^i = -N_i V_{0m} - \sum_k \left(D_{mk}^i \frac{\partial N_i}{\partial x_k} - F_{mk}^i e E_k N_0 \right), \tag{5.58}$$

where D_{mk}^i is a tensor whose elements are the coefficients of diffusion for an ion in a magnetic field, F_{mk}^i is the tensor mobility of the ions, and $\mathbf{E} = -\operatorname{grad}\varphi$ is the intensity of the electric field. The first term in (5.58) describes the motion of all the ions, as a whole, relative to the body. Now changing over in Eqs. (5.55) and (5.58) to Fourier components with respect to coordinates, we obtain

$$iqj_q^i = \tilde{I},$$

where, as always,

$$j_q^i = \int e^{-iqr} j(r)\, d^3r \tag{5.59}$$

*It is assumed that collisions of electrons and ions with neutral particles play the principal role.

†Since a magnetic field has no influence on the motion of neutral particles, their concentration reaches equilibrium much more rapidly than the concentrations of electrons and ions does, and therefore, one can neglect the motion of neutral particles.

and

$$j^{i}_{qm} = -N_{qi}V_{0m} - i\sum_{k}(D^{i}_{mk}q_k N_{qi} + F^{i}_{mk}q_k e\varphi_q N_0). \tag{5.60}$$

We now remark that the tensors D^{i}_{mk} and F^{i}_{mk} are not independent, as is well known. A relation, the so-called Einstein relation, exists between them. This relation is determined by the requirement that the current of ions in a fixed coordinate system must vanish in the state of equilibrium, i.e., when N_{qi} and φ_q are related by the formula of the Boltzmann distribution

$$N_{qi} = -\frac{eN_0}{\varkappa T}\varphi_q. \tag{5.61}$$

If we substitute (5.61) into (5.60) and equate, in virtue of what has been said above, the last two terms in (5.60) to zero, then

$$\sum_{k} F^{i}_{mk}q_k = \sum_{k} \varkappa T D^{i}_{mk}q_k. \tag{5.62}$$

As a result, using (5.62), (5.60), and (5.59) we find the following equation for N_{qi}:

$$\left[-iqV_0N_{qi} + \left(N_{qi} + \frac{eN_0}{\varkappa T}\varphi_q\right)\sum_{m,k} q_m q_k D^{i}_{mk}\right] = \tilde{I}. \tag{5.63}$$

The corresponding equation for N_{qe} naturally differs from (5.63) only by the sign of the charge and the replacement of D^{i}_{mk} by D^{e}_{mk}. The source function $\tilde{I}$ on the right side remains the same, since the number of electrons absorbed by the body in the stationary (steady-state) case is equal to the number of ions absorbed. Thus,

$$\left[-iqV_0N_{qe} + \left(N_{qe} - \frac{eN_0}{\varkappa T}\varphi_q\right)\sum_{m,k} q_m q_k D^{e}_{mk}\right] = \tilde{I}. \tag{5.64}$$

It is necessary to still add Poisson's equation for the potential to Eqs. (5.62) and (5.63). However, as we have repeatedly mentioned, when $Dq \ll 1$ it follows from Poisson's equation that $\delta N_i = \delta N_e$ and

$$N_{qi} = N_{qe} = N_q. \tag{5.65}$$

Therefore, eliminating φ_q from Eqs. (5.63) and (5.64), we find that

$$N_q = \frac{\tilde{I}}{-iqV_0 + D_a q^2}, \tag{5.66}$$

where D_a is defined by the formula

$$\frac{1}{D_a} = \frac{q^2}{2}\left(\frac{1}{\sum_{m,k} D^{i}_{mk}q_m q_k} + \frac{1}{\sum_{m,k} D^{e}_{mk}q_m q_k}\right). \tag{5.67}$$

It is natural to call the coefficient D_a, which depends on the direction of **q** relative to the magnetic field, the coefficient of ambipolar diffusion of ions and electrons in a magnetic field.*

*The question of the simultaneous diffusion of electrons and ions in a magnetic field was first rigorously investigated in [34].

It is obvious from symmetry considerations that the quadratic forms $\sum_{m,k} D^i_{mk} q_m q_k$ and $\sum_{m,k} D^e_{mk} q_m q_k$ contain the components of **q** only in the combinations $q_{\parallel}^2$ and $q_{\perp}^2$. Therefore,

$$\left.\begin{aligned} \sum_{m,k} D^i_{mk} q_m q_k &= D^i_{\parallel} q_{\parallel}^2 + D^i_{\perp} q_{\perp}^2 = q^2 (D^i_{\parallel} \sin^2\vartheta_1 + D^i_{\perp} \cos^2\vartheta_1), \\ \sum_{m,k} D^e_{mk} q_m q_k &= D^e_{\parallel} q_{\parallel}^2 + D^e_{\perp} q_{\perp}^2 = q^2 (D^e_{\parallel} \sin^2\vartheta_1 + D^e_{\perp} \cos^2\vartheta_1), \end{aligned}\right\} \tag{5.68}$$

where $D^i_{\parallel}$ and $D^e_{\parallel}$ are the coefficients for longitudinal diffusion of ions and electrons, $D^i_{\perp}$ and $D^e_{\perp}$ are the coefficients for transverse diffusion, and ϑ_1 is the angle between **q** and a plane perpendicular to the magnetic field. Thus,

$$\frac{1}{D_a(\vartheta_1)} = \frac{1}{2}\left(\frac{1}{D^i_{\parallel}\sin^2\vartheta_1 + D^i_{\perp}\cos^2\vartheta_1} + \frac{1}{D^e_{\parallel}\sin^2\vartheta_1 + D^e_{\perp}\cos^2\vartheta_1}\right). \tag{5.69}$$

If no magnetic field is present, then $D^i_{\parallel} = D^i_{\perp} = D^i$, $D^e_{\parallel} = D^e_{\perp} = D^e$, and $D^i \ll D^e$ so that $D_a \approx 2D^i$. Substituting this relation into (5.66), we obtain

$$N_q = \frac{\tilde{I}}{-iqV_0 + 2D^i q^2}.$$

This formula coincides, as it should, with formula (5.14) obtained from the kinetic equation.

In order of magnitude, one can express the coefficients $D_{\parallel}$ and $D_{\perp}$ in terms of the effective numbers of collisions of electrons and ions with neutral particles [34], i.e.,

$$\left.\begin{aligned} D^i_{\parallel} &\approx \frac{\varkappa T}{M_i \nu_{in}}, & D^i_{\perp} &\approx \frac{\varkappa T}{M_i \nu_{in}\left[1 + (\Omega/\nu_{in})^2\right]}, \\ D^e_{\parallel} &\approx \frac{\varkappa T}{m \nu_{en}}, & D^e_{\perp} &\approx \frac{\varkappa T}{m \nu_{en}\left[1 + (\omega_H/\nu_{en})^2\right]}. \end{aligned}\right\} \tag{5.70}$$

From here it is clear that the following inequality exists between these coefficients:

$$\frac{D^i_{\parallel}}{D^e_{\parallel}} \sim \sqrt{\frac{m}{M_i}} \ll 1, \quad \frac{D^i_{\perp}}{D^e_{\parallel}} \ll 1. \tag{5.71}$$

Formulas (5.66) and (5.69) solve the problem of the calculation of the Fourier components of the disturbances of the concentrations of ions and electrons. Now let us determine these disturbances in coordinate space. For this purpose, we take the inverse Fourier transform

$$\delta N(r) = \int N_q e^{iqr} \frac{d^3q}{(2\pi)^3} = \frac{\tilde{I}}{(2\pi)^3} \int \frac{e^{iqr}\, d^3q}{-iqV_0 + D_a(\vartheta_1)\, q^2}. \tag{5.72}$$

In order to evaluate the integral with respect to d^3q, we introduce a system of coordinates (x, y, z) with the z axis along $\mathbf{V}_0$ and the y axis perpendicular to the $\mathbf{V}_0$, $\mathbf{H}_0$ plane. Then

$$qV_0 = q_z V_0$$

and

$$\sin\vartheta_1 = \frac{q_z \cos\alpha + q_x \sin\alpha}{q},$$

where α is the angle between $\mathbf{V}_0$ and $\mathbf{H}_0$. Since the "trail" is greatly extended along the z axis, it is clear in advance that small values of q_z are important in the integral (5.72). Therefore, under the integral sign, one can set

$$\left.\begin{aligned} q^2 &\approx q_x^2 + q_y^2, \\ \sin\vartheta_1 &\approx \frac{q_x}{q}\sin\alpha. \end{aligned}\right\} \tag{5.73}$$

After this, the integration with respect to q_z reduces to an evaluation of the residue at the point

$$q_z = -i\,\frac{D_a(\vartheta_1)\,q^2}{V_0},$$

so that

$$\delta N(r) = \frac{1}{(2\pi)^2}\frac{\tilde{I}}{V_0}\int \exp\left[-z\,\frac{D_a(\vartheta_1)}{V_0}(q_x^2+q_y^2)\ i(q_x x + q_y y)\right] dq_x\,dq_y. \tag{5.74}$$

We shall evaluate the integral (5.74) in two cases: when the body moves exactly along the field and not too close to the direction of the magnetic field. First let

$$\sin^2\alpha \gg \frac{D_\perp^i}{D_\parallel^e},$$

i.e., the body does not move exactly in the direction of $\mathbf{H}_0$; then, it follows from (5.73) that

$$\vartheta_1^2 \sim \alpha^2 \gg \frac{D_\perp^e}{D_\parallel^i}. \tag{5.75}$$

Under condition (5.75), it is clear from inequality (5.71) that in (5.69) one can neglect the term containing the coefficients of diffusion of the electrons. Hence

$$q^2 D_a = q^2(D_\parallel^i \sin^2\vartheta_1 + D_\perp^i\cos^2\vartheta_1) = (D_\parallel^i\sin^2\alpha + D_\perp^i\cos^2\alpha)\,q_x^2 + D_\perp^i q_y^2. \tag{5.76}$$

Substituting (5.76) into (5.74) and evaluating the integrals with respect to dq_x and dq_y according to the formula

$$\int_{-\infty}^{\infty} e^{-\alpha t^2 + it}\,dt = \sqrt{\frac{\pi}{\alpha}}\,e^{-1/4\alpha},$$

we finally have

$$\delta N(r) = \frac{\tilde{I}}{8\pi z}\,\frac{\exp\left[-\frac{V_0}{8z}\left(\frac{x^2}{D_\parallel^i\sin^2\alpha + D_\perp^i\cos^2\alpha} + \frac{y^2}{D_\perp^i}\right)\right]}{\sqrt{D_\perp^i\left(D_\parallel^i\sin^2\alpha + D_\perp^i\cos^2\alpha\right)}}. \tag{5.77}$$

It follows from formula (5.77) that, thanks to the collisions, the "spreading out" of the trail behind the body is retarded. This is already evident from the fact that disturbances of the concentrations of electrons and ions fall off along the z axis like $1/z$ with increasing distance from the body, both in the presence of a magnetic field as well as in its absence (for $D_\parallel = D_\perp$), whereas without collisions in the absence of a magnetic field the concentration disturbances decrease much more rapidly, that is, like $1/z^2$. As before, the surfaces of constant density have the shape of a "petal," greatly elongated in the direction of the body's motion (the z axis) and flattened in the direction perpendicular to the V_0H_0 plane (the y axis). The cross sections of the "petal" in the xy plane are ellipsoids whose minor semiaxes are directed along the y axis, their major semiaxes are parallel to the z axis.

Now let us consider the case when $\alpha = 0$ or, more precisely, $\alpha^2 \ll D^i_\perp / D^e_\parallel$ and a body moves strictly in the direction of the magnetic field. If $\alpha = 0$, then $\vartheta_1 = 0$ [see Eq. (5.73)] and taking (5.71) into consideration, we have

$$q^2 D_a \approx 2(q_x^2 + q_y^2)\frac{D^i_\perp D^e_\perp}{D^i_\perp + D^e_\perp}.$$

Substituting the last relation into (5.74), we obtain

$$\delta N(r) = \frac{\tilde{I}}{8\pi z}\frac{D^i_\perp + D^e_\perp}{D^i_\perp D^e_\perp}\exp\left[-\frac{1}{8}\frac{(x^2+y^2)(D^i_\perp + D^e_\perp)V_0}{zD^i_\perp D^e_\perp}\right]. \tag{5.78}$$

If the number of electron collisions is sufficiently small so that

$$\nu_{en} \ll \omega_H, \tag{5.79}$$

then, as is evident from (5.70),

$$D^e_\perp \ll D^i_\perp$$

and therefore,

$$D_a = \frac{2D^i_\perp D^e_\perp}{D^i_\perp + D^e_\perp} \approx 2D^e_\perp.$$

In this case, Eq. (5.78) assumes the form

$$\delta N(r) = \frac{\tilde{I}}{8\pi z D^e_\perp}\exp\left[-\frac{1}{8}\frac{(x^2+y^2)V_0}{zD^e_\perp}\right]. \tag{5.80}$$

We see that only the coefficient for transverse diffusion of electrons enters into this formula. In formula (5.77), however, only the coefficients for the diffusion of ions appear.

It follows from formulas (5.78) and (5.80) that for $\alpha = 0$ the trail is even more strongly elongated in the direction of the body's motion. In fact, formula (5.80) leads to even larger values of δN than (5.77) since (5.80) contains the small quantity $D^e_\perp$ in the denominator. This is associated with the fact that it is now necessary that the particles move across the magnetic field in order to fill the rarefied region behind the body, but such motion is impeded. In addition, the very velocity of motion of the particles across the magnetic field in the case $\alpha = 0$ is strongly decreased in comparison with the velocity of motion of particles across the magnetic field in the case $\alpha \neq 0$. Whereas in the case $\alpha \neq 0$ the particles move with the velocity of transverse diffusion of ions, in the case $\alpha = 0$ under condition (5.79) they move with the velocity of transverse diffusion of electrons, i.e., approximately $(M_i/m)^{1/2}$ times more slowly (see Eq. (5.70)). Thanks to such a slow filling by the particles of the rarefied zone, the size of the trail for longitudinal motion of a body in the lower ionosphere is extremely large. For example, at an altitude of 100 km, $\nu_{en} \approx 10^5$, $T \approx 250°$, $\omega_H \approx 10^7$. Under these conditions, the disturbance behind a body of dimension $R_0 \sim 1$ m reaches 1% of N_0 at a distance $z = 10^2 m\omega_H^2 \pi R_0^2 V_0 / 8\varkappa T \nu_{en} \sim 20$ km and 0.1% of N_0 at a distance $z \sim 200$ km from the body.

Having eliminated N_q from Eqs. (5.63)-(5.65), one can determine φ_q, and having taken the inverse Fourier transform – one can determine $\varphi(\mathbf{r})$. We shall not carry out these calculations. We only note that for $\alpha = 0$, i.e., for motion along the field, the charge of the trail turns out to be positive, whereas it is usually negative. This is associated with the fact that ions are the more mobile particles for diffusion across the field, and in this case they are the first particles to fill the trail and they "drag" the electrons behind them.

In the present chapter we have only considered the influence of collisions on the particle distribution at large distances from a body. As stated in the introduction, collisions are also important in connection with a determination of the distribution function for the particles undergoing finite motion around a body. This question will be examined in Chapter VIII.

CHAPTER VI

DISTURBANCE OF THE MAGNETIC FIELD

§23. Calculation of the Disturbance of a Magnetic Field

In the preceding chapters we made it our aim to calculate the disturbances of the concentrations of charged particles and the disturbance of the electric field associated with these perturbations. In this connection, the magnetic field was assumed to be equal to the earth's field $\mathbf{H}_0$. However, moving space charge appears during the motion of a body in a plasma. The motion of charges creates an electric current which leads to the appearance of additional magnetic field associated with this current, i.e., it leads to a disturbance of the magnetic field. In the present chapter we shall investigate this disturbance. Such an investigation is of interest from various points of view. In the first place, sensitive instruments for measuring the magnetic field are installed in artificial satellites and rockets. Disturbances of the field can in principle distort the readings of these instruments. Therefore, it is important to have some idea about the possible magnitude of these distortions [68]. In the second place, the additional magnetic field may in turn influence the motion of ions and electrons, which leads to corrections in the formulas for the disturbances of the particle concentrations, and it is also necessary to estimate these corrections. In addition, an exact investigation of the structure of the magnetic field around a body may be used as a method for studying the disturbed zone.

In order to calculate the disturbances of the magnetic field, as well as for the solution of all the problems considered by us previously, it is necessary to use kinetic theory. Articles in which these disturbances are calculated on the basis of the equations of magnetohydrodynamics are, for the case of a body moving in a plasma, only of methodological value. In other cases, of course, the application of magnetohydrodynamics may be completely justified. We shall therefore not discuss these articles in detail, we only mention the comparatively recent articles by Reitz and Foldy [53] and by Dorman and Mikhailov [54]. The motion of a conducting sphere in a conducting liquid, located in a magnetic field, was considered in these articles. In this connection, in article [53] the disturbances of the magnetic field were calculated for the case when the sphere moves in the direction of the magnetic field, and in [54] they were calculated for an arbitrary direction of motion. The motion of the liquid was assumed to be potential, and the disturbances of the magnetic field were assumed to be small and to have no influence on this motion. The boundary layer near the body's surface was also considered in the last article. For motion of the body along the field, the disturbances H' of the magnetic field have the form

$$H'_r = \frac{3\pi\hat{\sigma}V_0R_0}{c^2}\left(\frac{R_0^4}{5r^4} - \frac{R_0^2}{3r^2}\right)(3\cos\vartheta - 1),\ H'_\vartheta = \frac{3\pi\hat{\sigma}V_0R_0}{5c^2}\,\frac{R_0^4}{r^4}\sin 2\vartheta,$$

where $\hat{\sigma}$ is the conductivity of the liquid, and the disturbances are written down in a spherical coordinate system with axis along the direction of motion. We see that at large distances from a body the disturbances fall off according to a $1/r^2$ law. This formula, however, is not suitable at very large distances from a body since in this region it is necessary to take the influence of H' on the motion of the liquid into consideration. Ludford and Murray [59] showed that the asymptotic law describing the decrease of the magnetic field with distance has the form

$$H'_r \sim H'_\vartheta \sim \exp\left[-\frac{2\pi\hat{\sigma}V_0R_0}{c^2}(1-\cos\vartheta)\right].$$

After these preliminary remarks, we go ahead to an evaluation of the magnetic field disturbances in the vicinity of a moving body.

In a coordinate system attached to the body, the density of the current created by the motion of electrons and ions can be expressed in terms of the distribution functions in the following manner:

$$\hat{j} = e \int v\,[f_i(v) - f_e(v)]\,d^3v. \tag{6.1}$$

Here, since the complete picture is stationary in this coordinate system, $\hat{\mathbf{j}}$ satisfies the equation of continuity

$$\operatorname{div} \hat{j} = 0. \tag{6.2}$$

The intensity of the additional magnetic field **H'** which appears can be determined in terms of $\hat{\mathbf{j}}$ with the aid of Maxwell's equation

$$\operatorname{rot} H' = \frac{4\pi}{c}\hat{j}. \tag{6.3}$$

Taking the rot of both sides of this equation, and remembering that rot rot = grad div − Δ, we bring Eq. (.63) to the more convenient form

$$\Delta H' = -\frac{4\pi}{c}\operatorname{rot}\hat{j}. \tag{6.4}$$

As in the calculation of the distribution function of the ions with account of an electric field, in the evaluation of the magnetic field disturbances it turns out to be possible to obtain closed expressions only for the Fourier components of these disturbances. The calculation of **H'** in coordinate space requires a numerical integration; however, even by themselves the formulas for the Fourier components of **H'** enable us to reach definite conclusions about the magnitude and nature of the disturbances.

Introducing the Fourier components of **H'** and $\hat{\mathbf{j}}$ in the form

$$H_q = \int e^{-iqr} H'\,d^3r, \qquad \hat{j}_q = \int e^{-iqr}\hat{j}\,d^3r,$$

instead of (6.1) we have

$$\hat{j}_q = e\int v\,[f_{qi}(v) - f_{qe}(v)]\,d^3v. \tag{6.5}$$

From Eq. (6.4) it follows that

$$H_q = i\frac{4\pi}{cq^2}[q\hat{j}_q]. \tag{6.6}$$

First let us calculate H_q for the case of motion of a body in a plasma, when no external magnetic field is present. For this purpose, we first determine the Fourier components of the current density of ions and electrons.

For the current density of the ions we have

$$\hat{j}_{qi} = e\int v f_{qi}\,d^3v = e\int u f_{qi}\,d^3v - eV_0\int f_{qi}\,d^3v = e\int u f_{qi}\,d^3u - eV_0 N_{qi}. \tag{6.7}$$

It is easy to evaluate the integral $\int u f_{qi}\,d^3u$ if we remember that f_{qi} [see Eq. (4.40)] only contains **u** in the combination **qu**. Therefore, it is clear that $\int u f_{qi}\,d^3u$ is a vector directed along **q**, so that

$$\left.\begin{aligned} \int u f_{qi}\, d^3u &= b\mathbf{q}, \\ \hat{\mathbf{j}}_{qi} &= e(b\mathbf{q} - \mathbf{V}_0 N_{qi}), \end{aligned}\right\} \tag{6.8}$$

where b is some coefficient. One can determine the coefficient b from the equation of continuity $\hat{\mathbf{j}}_i = e\tilde{I}\delta(\mathbf{r})$ without complicated calculations. Here $(-\tilde{I})$ is the total number of ions absorbed by the body per unit time [see Eq. (5.56)]:

$$\tilde{I} = \int I(u)\, d^3u \approx -\pi R_0^2 V_0 N_0.$$

In terms of Fourier components, the equation of continuity has the form

$$\mathbf{q}\hat{\mathbf{j}}_{qi} = -ie\tilde{I}.$$

Using (6.8) we have

$$\mathbf{q}\hat{\mathbf{j}}_{qi} = e(bq^2 - N_{qi}\mathbf{V}_0\mathbf{q}) = -ie\tilde{I},$$

therefore,

$$b = \frac{\mathbf{V}_0\mathbf{q}}{q^2} N_{qi} - \frac{i\tilde{I}}{q^2}.$$

Thus,

$$\hat{\mathbf{j}}_{qi} = eN_{qi}\left[\frac{(\mathbf{V}_0\mathbf{q})\,\mathbf{q}}{q^2} - \mathbf{V}_0\right] - ie\tilde{I}\frac{\mathbf{q}}{q^2} = eN_{qi}[(\mathbf{l}\mathbf{V}_0)\mathbf{l} - \mathbf{V}_0] - \frac{ie\tilde{I}\mathbf{l}}{q}, \tag{6.9}$$

where $\mathbf{l} = \mathbf{q}/q$ is a unit vector in the direction of $\mathbf{q}$. Similarly we find

$$\hat{\mathbf{j}}_{qe} = -eN_{qe}[(\mathbf{l}\mathbf{V}_0)\mathbf{l} - \mathbf{V}_0] + ie\tilde{I}\frac{\mathbf{l}}{q}.$$

Thus, finally,

$$\hat{\mathbf{j}}_q = e[(\mathbf{l}\mathbf{V}_0)\mathbf{l} - \mathbf{V}_0](N_{qi} - N_{qe}). \tag{6.10}$$

We note that in this case the current $\hat{\mathbf{j}}_q$ is proportional to the difference $(N_{qi} - N_{qe})$ which is a small quantity, as shown in § 14.

Now using Poisson's equation

$$-q^2\varphi_q = -4\pi e(N_{qi} - N_{qe}) \tag{6.11}$$

and substituting (6.11) into (6.10), we obtain

$$\hat{\mathbf{j}}_q = \frac{1}{4\pi}[(\mathbf{V}_0\mathbf{q})\mathbf{q} - \mathbf{V}_0 q^2]\,\varphi_q. \tag{6.12}$$

Then, from (6.12) and (6.6) it follows that

$$\mathbf{H}_q = -\frac{i}{c}[\mathbf{q}\mathbf{V}_0]\,\varphi_q = -\frac{1}{c}[\mathbf{V}_0\mathbf{E}_q], \tag{6.13}$$

where $\mathbf{E}_q$ is the Fourier component of the electric field intensity. We see that when no external magnetic field is present,

$$\mathbf{H}' = -\frac{1}{c}[\mathbf{V}_0\mathbf{E}]. \tag{6.14}$$

Since in the case under investigation the field **E** falls off according to a $1/r^3$ law at large distances from the body, the field **H'** also falls off according to the same law. We note that if a transformation to a fixed coordinate system is carried out according to the relativistic formulas for the transformation of fields, then it follows from (6.14) that **H'** = 0 in the fixed coordinate system.

Now let us go on to motion in an external magnetic field. Here many different cases occur, depending on the ratio of the components of **q** to the Debye radius, the dimensions of the body, and the Larmor radii of ions and electrons. We shall consider only the practically interesting case when $1/q$ is large in comparison with the Debye radius, the Larmor radius of electrons, and the dimensions of the body. In accordance with this, let us assume that

$$q \ll D^{-1}, \quad \varrho_{He}^{-1}, \quad R_0^{-1}, \tag{6.15}$$

$$q V_0 \ll \omega_H. \tag{6.16}$$

We note that for the calculation of the disturbances in coordinate space **r** according to the Fourier components of these disturbances, values of $q \sim 1/r$ are usually important in the inverse Fourier transform integral. In this sense, the formulas obtained below under conditions (6.15) and (6.16), i.e., for small values of q, describe the disturbances of a magnetic field at large distances from the body.

Under conditions (6.15) and (6.16), it is possible to obtain an expression for the current density created by electrons, almost without calculations. In fact, as shown in Chapter V, one can average the electron distribution function with respect to the directions of the vector $\mathbf{u}_\perp$. Therefore, $f_{qe} \approx f_{qe}(u_\parallel, u_\perp)$. This means, in particular, that the integral $\int u f_{qe}\, d^3u$ is directed along the external magnetic field

$$\int u f_{qe}\, d^3u = b_1 l_1,$$

where $l_1 = H_0/H_0$ is a unit vector in the direction of the external magnetic field.

Thus,

$$\hat{j}_{qe} = -e \int (u - V_0) f_{qe}\, d^3u = -e b_1 l_1 + e V_0 N_{qe}. \tag{6.17}$$

We again determine the coefficient b_1 from the condition $\mathbf{q}\hat{\mathbf{j}}_{qe} = ie\tilde{\mathbf{I}}$ or the condition

$$-e b_1 (l_1 q) + e (V_0 q) N_{q0} = ie\tilde{I},$$

so that

$$b_1 = \frac{V_0 q}{q_\parallel} N_{qe} + \frac{i\tilde{I}}{q_\parallel}.$$

Finally

$$\hat{j}_{qe} = e\left[V_0 - \frac{l_1 (V_0 q)}{q_\parallel}\right] N_{qe} + ie\tilde{I}\frac{l_1}{q_\parallel}. \tag{6.18}$$

The formulas obtained for $\hat{\mathbf{j}}_{qi}$ are, of course, different depending on whether the values of $1/q$ considered are large or small in comparison with the Larmor radius of the ions. First let

$$q \gg \varrho_{Hi}^{-1}, \tag{6.19}$$

where ρ_{Hi} is the Larmor radius of the ions. Under such conditions, a constant magnetic field has little influence on the motion of the ions, and we may directly use for $\hat{\mathbf{j}}_{qi}$ the expression (6.9) obtained earlier. In this connection, one can set $N_{qi} = N_{qe} = N_q$, neglecting the small difference between the concentration of ions and the concentration of electrons. The Fourier component $\hat{\mathbf{j}}_q$ of the total current density is therefore given by

$$\hat{j}_q = \hat{j}_{qi} + \hat{j}_{qe} = e\left(\frac{l}{q} - \frac{l_1}{q_{\|}}\right)[(q V_0) N q - i\widetilde{I}]. \tag{6.20}$$

Therefore,

$$H_q = \frac{i4\pi e}{cq^2}\,[q V_0 - i\widetilde{I}]\,\frac{[ql_1]}{q_{\|}}\,N_q. \tag{6.21}$$

We see that **H'** divides naturally into two terms: $\mathbf{H}' = \mathbf{H}'_1 + \mathbf{H}'_2$, where

$$H'_{1q} = i\,\frac{4\pi e\,(qV_0)}{cq^2}\,N_q\,\frac{[ql_1]}{q_{\|}}, \qquad H'_{2q} = \frac{4\pi e\widetilde{I}}{cq^2}\,\frac{[ql_1]}{q_{\|}}. \tag{6.22}$$

It is easy to convince oneself that both terms are, in general, of the same order. When the body is moving in a direction close to the direction of the external magnetic field, then one can set $(\mathbf{q}\mathbf{V}_0)l_1 = q_{\|}\mathbf{V}_0$ in (6.22). After this, using the inverse Fourier transformation, (6.22) can be rewritten in the form

$$\Delta H'_1 = \frac{4\pi e}{c}\,[\nabla(\delta N)\,V_0]. \tag{6.23}$$

Now let us represent Eq. (6.23) in another form by introducing a scalar function of the coordinates ψ, which satisfies the equation

$$\Delta\psi = -4\pi e\delta N.$$

Here it is necessary to emphasize that ψ of course is not the potential of the electric field around the body. In the equation for the potential, the expression $(\delta N_i - \delta N_e)$ would stand to the right.

From the last formula it follows that H' equals

$$H'_1 = \frac{1}{c}\,[V_0\nabla\psi].$$

It is clear that in the case under investigation, because of the axial symmetry of δN (and consequently also of ψ) the lines of magnetic force are concentric circles around the direction $\mathbf{V}_0$. If the body moves with a velocity much greater than the thermal velocity of the ions, one can simplify the expression for **H'** still more. For this we introduce the vector $\mathbf{p} = -\nabla\psi$. The vector **p** obviously lies in the plane $\mathbf{V}_0$. The absolute magnitude of **H'** is related to **p** by the equation

$$H'_1 = \frac{1}{c}\,V_0 p_{\perp},$$

where $p_{\perp}$ is the projection of **p** on a plane perpendicular to $\mathbf{V}_0$. It is clear that the vector **p** satisfies the equation $\operatorname{div}\mathbf{p} = 4\pi e\delta N$. If, however, the value of $\mathbf{V}_0$ is sufficiently large so that $\mathbf{V}_0 \gg (\varkappa T/M)^{1/2}$, the body's trail is strongly elongated, and the derivatives of **p** with respect to z are small (z is the coordinate along $\mathbf{V}_0$). In this case one can write div **p** in the form $\nabla_\rho\,\mathbf{p}_{\perp} = 4\pi e\delta N$, where $\boldsymbol{\rho}$ is the vector in the plane perpendicular to $\mathbf{V}_0$, and ∇_ρ is the gradient operator with respect to this radius vector. Now integrating the last expression over the area of a circle of radius ρ and applying Gauss' Theorem, we obtain

$$p_{\perp}(\varrho, z) = \frac{4\pi e}{\varrho}\int_0^{\rho}\delta N(\varrho', z)\,\varrho'\,d\varrho' \tag{6.24}$$

or finally

$$H'_1(\varrho, z) = \frac{4\pi e V_0[\varrho V_0]}{c\varrho}\int_0^{\rho}\delta N(\varrho', z)\,\varrho'\,d\varrho'.$$

Using the expressions for $\delta N(\rho, z)$ obtained above in §5 and in Chapter III, one can now calculate the magnetic field disturbance H' by numerically integrating (6.24). It is easy, however, to show without special calculations that the value of H' is negligible over the entire region covered by formula (6.24) and reaches in the ionosphere a value of only a few times γ ($\gamma \sim 10^{-5}$ oersted). In fact, the maximum value of the integral appearing in formula (6.24) corresponds to the value $\delta N/N = 1$, and is equal to $\rho^2/2$. Therefore, the maximum possible value of the disturbance of the magnetic field is

$$H_1' \sim \frac{4\pi e V_0 N}{c} \frac{\varrho}{2} \approx 8 \cdot 10^{-14} N\varrho \text{ Oe}. \tag{6.24a}$$

H_1' is equal to zero along the axis coinciding with the direction of the vector $\mathbf{V}_0$. On the other hand, with increasing ρ, $\delta N \to 0$. Therefore, H' has a maximum in a certain intermediate range of values of ρ, where δN is maximal. From the data presented in §5 and in Chapter III it is clear that this range of values of ρ is on the order of the dimensions of the body itself. Since the maximal value of N in the ionosphere is $\sim 5 \cdot 10^6$, and the dimensions of artificial satellites are on the order of 10^2 cm, we find that $H'_{1\,max} \sim 4\gamma$.

In order to calculate H_2', it is necessary to know the rule for going around the pole during the integration with respect to $q_{\parallel}$. One can show that this rule does not have a universal form, but it depends on the specific shape of the body. We shall not dwell on this question. The final answer for the case of a flat disk has the form

$$H_2' = \frac{\pm 2e\pi R_0^2 V_0 N_0 \, [l_1 r]}{r_\perp^2}.$$

The sign in this formula coincides with the sign of $(\mathbf{l}_1 \mathbf{V}_0)$. We see that $\mathbf{H}_2'$ is the field of a line current of strength $e\pi R_0^2 V_0$ flowing from the body along the magnetic field $\mathbf{H}_0$.

Now let us consider the case of very small values of q, when

$$q \ll \varrho_{Hi}^{-1}, \quad qV_0 \ll \Omega. \tag{6.25}$$

As before, for $\hat{j}_{qe}$ we may use expression (6.18). The same expression with replacement of $-e$ is also approximately valid under conditions (6.25) for $\hat{j}_{qi}$. If, however, we take $\hat{j}_{qi}$ in that approximation, the total current density turns out to vanish. This means that we must calculate $\hat{j}_{qi}$ correct to terms of order $(q_\perp \rho_{Hi})$. For this we use the fact that (see the footnote on p. 110) under conditions (6.25) one can represent the distribution function of the ions in the form

$$f_{qi} = -f_{0i} \frac{e}{\varkappa T} \varphi_q + \exp\left(\frac{iq_\perp u_\perp}{\Omega} \sin\beta\right) g_q(u_\parallel, u_\perp)$$

$$\approx -f_{0i} \frac{e}{\varkappa T} \varphi_q + \left(1 + \frac{iq_\perp u_\perp}{\Omega} \sin\beta\right) g_q(u_\parallel, u_\perp) = -f_{0i} \frac{e}{\varkappa T} \varphi_q + \left(1 + iq_\perp \frac{u_\perp l_3}{\Omega}\right) g_q(u_\parallel, u_\perp), \tag{6.26}$$

where l_3 is a unit vector perpendicular to $\mathbf{H}_0$ and $\mathbf{q}_\perp$. Now having multiplied (6.26) by $e(\mathbf{V}_0 + \mathbf{u})$ and integrating with respect to d^3u, we find that the first two terms of the result give $\hat{j}_{qe}$. Thus,

$$\hat{j}_{qi} = -\hat{j}_{qe} + \frac{e}{\Omega} i \int q_\perp (u_\perp l_3) u_\perp g_q(u_\parallel, u_\perp) \, d^3u = -\hat{j}_{qe} + il_3 \frac{eq_\perp}{2\Omega} \int u_\perp^2 g_q(u_\parallel, u_\perp) \, d^3u$$

and

$$\hat{j}_q = \hat{j}_{qe} + \hat{j}_{qi} = \frac{il_3 eq_\perp}{2\Omega} \int u_\perp^2 g_q(u_\parallel, u_\perp) \, d^3u. \tag{6.27}$$

The function $g_q(u_\parallel, u_\perp)$ is determined by the collision integral $I(\mathbf{u})$ (see Eq. (5.23)).* If we take, as usual,

$$I(u) = -\pi R_0^2 V_0 f_{0i},$$

then $g_q(u_\parallel, u_\perp)$ will depend on $\mathbf{u}_\perp$ according to the law

$$\exp\left(-\frac{M_i u_\perp^2}{2\varkappa T}\right).$$

In this case, it is possible to rewrite (6.27) in the form

$$\hat{j}_q = \frac{l_3 i e q_\perp \varkappa T}{M_i \Omega} \int g_q(u_\parallel, u_\perp)\, d^3u. \tag{6.28}$$

In order to determine $\int g_q\, d^3u$, we integrate formula (6.26) with respect to d^3u, here omitting small terms of order $q_\perp \rho_{Hi}$. We obtain

$$\int f_{qi}\, d^3u = N_q = -\frac{eN_0}{\varkappa T}\varphi_q + \int g_q(u_\parallel, u_\perp)\, d^3u$$

or

$$\int g_q(u_\parallel, u_\perp)\, d^3u = N_q + \frac{eN_0}{\varkappa T}\varphi_q = \frac{2eN_0}{\varkappa T}\varphi_q. \tag{6.29}$$

Here we omit the index "i" on N_q since the calculation is carried out in the approximation $N_{qi} = N_{qe} = N_q$. Substituting (6.29) into (6.28), we finally have

$$\hat{j}_q = \frac{l_3 i 2e^2 N_0 q_\perp}{M_i \Omega}\varphi_q = \frac{i}{2\pi D^2}\frac{\varkappa T q_\perp l_3 \varphi_q}{M_i \Omega} = \frac{\varkappa T}{2\pi D^2 M_i \Omega}[\mathbf{E}_q \mathbf{l}_1]. \tag{6.30}$$

Using (6.30) we find that the Fourier components of the magnetic field are given by

$$\mathbf{H}_q = \frac{2\varkappa T}{c(Dq)^2 M_i \Omega}[\mathbf{E}_q q_\parallel - \mathbf{l}_1(\mathbf{q}\mathbf{E}_q)]. \tag{6.31}$$

It is evident from (6.31) that the magnetic field intensity in the $\mathbf{V}_0\mathbf{H}_0$ plane falls off according to the same law as the electric field intensity, i.e., like $1/r^2$ for increasing distance from the body. To also obtain formulas for the Fourier components of the magnetic field for arbitrary values of $\mathbf{q}$ does not present any major difficulties. However, we shall not present these extremely unwieldy formulas, since in order to obtain $\mathbf{H}'(\mathbf{r})$ from them, i.e., to take the inverse Fourier transform, requires laborious numerical calculations.

§ 24. Influence of Disturbances of the Magnetic Field on the Motion of Ions and Electrons

Up to now the disturbances of the magnetic field were considered for a given motion of the ions, i.e., the "feedback" effect of the magnetic disturbances on the motion of the ions was neglected. Of course, such a calculation leads to approximate results. For a rigorous investigation of this problem, in the kinetic equation it is necessary to add the additional force acting on a unit charge in the form

$$\frac{1}{c}[\mathbf{v}\mathbf{H}'] = \frac{1}{c}[\mathbf{u} - \mathbf{V}_0 \mathbf{H}'].$$

*Here we note that collisions between particles are neglected everywhere in this chapter, i.e., we take $\Upsilon = 0$.

One can, however, regard this term as small if only because it is proportional to δN; in addition, it contains the velocity of light in the denominator. Therefore, in the equations it is necessary to leave only its product by the factor

$$\frac{\partial f_{0i}}{\partial u} = -\frac{M_i u}{\varkappa T} f_{0i} \quad \text{or} \quad \frac{\partial f_{0e}}{\partial u} = -\frac{mu}{\varkappa T} f_{0e}.$$

Here the term containing the vector product [uH'] vanishes. Finally, we arrive at the result that a calculation of the inverse influence of the magnetic field disturbances reduces to the replacement of the product $-iq\varphi_q$ in the equations for the distribution functions of electrons and ions by

$$-iq\varphi_q - \frac{1}{c}[V_0 H_q]. \tag{6.32}$$

From here, it is clear that one can neglect the "feedback" effect of the magnetic field disturbances only in the case when the first term in this formula is much larger than the second. Under condition (6.19), as is clear from (6.22),

$$H_q \sim \frac{V_0}{cD^2 q}\frac{E_q}{q_{\|}}$$

and the condition, enabling us to neglect the influence of magnetic disturbances, has in this range of values of q the form

$$\frac{V_0^2}{c^2 D^2 q q_{\|}} \ll 1 \quad \left(\varrho_{He} \ll \frac{1}{q} \ll \varrho_{Hi}\right). \tag{6.33}$$

Upon fulfillment of inequality (6.25), from (6.31) we obtain

$$H_q \sim \frac{\varkappa T}{M_i D^2 q \Omega} E_q.$$

As a result, the desired (sought for) condition in this case has the form

$$\frac{\varkappa T q V_0}{M_i c^2 \Omega (Dq)^2} \ll 1 \quad \left(\frac{1}{q} \gg \varrho_{Hi}\right). \tag{6.34}$$

One can show, however, that in that approximation, in which formula (6.30) is obtained, the disturbances of the magnetic field only influence the current distribution in the plasma and do not influence the distribution of the concentrations of ions and electrons. Therefore the corrections to N_q turn out to be only on the order of

$$\frac{\varkappa T q V_0}{M_i c^2 \Omega (Dq)^2} \sqrt{\frac{\varkappa T q^2}{M_i \Omega^2}}. \tag{6.35}$$

The formulas obtained, free from the limitations (6.33) and (6.34), i.e., taking the influence of the magnetic field disturbances on the motion of the particles into consideration, are associated with rather complicated calculations. We shall carry out these calculations only for case (6.19), when the Larmor radius of the ions is much larger than 1/q. The case of arbitrary values of q is considered in [66].

First of all we note that expression (6.21) for the magnetic field disturbances is valid also upon account of its inverse influence on the concentration of charges. This is clear from the fact that only the law of conservation of charge and very general properties of the equations for the particle distribution function were used for the derivation of the indicated expression. With the aid of Eq. (6.21), one can represent the force acting on a unit charge in the form

$$F = -i\left(q\varphi_q + f\frac{N_q \varkappa T}{N_0 e}\right) - f\frac{(q\varkappa T\tilde{I})}{eN_0(qV_0)}, \tag{6.36}$$

where

$$f = \frac{qV_0}{c^2(Dq)^2}\frac{[V_0[ql_1]]}{qq_{\parallel}}. \tag{6.37}$$

These relations enable one to easily relate N_q with φ_q. For this, first we write the equation for the electron distribution function, having averaged it beforehand over the angle β. We can directly obtain this equation from (4.66), having set $\mathbf{I} = 0$ and having replaced e by $-e$ and $-i\mathbf{q}\varphi_q$ by $\mathbf{F}$. In addition, we may set $\mathbf{qV}_0 = 0$, since $\mathbf{V}_0$ is much smaller than the velocity of motion of the electrons. Finally the equation has the form

$$iq_{\parallel}u_{\parallel}f_{qe} = -\frac{e}{\varkappa T}f_{0e}\overline{uF}. \tag{6.38}$$

From here it is seen that

$$N_q = \int f_{qe}\,d^3u = -\frac{eN_0}{\varkappa T}\frac{(l_1F)}{iq_{\parallel}}$$

and

$$\varphi_q = \frac{\varkappa T}{eN_0}\left[1 - \frac{fl_1}{q_{\parallel}}\right]N_q + \frac{i\varkappa\widetilde{I}}{eN_0(qV_0)}(fl_1). \tag{6.39}$$

Relation (6.39) replaces for electrons the Boltzmann distribution formula which we previously used everywhere. For the distribution function of the ions we now have the equation

$$i(qu - qV_0)f_{qi} = \frac{e}{\varkappa T}f_{0i}uF + I(u),$$

from here it follows that

$$N_q = \int \frac{I}{i(qu - qV_0)}d^3u + \frac{e}{\varkappa T}\int \frac{f_{0i}(lu)}{i(qu - qV_0)}d^3u\,(lF). \tag{6.40}$$

Simultaneously solving Eqs. (6.40), (6.39), and (6.37), we finally obtain

$$N_q = \frac{\pi R_0^2 N_0}{q}\,\frac{\left(\frac{M_iV_0^2}{2\varkappa T}\right)^{1/2}Q(a) + \gamma\,[aQ(a) - i]\left(\frac{qV_0}{qV_0}\right)}{1 + (1-\gamma)\,[1 + iaQ(a)]}, \tag{6.41}$$

where

$$\gamma = \frac{[(qV_0)(q_{\perp}V_0)(q_{\perp}q)]}{[(cDq)^2q^2q_{\parallel}]}. \tag{6.42}$$

Formula (6.42) goes over into formula (4.41) obtained in Chapter IV for $\gamma \ll 1$. This condition, as it should, coincides with condition (6.33).

CHAPTER VII

SCATTERING OF ELECTROMAGNETIC WAVES BY THE "TRAIL" OF A RAPIDLY MOVING BODY

§ 25. Effective Cross Section for the Scattering in the Approximation of Perturbation Theory

Disturbance of the electron concentration leads to a change of the dielectric permittivity of the plasma in the vicinity of the body. The region of altered dielectric permittivity formed because of this "trail" causes the scattering of electromagnetic waves incident on the body. Of course, electromagnetic waves will also be scattered by the body itself, for example, by a metallic sphere. However, the scattering on the body is described by the usual formulas of diffraction theory, and is not of interest to us here. Perturbation theory (see, for example, p. 486 of [33]) will be used for the calculation of scattering by the trail.

We shall assume that the frequency of the scattered waves is much larger than the Larmor frequency ω_H of the electrons:

$$\omega \gg \omega_H = \frac{eH_0}{mc}, \tag{7.1}$$

so that the dielectric permittivity of the plasma, $\varepsilon(\omega, \mathbf{r})$, is given by the formula

$$\varepsilon(\omega, r) = 1 - \frac{4\pi e^2}{m\omega^2} N(r). \tag{7.2}$$

Let us use, then, Maxwell's equations:

$$\left.\begin{aligned} \operatorname{rot} E &= i\frac{\omega}{c} H, \\ \operatorname{rot} H &= -i\frac{\omega}{c} D. \end{aligned}\right\} \tag{7.3}$$

Eliminating **H** from (7.3), we obtain

$$\operatorname{rot}\operatorname{rot} E = \frac{\omega^2}{c^2} D, \tag{7.4}$$

where $\mathbf{D} = \varepsilon(\omega, \mathbf{r})\,\mathbf{E}$ is the electric displacement vector. Now let us represent ε in the form

$$\varepsilon(\omega, r) = \bar{\varepsilon}(\omega) + \delta\varepsilon(\omega, r), \tag{7.5}$$

where $\bar{\varepsilon}(\omega)$ is the dielectric permittivity of the undisturbed plasma, i.e.,

$$\bar{\varepsilon}(\omega) = 1 - \frac{4\pi N_0 e^2}{m\omega^2} = 1 - \left(\frac{\omega_0}{\omega}\right)^2. \tag{7.6}$$

Here ω_0 is the plasma frequency of the ionosphere at the altitude where the body is located. Taking into consideration that rot rot = grad div − Δ, and substituting (7.5) into (7.4), we bring the equation for **D** to the form

$$\Delta \boldsymbol{D}+k^2\boldsymbol{D}= -\operatorname{rot}\operatorname{rot}\frac{\delta\varepsilon}{\varepsilon}\boldsymbol{D}, \tag{7.7}$$

where

$$k^2=\bar{\varepsilon}\frac{\omega^2}{c^2}.$$

We shall represent the electric field **E** in the form

$$\boldsymbol{E}=\boldsymbol{E}_1+\boldsymbol{E}_2, \tag{7.8}$$

where $\mathbf{E}_1 = \mathbf{E}_0 e^{i(\mathbf{k}\mathbf{r}-\omega t)}$ is the field of the incident wave, $\mathbf{E}_2$ is the field of the scattered wave, and **k** is the wave vector of the incident wave ($|\mathbf{k}| = k$). We also write the displacement vector **D** in the form

$$\boldsymbol{D}=\boldsymbol{D}_1+\boldsymbol{D}_2=\bar{\varepsilon}\boldsymbol{E}_1+\boldsymbol{D}_2.$$

The approximation of perturbation theory is that $\delta\varepsilon$, $\mathbf{E}_2$, and $\mathbf{D}_2$ are assumed to be small, which enables us to neglect their products. By virtue of this, we have

$$\frac{\delta\varepsilon}{\varepsilon}\boldsymbol{D}=\delta\varepsilon\boldsymbol{E}\approx\delta\varepsilon\boldsymbol{E}_1. \tag{7.9}$$

Substituting (7.9) into (7.7), we bring the latter equation to the form

$$\Delta\boldsymbol{D}_2+k^2\boldsymbol{D}_2= -\operatorname{rot}\operatorname{rot}(\boldsymbol{E}_1\delta\varepsilon). \tag{7.10}$$

Thus $\mathbf{D}_2$ is described by a linear inhomogeneous equation with a known function appearing on the right side. As is well known, its solution has the form

$$\boldsymbol{D}_2(\boldsymbol{r})=\frac{1}{4\pi}\int\frac{e^{ik|\boldsymbol{r}-\boldsymbol{r}'|}}{|\boldsymbol{r}-\boldsymbol{r}'|}\{\operatorname{rot}\operatorname{rot}\delta\varepsilon(\boldsymbol{r}')\boldsymbol{E}_1(\boldsymbol{r}')\}\,d^3r'=\frac{1}{4\pi}\operatorname{rot}\operatorname{rot}\int\frac{e^{ik|\boldsymbol{r}-\boldsymbol{r}'|}}{|\boldsymbol{r}-\boldsymbol{r}'|}\delta\varepsilon(\boldsymbol{r}')\boldsymbol{E}_1(\boldsymbol{r}')\,d^3r'. \tag{7.11}$$

At large distances from the body, where $kr \gg 1$, one can set

$$k|\boldsymbol{r}-\boldsymbol{r}'|\approx kr-\frac{k\boldsymbol{r}\boldsymbol{r}_1}{r}=kr-\boldsymbol{k}'\boldsymbol{r}',$$

where **k'** is the wave vector of the scattered wave; at large distances from the body the direction of **k'** coincides with the direction of **r**. As a result

$$\boldsymbol{D}_2=\operatorname{rot}\operatorname{rot}\frac{e^{ikr}}{4\pi r}\int e^{-i\boldsymbol{k}'\boldsymbol{r}'}\boldsymbol{E}_1(\boldsymbol{r}')\,\delta\varepsilon(\boldsymbol{r}')\,d^3r'=-\frac{e^{ikr}}{4\pi r}\left[\boldsymbol{k}'\left[\boldsymbol{k}'\int e^{-i\boldsymbol{k}'\boldsymbol{r}'}\boldsymbol{E}_1(\boldsymbol{r}')\,\delta\varepsilon(\boldsymbol{r}')\,d^3r'\right]\right], \tag{7.12}$$

if the terms $\sim 1/r^2$ are omitted upon differentiation. It is clear from Eq. (7.2) that

$$\delta\varepsilon= -\frac{4\pi e^2}{m\omega^2}\delta N= -\left(\frac{\omega_0}{\omega}\right)^2\frac{\delta N}{N_0}\qquad(\delta N=N-N_0). \tag{7.13}$$

It is natural that at large distances from the body ($r\to\infty$)

$$\delta\varepsilon\to 0,\quad \boldsymbol{D}_2=\bar{\varepsilon}\boldsymbol{E}_2.$$

Substituting (7.13) into (7.12), we finally have

$$E_2 = \frac{c^2}{m\omega^2\varepsilon} \frac{e^{ikr}}{r} [k' [k'E_0] N_q], \tag{7.14}$$

where N_q is the Fourier component of the disturbance of the electron density, that is

$$N_q = \int \delta N(r) e^{-iqr} d^3r, \tag{7.15}$$

$$q = k' - k, \quad |q| = 2k \sin\frac{\psi}{2} = 2\frac{\omega}{c}\sqrt{1 - \left(\frac{\omega_0}{\omega}\right)^2} \sin\frac{\psi}{2}. \tag{7.16}$$

Here ψ is the scattering angle, i.e., the angle between the wave vectors of the incident and scattered waves.

The quantity $\frac{|E_2|^2}{|E_1|^2} r^2 do$ determines what part of the energy of a wave incident per unit area behind the body is scattered into the element of solid angle do. This quantity is the so-called effective differential scattering cross section $d\sigma$. Thus,

$$d\sigma = \frac{|E_2|^2}{|E_1|^2} r^2 do = \frac{1}{16\pi^2}\left(\frac{\omega_0}{c}\right)^4 \frac{|N_q|^2}{N_0^2} \sin^2\psi_1 do, \tag{7.17}$$

where ψ_1 is the angle between $\mathbf{E}_0$ and $\mathbf{k}'$.

In order for the approximations of perturbation theory used here to be applicable, the following inequality must be satisfied at those distances which give the major contribution to the integrals in formulas (7.11) or (7.15):

$$|E_2| \ll |E_0|. \tag{7.18}$$

Verification of the fulfillment of this inequality requires a rather detailed investigation of expression (7.11) with a specific function $\delta\varepsilon$. In conclusion, we remark that within the limits of applicability of the approximation used here, the electric fields formed because of scattering by the body itself and by its trail are simply to be added together. In the general case, this is by no means the case.

§ 26. The Scattering Function

It is clear from formula (7.17) of the preceding section that the effective cross section for the scattering of electromagnetic waves is determined by the Fourier component N_q of the disturbance of the electron concentration. Formulas for N_q, based on different approximations and valid in different ranges of values of the parameters, were obtained by us in §§8, 13, and others. All of these formulas, however, are rather complicated. Therefore, for numerical calculation of the effective cross section $d\sigma$ it is necessary to resort to numerical integration. It is more convenient to carry out such numerical calculations on the basis of the most general formula, which gives the value of N_q over a wide region of variation of **q** and of the other parameters. Formula (5.6), which was obtained by taking the magnetic field and collisions into consideration, is such a formula. Here, of course, one has to put up with the interpolative nature of this formula, which was mentioned at the end of §21.

Equation (5.6) was written down for the case $qR_0 \ll 1$. In order to take the finite size of the body into account, it is necessary to multiply the right side of (5.6) by the function $G(qR_0, \chi)$, defined by Eq. (2.88). Now calculating the absolute value squared of the right side of (5.6) [in order to do this, it is necessary to separate real and imaginary parts in the numerator and denominator of (5.6)], multiplying it by $|G|^2$, and substituting into the right side of (7.17), we bring $d\sigma$ to the form

$$d\sigma = \frac{1}{16}\left(\frac{\omega_0}{c}\right)^4 \frac{R_0^4 V_0^2}{\Omega^2} \sin^2 \psi_1 F_3(\tau, \eta, \zeta, \delta) \mid G(qR_0, \chi) \mid^2 do. \tag{7.19}$$

In this formula

$$F_3(\tau, \eta, \zeta, \delta) = \frac{F_1^2 + F_2^2}{(2 - 2\eta F_1 - \tau F_2)^2 + (\tau F_1 - 2\eta F_2)^2}, \tag{7.20}$$

$$F_1 = \exp(-\delta) \int_0^\infty \cos(\tau x) \exp(-\eta x - \zeta x^2 + \delta \cos x)\, dx, \tag{7.21}$$

$$F_2 = \exp(-\delta) \int_0^\infty \sin(\tau x) \exp(-\eta x - \zeta x^2 + \delta \cos x)\, dx. \tag{7.22}$$

The dimensionless parameters η, τ, ζ, and δ are expressed in terms of physical quantities by the relationships

$$\tau = \frac{qV_0}{\Omega} = \frac{qV_0}{\Omega} \cos\chi, \quad \eta = \frac{\nu}{\Omega}, \quad \zeta = \frac{\varkappa T}{2M_i\Omega^2} q_{\|}^2, \quad \delta = \frac{\varkappa T}{M_i\Omega^2} q_{\perp}^2, \tag{7.23}$$

where χ is the angle between $\mathbf{q}$ and $\mathbf{V}_0$.

We shall define the direction of the vector $\mathbf{q} = \mathbf{k}' - \mathbf{k}$, determining the change of the incident wave's wave vector upon scattering, with the aid of the angle ϑ_1 formed by the vector $\mathbf{q}$ with the plane perpendicular to $\mathbf{H}_0$, and the angle φ between the planes $\mathbf{q}\mathbf{H}_0$ and $\mathbf{V}_0\mathbf{H}_0$. These angles are shown in Fig. 31. The point S on this figure is the point of radiation, the point E is the point of observation, O is the point where the body is located. The wave vectors $\mathbf{k}$ and $\mathbf{k}'$ are, respectively, directed along the rays SO and OE, and the vector $\mathbf{q}$ is directed along the line which bisects the angle SOE. The angle ϑ_1 is regarded as positive if $\mathbf{q}$ and $\mathbf{H}_0$ form an acute angle with each other. The parameters τ, ζ, and δ can be expressed in terms of the angles ϑ_1, φ, and α between $\mathbf{V}_0$ and $\mathbf{H}_0$ by formulas which can be immediately obtained from (7.23), namely:

$$\tau = \frac{qV_0}{\Omega}(\cos\alpha \sin\vartheta_1 + \sin\alpha \cos\vartheta_1 \cos\varphi), \quad \zeta = \frac{\varkappa T q^2}{2M_i\Omega^2} \sin^2\vartheta_1, \quad \delta = \frac{\varkappa T q^2}{M_i\Omega^2} \cos^2\vartheta_1. \tag{7.24}$$

The integrals (7.21) and (7.22) cannot be expressed in terms of well known special functions, and it is necessary to evaluate them by numerical methods. Similarly the function $G(qR_0, \cos\chi)$ also cannot be expressed in terms of special functions.

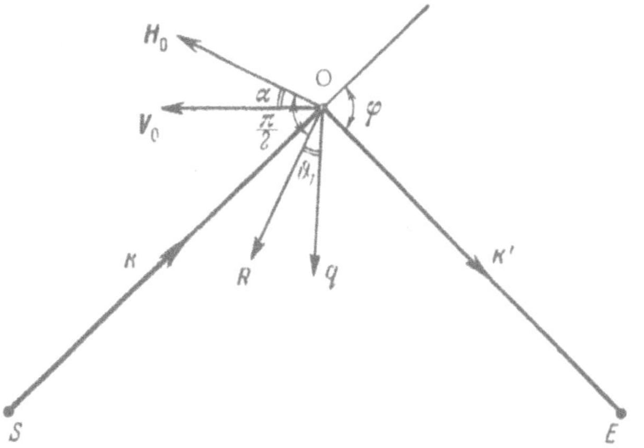

Fig. 31. Arrangement of the vectors in space.

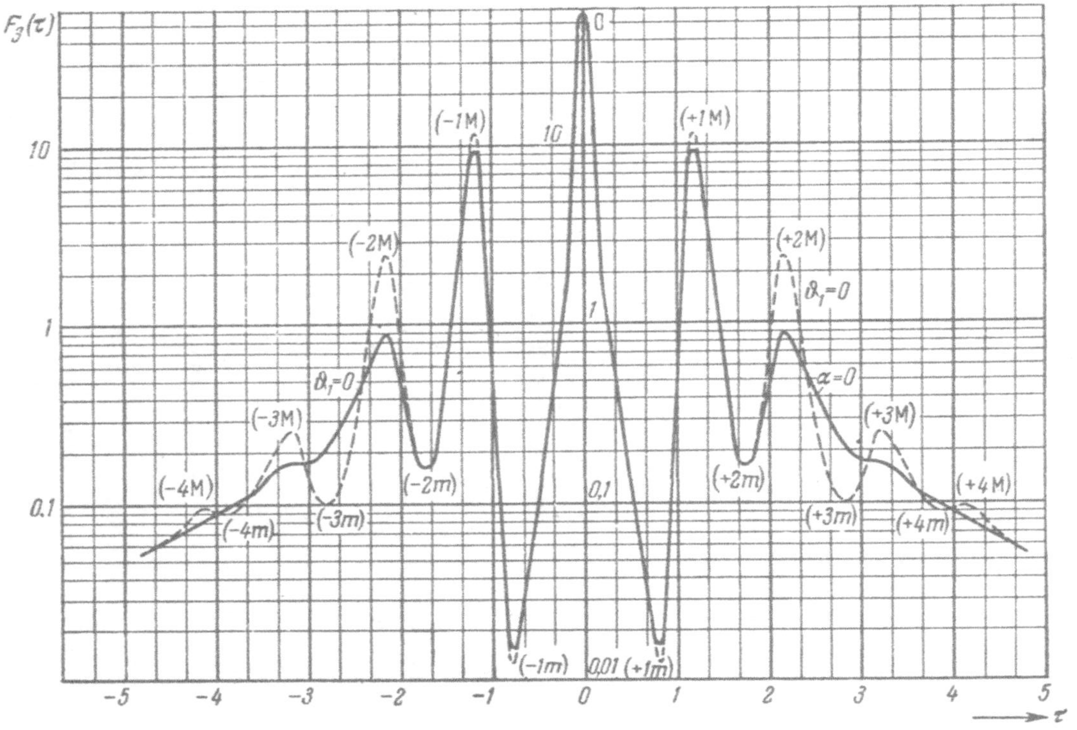

Fig. 32. Behavior of the function $F_3(\tau)$ for the case $\alpha = 0$, $\vartheta_1 \neq 0$ and the case $\alpha \neq 0$, $\vartheta_1 = 0$. The maxima and minima of different order (0), (±1 M), (±1 m), (±2 M), (±2 m), etc., are indicated on the figure.

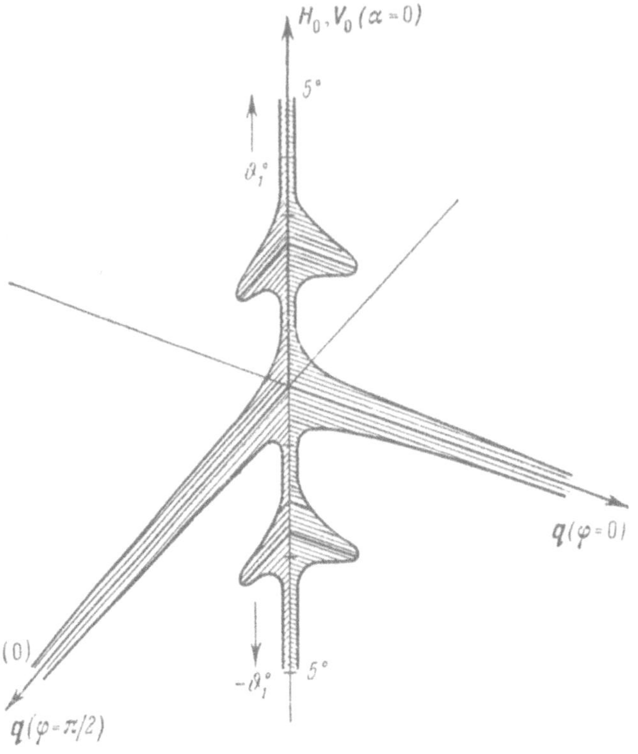

Fig. 33. The surface $F_3 = F_3(\vartheta_1, \varphi)$ in cylindrical coordinates when $\mathbf{V}_0 \parallel \mathbf{H}_0$ ($\alpha = 0$).

Tabulation of the function $F_3(\tau, \eta, \zeta, \delta)$, which we shall hereafter refer to as the scattering function, for different values of τ, η, ζ, and δ have made it possible to ascertain its behavior as a function of altitude in the ionosphere, wavelength, temperature, and direction of motion of the body. The corresponding results are presented in detail in the following sections. Here it is appropriate to discuss certain general properties of F_3, necessary for what follows.

The basic property of the function $F_3(\tau, \eta, \zeta, \delta)$ is its oscillating character. The maxima and minima of F_3 correspond to definite values of τ. For the values of the parameters used by us, F_3 displays in addition to the principal maximum another six or eight maxima and six or eight minima, located symmetrically about the principal maximum when $\alpha = 0$. The principal maximum of F_3 (which we also call the maximum of zero order (0)) occurs at

$$\tau = 0. \tag{7.25}$$

The side maxima M and minima m [we denote them, respectively by (±1 M), (±2 M), (±3 M), (±4 M) ... and (±1 m), (± 2 m), (±3 m), and (± 4 m) ...] for $\alpha = 0$ are defined by the values

$$\left.\begin{array}{lllll} \tau_{max} \approx \pm 1.22, & \pm 2.18, & \pm 3.15, & \pm 4.23, \\ \tau_{min} \approx \pm 0.73, & \pm 1.70, & \pm 2.91, & \pm 3.86. \end{array}\right\} \tag{7.26}$$

The principal maximum of F_3 ($\tau = 0$) has the largest value when $\alpha = 0$ or $\vartheta_1 = 0$. For other values of α or ϑ_1 the value of F_3 for $\tau = 0$ can be smaller than the first-order maximum of F_3 ($\tau = \pm 1.22$). In the case when $\alpha = 0$, that is, when the velocity of the body $\mathbf{V}_0$ and the magnetic field $\mathbf{H}_0$ are collinear, the value of F_3 when $\tau = 0$, that is, the principal maximum is always much larger than the side maxima, whose values decrease as their number increases. The number of side maxima depend essentially on η and ζ, which determine the rate of convergence of integrals (7.21) and (7.22). Naturally the value of ζ, the coefficient of the

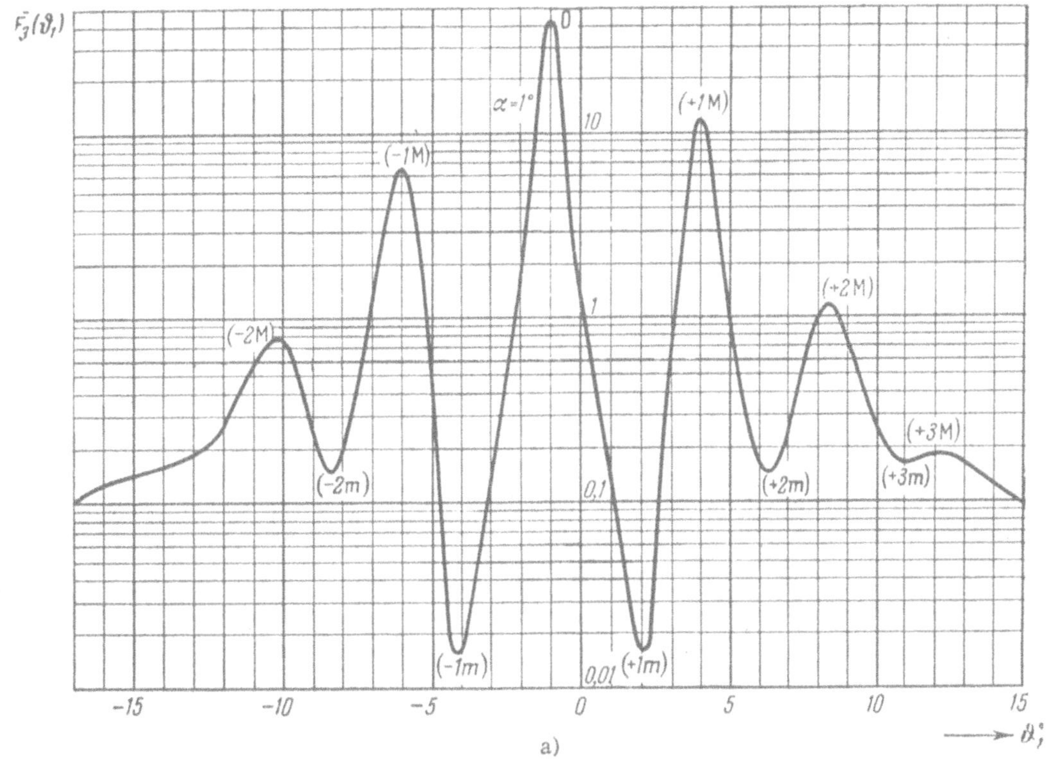

a)

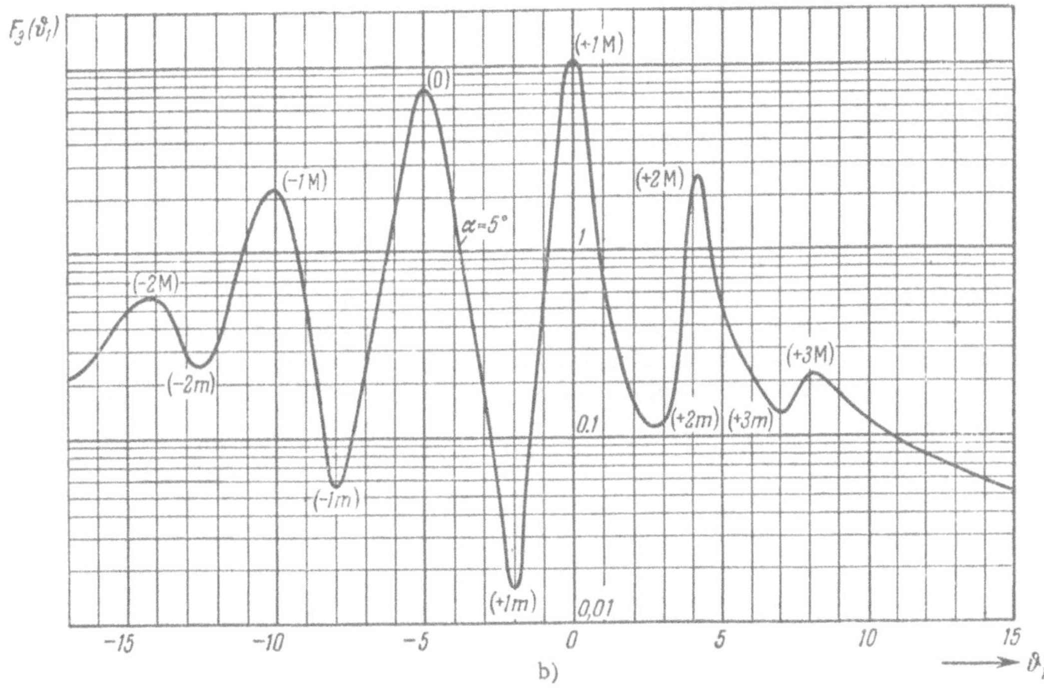

b)

Fig. 34. Behavior of the function F_3 (ϑ_1, $\varphi = 0$) for $\alpha = 1°$ and 5°. The maxima and minima of different order are indicated on the figure: (0), (±1M), (±1m), (±2M), (±2m), etc.

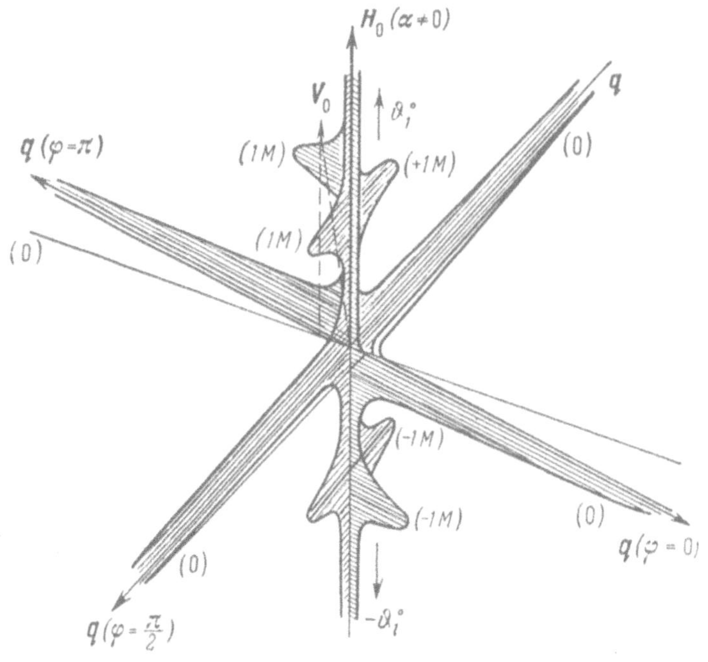

Fig. 35. Sections of the surface $F_3(\vartheta_1, \varphi)$ in the plane of $\varphi = 0$ and $\varphi = \pi/2$ when $\alpha \neq 0$.

quadratic term in the exponents of the integrands, plays a decisive role in the convergence of these integrals. We shall show, for example, that for $\varkappa Tq^2/M_i\Omega^2 \approx 1$ the maximum of third order ($\tau \approx 3.5$, $\vartheta_1 = 17.5°$ when $\alpha = 0$) already practically disappears. If $\vartheta_1 = 0$, then in this case the maxima of third and fourth order are still noticeable. On Fig. 32 are shown for illustration two curves of $F_3(\tau)$, calculated for $\varkappa Tq^2/M_i\Omega = 1$, $\eta \approx 0.06$, and $qV_0/\Omega \approx 14$. One can find more detailed results of the corresponding calculations in the tables of the following section.

If $\alpha = 0$ then, as is clear from (7.24) the value of τ, and therefore the value of $F_3(\tau)$ also, does not depend on φ. The surface $F_3 = F_3(\vartheta_1, \varphi)$ when $\alpha = 0$ is shown in Fig. 33 in cylindrical coordinates, where the angle ϑ_1 is laid off along the vertical axis, the radius vector of the cylindrical coordinate system is proportional to the magnitude of F_3, and the angle φ is chosen as the azimuthal angle. Since when $\alpha = 0$, F_3 does not depend on φ, the surface $F_3 = F_3(\vartheta_1, \varphi)$ is a surface of revolution about the vertical axis. The intersections of this surface with the planes $\varphi = 0$ and $\varphi = \pi/2$ are shown in the figures. It is clear from Figs. 32 and 33 that F_3 rapidly varies as a function of ϑ_1 (or as a function of the angle α for a fixed value of ϑ_1). Here the width of the principal and side maxima is of the order of a fraction of a degree or of one to two degrees. With increasing intensity (magnitude) of the principal maximum, which corresponds to an increase in T and a decrease in ν (an increase in the altitude at which the body is moving, see the tables of § 1), its width decreases and simultaneously its ratio to the maximum of the first and other orders increases. In Fig. 32, where the curves $F_3(\tau)$ are plotted on a logarithmic scale, the rapid oscillations of the scattering function are strongly smoothed out.

The symmetry of the function $F_3(\tau)$ relative to the angle φ is violated when the direction of the velocity $\mathbf{V}_0$ does not coincide with $\mathbf{H}_0$ ($\alpha \neq 0$). In this case τ depends on the angle φ between the planes $\mathbf{V}_0\mathbf{H}_0$ and $\mathbf{qH}_0$. If $\alpha > 0$ and the angle φ lies in the interval $\varphi < \pi/2$ or $3\pi/2 < \varphi$, then the condition $\tau = 0$ is satisfied for negative values of ϑ_1. Therefore, if the surface $F_3 = F_3(\vartheta_1, \varphi)$ is constructed as in Fig. 33, then the principal maximum affiliated with those same values of φ lies below the plane $\vartheta_1 = 0$. The line formed on the surface $F_3(\vartheta_1, \varphi)$ by the principal maxima is no longer a circle lying in the plane $\vartheta_1 = 0$, as in the case $\alpha = 0$, but is a nonplanar curve of elliptic type crossing the plane $\vartheta_1 = 0$ when $\varphi = \pi/2$ and $\varphi = 3\pi/2$. At these two points the zero-order maximum has the same values as in the case when $\alpha = 0$. The character of the

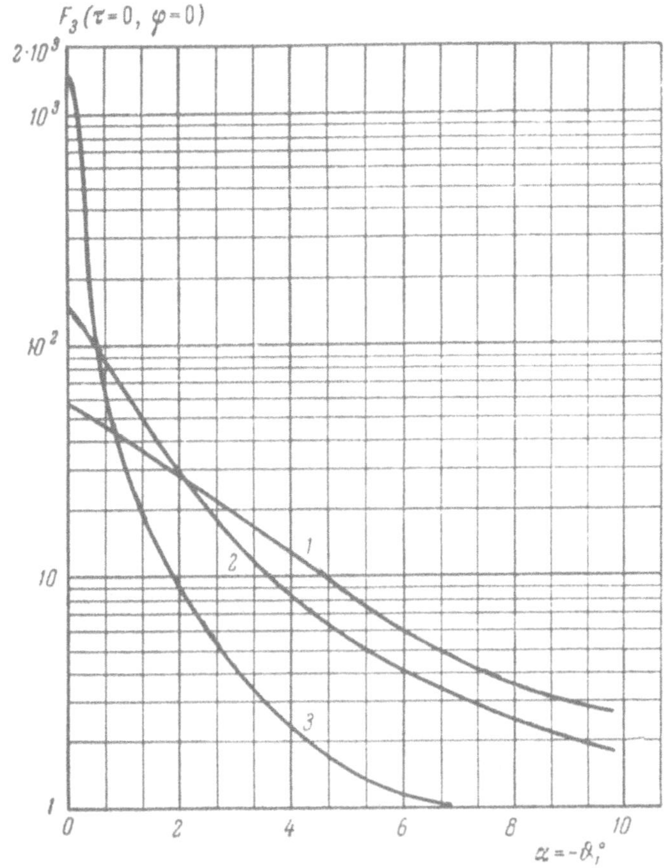

Fig. 36. Dependence of the principal maximum $F_3(\tau=0)$ on α for $\alpha = -\vartheta_1$ and $\varphi = 0$ for h = 300 km (curve 1), h = 400 km (curve 2), and h = 700 km (curve 3).

variation of $F_3(\vartheta_1, \varphi = 0)$ when $\alpha \neq 0$ can be traced in Figs. 34a and 34b, where for $\varkappa Tq^2/M_i\Omega^2 = 1$, $qV_0/\Omega = 14$, $\eta = 0.06$, these curves are shown for $\alpha = 1°$ and $5°$. The maxima and minima of the corresponding order are indicated on Fig. 34. Curves of $F_3(\vartheta_1)$ for $\varphi = 0, \pi/2$ are shown in Fig. 35. The dependence of the intensity (magnitude) of the zero-order maximum $F_3(\tau = 0)$ on $\alpha = -\vartheta_1$ when $\varphi = 0$ and on φ when $\alpha = 5°$ is shown in Figs. 36 and 37. Inside parentheses near the values of φ (Fig. 37) are written the values of ϑ_1 corresponding to them for $\alpha = 5°$. It is clear that $F_3(\tau = 0)$ increases as φ increases from 0 to $\pi/2$. As already mentioned, in this case the value of ϑ_1 for $\tau = 0$ negative. When $\varphi = \pi/2$ the function $F_3(\tau = 0)$ has a maximum equal to the maximum of F_3 when $\alpha = 0$; here the maximum lies in the plane $\vartheta_1 = 0$. The maximum $F_3(\tau = 0)$ is decreasing in the interval $\pi/2 < \varphi < \pi$. The quantity τ now assumes a value equal to zero for positive values of ϑ_1. In this connection, the same values of $F_3(\tau = 0)$ correspond in pairs to the values $\varphi = (\pi/2) + \varphi_M$, $\vartheta_1 = \vartheta_{1M}$ and $\varphi = (\pi/2) - \varphi_M$, $\vartheta_1 = -\vartheta_{1M}$. Or course, for a given value of φ_M the value of ϑ_{1M} is determined from the condition $\tau = 0$. The position of the higher order maxima when $\alpha \neq 0$ varies in a similar fashion. It is easy to determine their angular coordinates ϑ and φ for a given value of α with the aid of formulas (7.26) and (7.24).

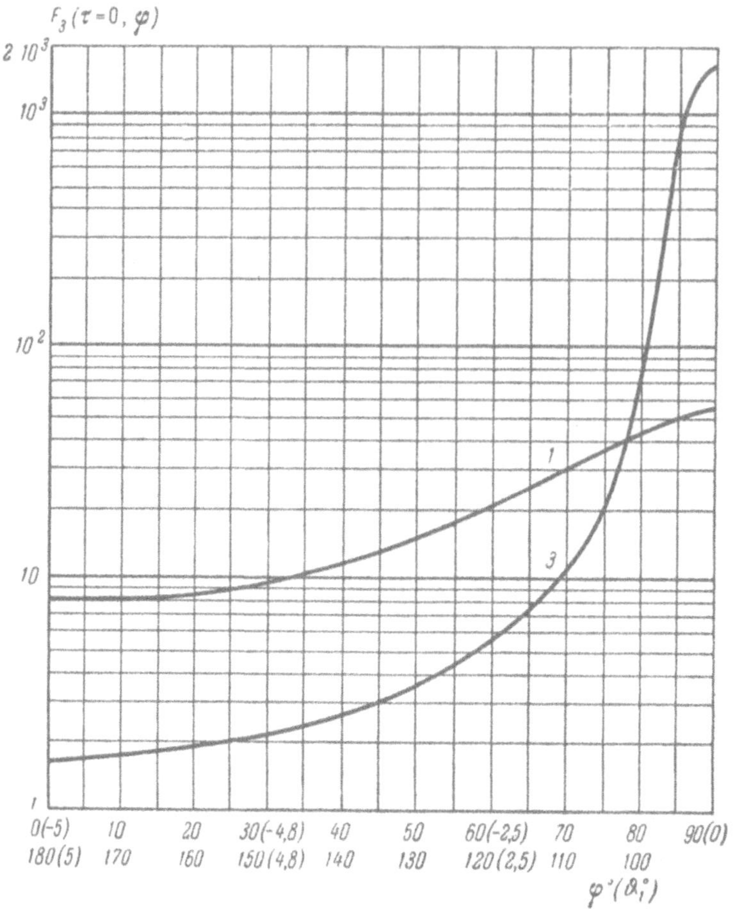

Fig. 37. Dependence of the value of the principal maximum $F_3(\tau = 0)$ on the angle φ between the planes $\mathbf{V}_0\mathbf{H}_0$ and $\mathbf{q}\mathbf{H}_0$ when $\alpha = 5°$ for h = 300 km (curve 1) and h = 700 km (curve 3).

TABLE X

Values of $\left\{\frac{1}{16}\left(\frac{\omega_0}{c}\right)^4 \frac{R_0^4 V_0^2}{\Omega^2}\right\}$

	Day			Night		
R_0, m \ h, km	300	400	700	300	400	700
0.5	$1.8\cdot10^{-2}$	$1.2\cdot10^{-2}$	$1.2\cdot10^{-3}$	$6\cdot10^{-5}$	$1.2\cdot10^{-3}$	$1.8\cdot10^{-4}$
1	0.3	0.2	$2\cdot10^{-2}$	10^{3}	$2\cdot10^{-2}$	$3\cdot10^{-3}$
2	4.8	3.2	0.32	$1.6\cdot10^{-2}$	0.32	$4.8\cdot10^{-2}$
3	24	16	1.6	$8\cdot10^{-2}$	1.6	0.24

TABLE XI. Values of the Function $|G(qR_0, \pi/2)|^2$

qR_0	0	0.05	0.1	0.2	0.3	0.4	0.5	0.6
$\|G\|^2$	1.00	0.999	0.995	0.99	0.97	0.96	0.94	0.91
qR_0	0.7	0.8	0.9	1.0	1.5	2.0	2.5	3
$\|G\|^2$	0.88	0.84	0.81	0.77	0.55	0.33	0.16	0.05
qR_0	3.5	3.83	4	4.5	5	5.5	5.54	6
$\|G\|^2$	$6\cdot10^{-3}$	—	$1.2\cdot10^{-3}$	$1.1\cdot10^{-2}$	$1.7\cdot10^{-2}$	$1.6\cdot10^{-2}$	$1.6\cdot10^{-2}$	$8\cdot10^{-3}$
qR_0	6.5	7.01	7.5	8	8.5	8.64	9	9.5
$\|G\|^2$	$2.2\cdot10^{-3}$	—	$1.3\cdot10^{-3}$	$3.4\cdot10^{-3}$	$4.1\cdot10^{-3}$	$4\cdot10^{-3}$	$3\cdot10^{-3}$	$1.2\cdot10^{-3}$

§ 27. Dependence of the Effective Scattering Cross Section on Altitude and on Wavelength

From the preceding, it follows that the effective differential cross section $d\sigma$, determining the intensity of electromagnetic waves scattered by the trail of a body, can be represented in a natural way in the form of a product of three factors, namely

$$d\sigma = \left\{\frac{1}{16}\left(\frac{\omega_0}{c}\right)^4 \frac{R_0^4 V_0^2}{\Omega^2}\right\} F_3(\tau, \eta, \zeta, \delta)\, |G(qR_0, \chi)|^2 \sin^2\psi_1\, do, \tag{7.27}$$

where ψ_1 is the angle between $\mathbf{E}_0$ and $\mathbf{k}'$. For an analysis of the properties of $d\sigma$ it is necessary to tabulate two of these factors, i.e., the functions F_3 and G. The appropriate calculations were carried out for three heights in the ionosphere, h = 300, 400, and 700 km.

For these altitudes, for $V_0 = 8$ km/sec, and the values of the ionosphere's parameters listed in Tables I and III, the values of the factor appearing inside the curly brackets in formula (7.27) are given in Table X.

It is easy to calculate the third factor in (7.27) if the fact that it depends rather smoothly on the angle χ is taken into consideration. We are mainly interested in small values of τ, i.e., values of χ close to $\pi/2$. Therefore, it is sufficient for us to know $G(qR_0, \pi/2)$. In this case one can use formula (2.88), according to which

$$|G(qR_0, \pi/2)|^2 = \left(\frac{J_1(qR_0)}{qR_0}\right)^2. \tag{7.28}$$

Values of the function (7.28) are given in Table XI. As far as the principal factor in formula (7.27) is concerned, that is, the scattering function F_3, its dependence on the angles α, ϑ_1, and φ were calculated in sufficient detail for the indicated altitudes and different wavelengths with the aid of an electronic computer.

We note first of all that when $\mathbf{V}_0 \parallel \mathbf{H}_0$, that is, when $\alpha = 0$, the principal maximum of the effective scattering cross section lies in the direction of "specular reflection" of the wave from the direction of the earth's magnetic field (this is clear from the foregoing analysis). In this case the bisector of the angle $(\mathbf{k}, \mathbf{k}')$, namely the vector $\mathbf{q}$, coincides with the normal to $\mathbf{H}_0$. On the other hand, if $\alpha \neq 0$, then the vector $\mathbf{q}$ is turned relative to the normal to $\mathbf{H}_0$ by an angle ϑ_1, determined from the equation $\tau(\alpha, \vartheta_1, \varphi) = 0$ for specified values of α and φ. Thus, in this case $\mathbf{q}$ makes an angle $(\pi/2) - \vartheta_1$ with $\mathbf{H}_0$. For example, when $\alpha = 5°$ and $\varphi = 0$, the maximum scattering will occur along the vector $\mathbf{k}'$, chosen such that the vector $\mathbf{q}$ makes an angle $(\pi/2) - 5°$ with $\mathbf{H}_0$. The position of the maxima of higher order, that is, the angles ϑ_1 through which they are turned, is determined from the corresponding values of τ_{max} [see (7.26) and formula (7.24)]. For example, when $\varphi = 0$, $\alpha = 0$, and $qV_0/\Omega = 14$ as used in the calculations, the principal maximum and the maxima of orders ± 1, 2, 3 correspond to values of the angles

$$\vartheta_{1\,max} \approx 0°;\ \pm 5°;\ \pm 9°;\ \pm 13°, \tag{7.29}$$

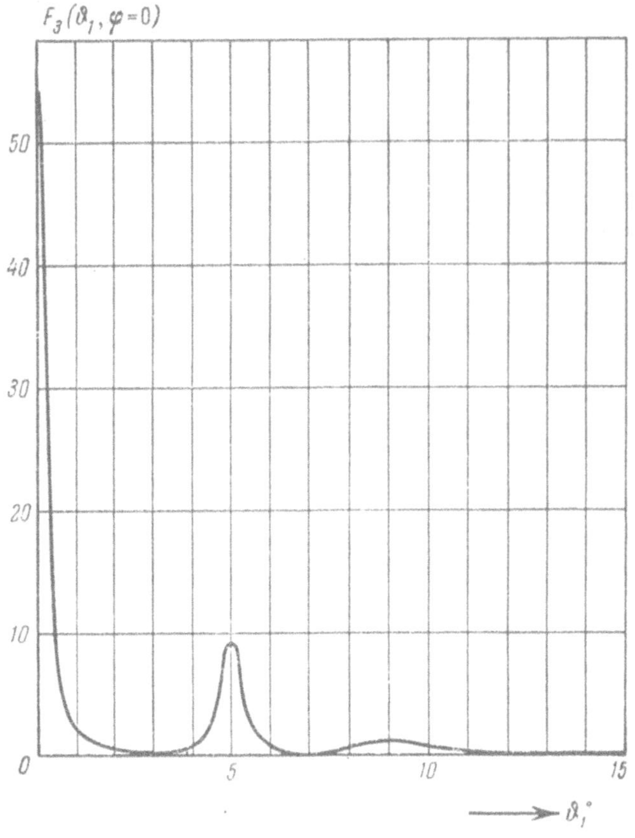

Fig. 38. Dependence of $F_3(\vartheta_1, \varphi = 0)$ on ϑ_1 for $\lambda = 30$ m, $\alpha = 0$, and $h = 300$ km.

and when $\alpha = 1°$, $\varphi = 0$ the maxima of the same orders correspond to the angles

$$\vartheta_{1\ \max} \approx 1°; \quad (+4°, \ -6°); \quad (+8°, \ -10°); \quad (+12°, \ -14°). \tag{7.30}$$

One can obtain a detailed idea about the character of the variations of F_3 from Tables XII through XVI. Curves of $F_3(\vartheta_1, \varphi = 0)$ are shown in Figs. 38, 39, and 40 for a wavelength $\lambda = 30$ m and heights $h = 300$, 400, and 700 km respectively for $\alpha = 0$, and in Fig. 41 for $h = 300$ km and $\alpha = 5°$. These curves of $F_3(\vartheta_1, \varphi = 0)$, in contrast to the curves of F_3 presented in Figs. 32 and 34, are constructed on a linear scale, and give an idea about the width of the individual maxima. The dependence of the principal maximum on the wavelength λ is shown in Fig. 42. One can estimate the width of the maximum for different values of λ according to the data in Table XIII. Here it is everywhere assumed that $\psi = \pi$, so that

$$q = \frac{2\omega}{c} \sqrt{1 - \left(\frac{\omega_0}{\omega}\right)^2} = \frac{4\pi}{\lambda}. \tag{7.31}$$

A change in the temperature, T, and accordingly in the number of collisions ν_{ii} naturally leads to a change in η, ζ, and δ [see Eqs. (7.23) and (7.24)]. Some of the data in Table XVII gives an idea of the magnitude of the changes in F_3 due to the change in T.

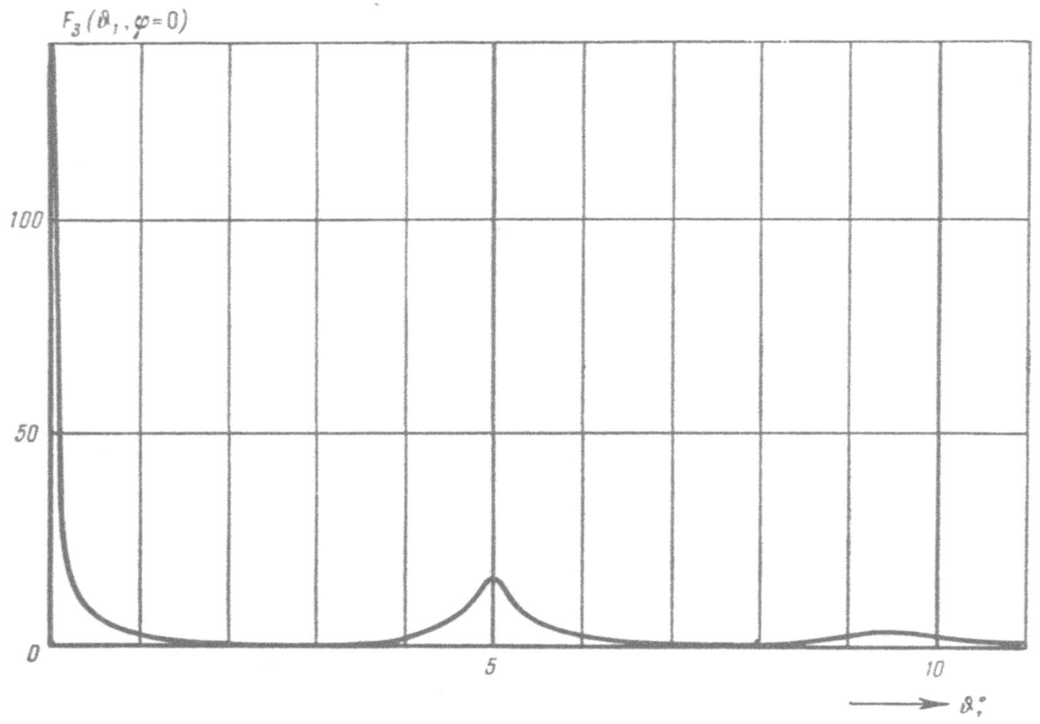

Fig. 39. Dependence of $F_3(\vartheta_1, \varphi = 0)$ on ϑ_1 for $\lambda = 30$ m, $\alpha = 0$, and $h = 400$ km.

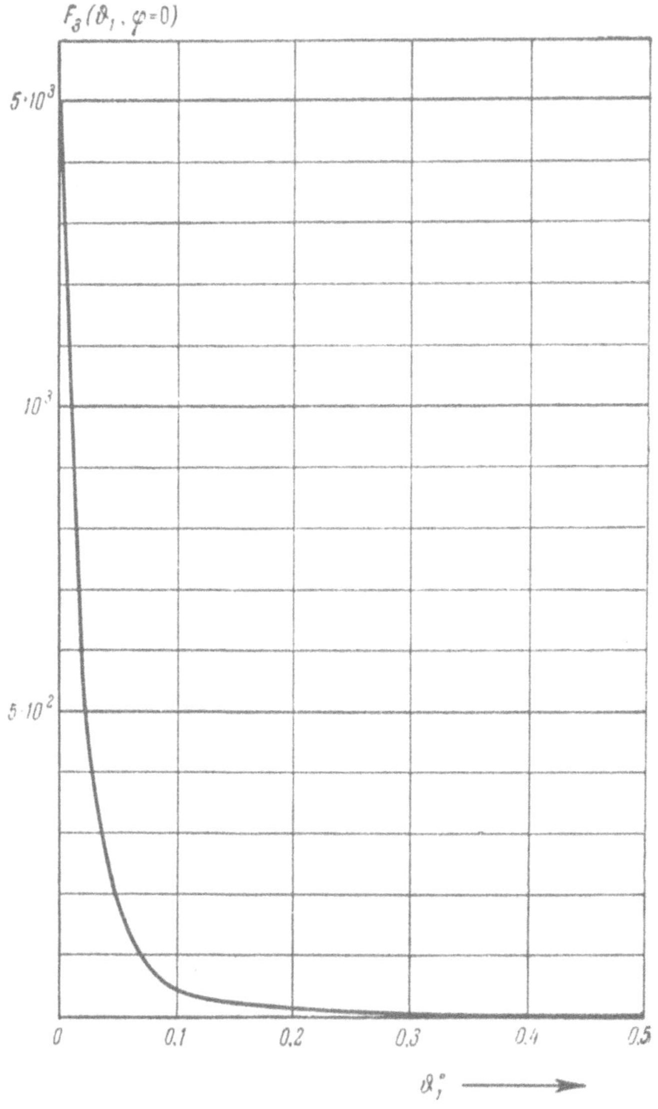

Fig. 40. Dependence of $F_3(\vartheta_1, \varphi=0)$ on ϑ_1 for $\lambda=30$ m, $\alpha=0$, and $h=700$ km.

TABLE XII. Values of $F_3(\alpha, \vartheta_1)$ for $\lambda = 30$ m, $\varphi = 0$ at Various Heights h

α, ϑ_1, deg. \ h, km	$\vartheta_1=0$			$\alpha=0$			$\alpha=-\vartheta_1$		
	300	400	700	300	400	700	300	400	700
0	53.46	134.4	1535	53.46	134.4	1535	53.46	134.4	1535
0.02	—	—	613.9	—	—	—	—	—	—
0.03	—	—	355.1	—	—	—	—	—	—
0.05	49.20	103.2	150.7	—	—	—	—	—	—
0.1	39.72	60.86	40.8	—	—	—	—	—	—
0.2	22.44	23.10	10.31	—	23.07	10.0	—	—	—
0.3	12.98	11.30	4.56	—	—	4.42	—	—	—
0.5	5.44	4.20	1.58	5.39	4.14	1.53	49.62	101.1	109.6
1.0	1.32	0.94	0.32	1.30	0.92	0.31	41.25	61.65	32.14
1.5	0.47	0.31	—	—	—	—	—	—	—
2.0	0.17	0.10	0.019	0.17	0.999	0.018	—	—	9.30
3.0	0.015	0.011	0.037	0.013	$7.8 \cdot 10^{-3}$	0.023	—	14.01	3.99
3.5	0.15	0.18	—	—	—	—	—	—	—
4.0	0.85	0.67	0.56	0.98	0.69	0.68	—	—	—
4.5	2.48	2.22	1.30	—	—	—	—	—	—
5.0	9.39	13.18	3.18	10.26	12.61	3.41	7.95	5.85	1.48
5.5	2.58	2.71	1.29	—	—	—	—	—	—
6.0	0.88	1.44	0.28	0.77	0.61	0.19	5.96	4.23	—
7	0.18	0.10	0.025	0.18	0.11	0.010	4.63	3.20	—
8	0.22	0.18	0.14	0.26	0.053	0.10	3.71	2.52	—
9	0.89	0.98	0.45	2.48	5.31	28.9	3.04	2.03	—
9.5	0.83	1.15	0.62	0.78	1.05	1.57	—	—	—
10	0.53	0.70	0.53	0.45	0.52	0.47	2.54	1.68	0.39
11	0.25	0.25	0.21	0.23	0.22	—	—	—	—
12	0.19	0.19	0.18	0.10	0.046	—	—	—	—
12.5	0.187	0.20	—	0.32	0.89	—	—	—	—
13	0.18	0.21	0.24	0.28	0.47	—	—	—	—
13.5	0.17	0.21	0.25	0.19	0.25	—	—	—	—
14	0.16	0.19	0.25	0.16	0.19	—	—	—	—
15	0.12	0.14	0.20	0.12	0.13	—	—	—	—
16	0.10	0.11	—	0.088	0.086	—	—	—	—
16.5	—	—	—	—	0.058	—	—	—	—
17	—	0.091	—	0.091	0.16	—	—	—	—
17.5	0.077	—	—	0.080	0.010	—	—	—	—
18	0.071	0.078	—	0.072	0.16	—	—	—	—
18.5	0.067	—	—	0.067	—	—	—	—	—
19	0.062	0.067	—	0.062	0.070	—	—	—	—
20	0.054	0.058	—	0.054	0.057	—	—	—	—
21	—	—	—	0.048	—	—	—	—	—
22	—	—	—	0.044	—	—	—	—	—
23	—	—	—	0.039	—	—	—	—	—
24	—	—	—	0.036	—	—	—	—	—

TABLE XIII. Values of $F_3(\vartheta_1)$ in the Vicinity of the Principal Maximum for $\alpha = 0$, $\varphi = 0$, Different Values of the Wavelength λ, and Various Altitudes h

h, km	300				400				700		
λ, m / ϑ_1, deg.	30	20	15	10	30	20	15	10	30	20	15
0	53.46	11.07	4.82	1.72	134.4	31.05	14.10	5.20	1535	479.3	241.3
0.02	—	—	—	—	—	—	—	—	613.9	—	—
0.03	—	—	—	0.080	—	—	—	0.063	355.1	—	—
0.05	49.20	9.75	3.96	—	103.2	20.54	7.71	—	150.7	25.3	8.67
0.1	39.72	7.16	2.58	—	60.86	10.2	3.27	—	40.8	6.58	2.20
0.2	22.44	3.46	—	—	23.10	3.36	—	—	10.3	1.63	—
0.3	12.98	1.83	0.51	—	11.30	1.55	0.42	—	4.56	0.70	—
0.5	5.44	0.68	0.16	0.016	4.20	0.52	0.12	0.011	1.58	0.22	0.056
1.0	1.32	0.11	0.01	0.020	0.94	0.074	0.005	0.017	0.32	0.027	0.001
1.5	0.47	—	0.03	—	0.31	—	0.027	—	—	$8 \cdot 10^{-5}$	0.023
2	0.17	0.031	0.48	0.016	0.10	$3.6 \cdot 10^{-3}$	—	0.010	0.019	0.027	—

TABLE XIV. Values of the Principal Maximum $F_3(\tau = 0)$ and of the Angles ϑ' for Different Angles φ Between the Planes $\mathbf{V_0H_0}$ and $\mathbf{qH_0}$

φ, deg.	0	15	30	60	75	90	105	150	180
ϑ_1, deg. / h, km	−5	−4.8	−4.3	−2.5	−1.3	0	1.3	4.3	5
300	7.95	8.38	9.88	20.38	35.98	53.46	35.98	9.88	7.95
700	1.48	—	—	5.67	19.74	1535	19.74	—	1.48

TABLE XV. Dependence of $F_3(\vartheta_1)$ for Different Values of α, $\varphi = 0$, and $\lambda = 30$ m

h, km / ϑ_1, deg.		$\alpha=5°$						$\alpha=1°$				$\alpha=0.5°$	
		300		400		700		300		700		700	
0		10.26		12,61		3.41		1.30		0.31		1531	
+0.5,	−0.5	1,94	6,24	1.68	7.63	0.57	1.44	—	5.44	0.090	1.58	0,31	109.6
+1	−1	0.77	0.75	0.61	0.72	0.19	0.75	0.17	41.26	0.019	32.14	0.091	1.77
+2,	−2	0,18	0.014	0.11	0.009	0.01	0.027	0.014	1.37	0.027	0,36	—	0.10
+3,	−3	0.12	0.18	0.10	0.10	0.14	0.020	0.68	0.18	0.64	0,020	0.22	—
+4,	−4	2,18	1.62	6.50	1.29	2.70	0.59	11.69	0.018	7.45	0.051	1.55	0.22
+5,	−5	0,47	7.95	0.56	5.85	0,69	1.48	0.86	0.64	0.25	0.48	1.24	1.10
+6,	−6	0.22	1.93	0.21	1.59	0.084	0.55	0.18	6.70	0.013	1.73	0.056	1.26
+7,	−7	0.15	0.21	0.14	0.13	0.09	0.032	0.21	0.94	0.14	0.31	0.058	0.066
+8,	−8	0.22	0.058	0.28	0.057	0.34	0.078	1.05	0.19	0.56	0,041	0.28	0.070
+9,	−9	0.16	0.53	0.20	0.48	0.41	0.28	0.53	0.24	0.64	0.15	0.71	0.26
+10,	−10	0,12	2.09	0.13	1.66	0.19	0.45	0.24	0.77	0.18	0.38	0.32	—
+11,	−11	0.094	1.28	—	1.03	—	—	0.18	0.53	—	—	—	—
+12,	−12	0.082	0.31	—	0.23	—	—	0.19	0.26	—	—	—	—
+13,	−13	0,072	0.30	—	0.26	—	—	0.16	0.19	—	—	—	—
+14,	−14	0.062	0.53	—	0.51	—	—	0.12	0.18	—	—	—	—
+15,	−15	0.054	0.49	—	0.54	—	—	0.096	0.15	—	—	—	—
+16,	−16	0.045	0.29	—	0.32	—	—	0.082	0.12	—	—	—	—
+17,	−17	0.043	0.21	—	0.046	—	—	0.072	0.097	—	—	—	—
+18,	−18	0.039	0.18	—	0.21	—	—	0.062	0.082	—	—	—	—
+19,	−19	0.036	0.15	—	0.17	—	—	0.054	0.071	—	—	—	—
+20,	−20	0.032	0.12	—	0.14	—	—	0.048	0.062	—	—	—	—

TABLE XVI. Dependence of $F_3(\vartheta_1)$ in the Vicinity of the Principal Maximum for $\lambda = 30$ m and Different Values of the Angle φ Between the Planes $\mathbf{V}_0\mathbf{H}_0$ and $\mathbf{qH}_0$

			$h=300$ km $\varphi=0°$, $\alpha=5°$				
ϑ_1^0		−3.5	−4	−5	−6	−6.5	
F_3		0.51	1.62	7.95	1.93	0.64	
			$h=300$ km $\varphi=15°$, $\alpha=4.8°$				
ϑ_1^0	−3.3	−3.8	−4.3	−4.8	−5.3	−5.8	−6.3
F_3	0.47	1.61	6.02	8.38	5.18	1.90	0.67
			$h=300$ km $\varphi=30°$, $\alpha=4.3°$				
ϑ_1^0	−2.8	−3.3	−3.8	−4.3	−4.8	−5.3	−5.8
F_3	0.46	1.52	6.28	9.88	5.61	1.85	0.65
			$h=700$ km $\varphi=0°$, $\alpha=5°$				
ϑ_1^0	−3.5	−4	−4.5	−5	−5.5	−6	−6.5
F_3	0.11	1.29	1.47	5.85	1.06	0.54	0.18
			$h=700$ km $\varphi=15°$, $\alpha=4.8°$				
ϑ_1^0	−3.3	−3.8	−4.3	−4.8	−5.3	−5.8	−6.3
F_3	0.10	0.59	1.47	1.58	1.12	0.56	0,20
			$h=700$ km $\varphi=30°$, $\alpha=4.3°$				
ϑ_1^0	−7.8	−3.3	−3.8	−4.3	−4.8	−5.3	−5.8
F_3	0.093	0.53	1.87	1.98	1.32	0.59	0.20
			$h=700$ km $\varphi=60°$, $\alpha=2.5°$				
ϑ_1^0	—	−1.5	−2	−2.5	−3	−3.5	—
F_3	—	0.34	2.68	2.67	2.5	0.56	—
			$h=700$ km $\varphi=75°$, $\alpha=1.3°$				
ϑ_1^0	—	−0.3	−0.8	−1.3	−1.8	−2.3	−2.8
F_3	—	0.32	11.70	19.74	2.57	0.39	0.10

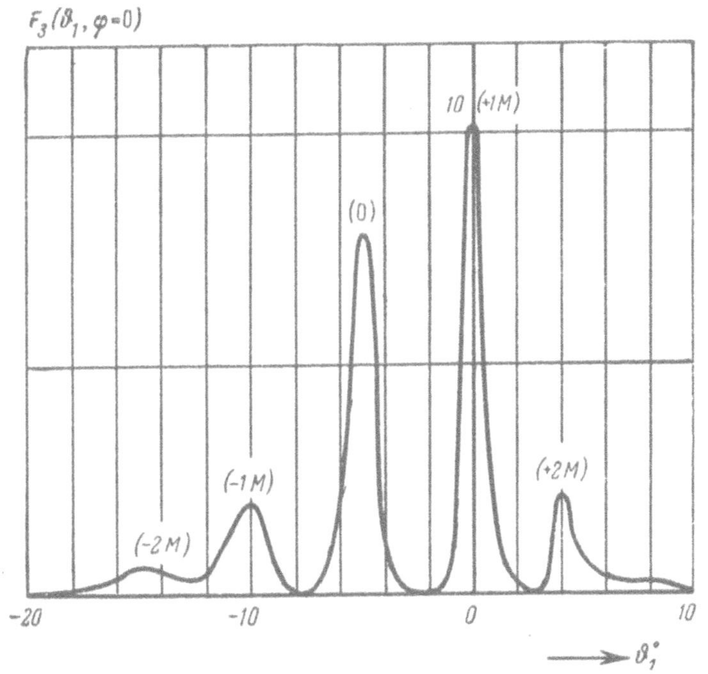

Fig. 41. Dependence of $F_3(\vartheta_1, \varphi=0)$ on ϑ_1 for $\lambda=30$ m, $\alpha=5°$, and h = 300 km.

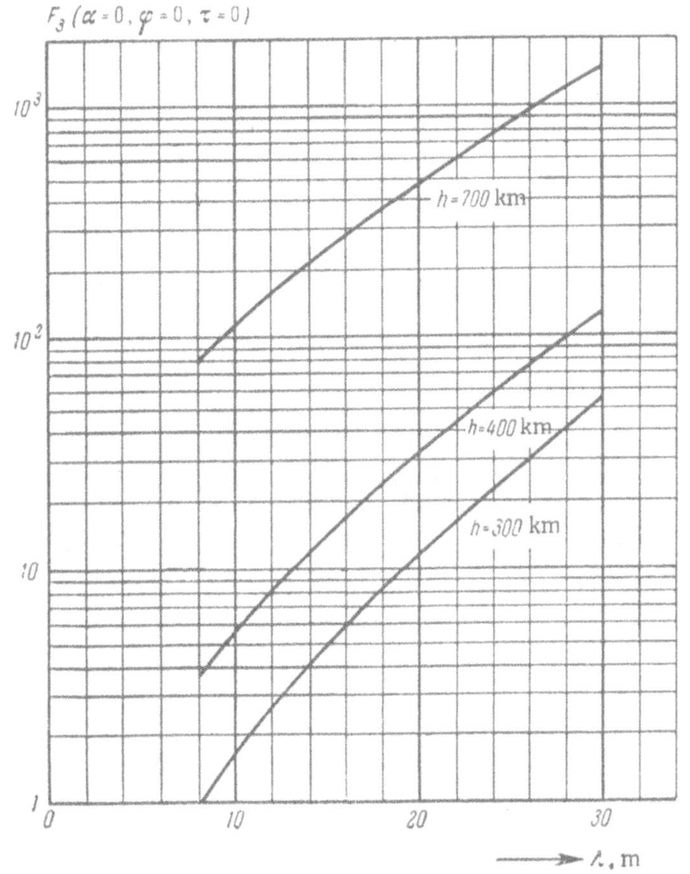

Fig. 42. Dependence of $F_3(\vartheta_1=0, \varphi=0)$ on the wavelength λ for $\alpha=0$ and h = 300, 400, and 700 km.

TABLE XVII. Values of the Principal Maximum of $F_3(\alpha, \vartheta_1 = 0)$ for Different Values of the Temperature T at $\lambda = 30$ m and h = 300 and 400 km

$\alpha°$ \ $T°$	800	1600	2000	$\alpha°$ \ $T°$	1000	2000
	$h=300$ km				$h=400$ km	
0°	53.46	14.04	9.67	0°	134.4	38.14
5°	10.26	4.12	—	5°	12.61	3.16

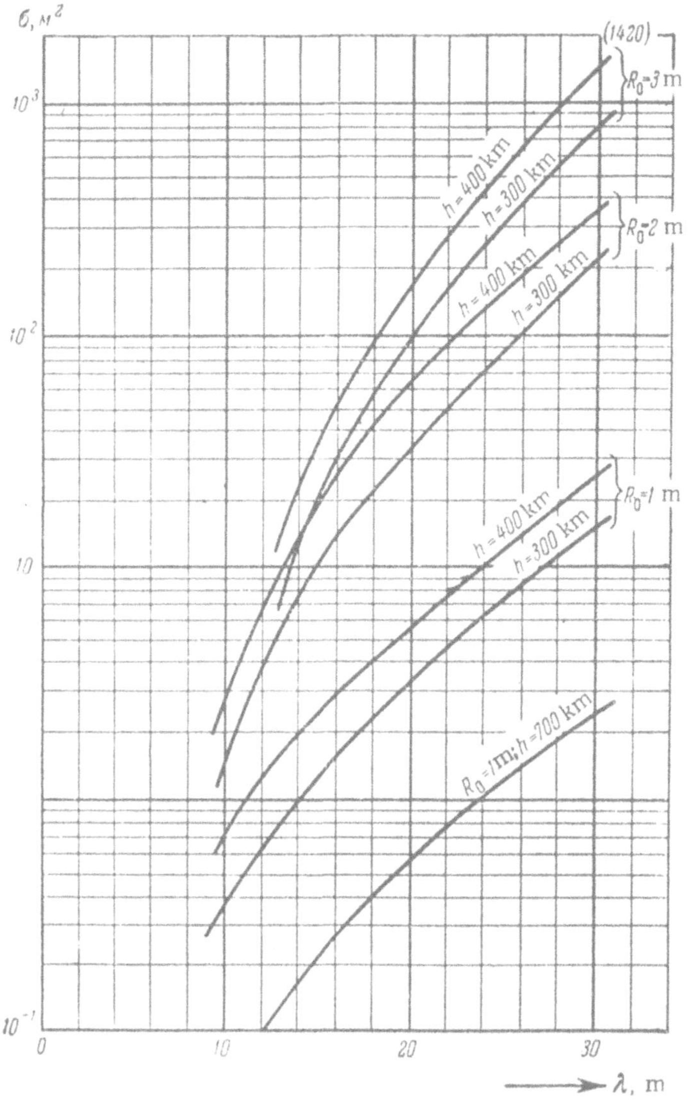

Fig. 43. Dependence of the effective cross section dσ/do during the day on the wavelength λ for altitudes h = 300, 400, and 700 km and for sphere radii R_0 = 1, 2, and 3 meters.

TABLE XVIII. Effective Differential Cross Section $d\sigma/do$ of the "Trail" in the Direction of the Principal Maximum; $\beta_0 = (1/4\pi)(d\sigma/d\sigma_0)$, $d\sigma_0$ is the Effective Cross Section of a Metallic Sphere

	h, km	300			400			700		
	R_0, m \ λ, m	30	20	15	30	20	15	30	20	15
Day	0.5	$\frac{d\sigma}{do}$ 0.97 β_0 1380	0,20 47	0,08 7	1.60 2290	0.36 90	0.2 14	1.8 2400	0,6 160	0.3 22
	1	$\frac{d\sigma}{do}$ 15 β_0 350	3 14	1 1,6	26 590	5.6 25	2 3	30 630	8 36	4 5
	2	$\frac{d\sigma}{do}$ 210 β_0 72	35 3	10 0,3	350 120	65 5	20 0.6	400 134	100 79	34 1,2
	3	$\frac{d\sigma}{do}$ 810 β_0 28	98 1	18 0,30	1420 50	184 2	35 0.5	1620 58	280 2,8	60 0.83
Night	0.5	$\frac{d\sigma}{do}$ $3\cdot10^{-3}$ β_0 4.3	$7\cdot10^{-4}$ 0.2	$2\cdot10^{-4}$ 0,02	0,2 229	0,04 9	0,02 1.4	0.3 373	$8\cdot10^{-2}$ 22	$4\cdot10^{-2}$ 3.61
	1	$\frac{d\sigma}{do}$ 0.05 β_0 1	10^{-2} 0,05	$3\cdot10^{-3}$ 0,014	2.6 60	0,61 2.5	0,2 0,3	4,6 96	1.3 57	0,6 0,8
	2	$\frac{d\sigma_0}{do}$ 0.7 β_0 0.24	0.1 $3\cdot10^{-3}$	0.03 $2.4\cdot10^{-3}$	35 12	6.5 0.5	2.0 0.06	61 20	15 1,2	5,3 0.2
	3	$\frac{d\sigma_0}{do}$ 2.7 β_0 0.1	0,3 $4\cdot10^{-3}$	0.06 $7\cdot10^{-4}$	140 5	18 0.2	3,5 0,05	240 8,5	51 0.6	9 0.8

With the aid of Tables X through XVII the effective cross section $d\sigma$ for a sphere of radius $R_0 = 0.5$, 1, 2, and 3 m is calculated for the above chosen values of h and for different wavelengths. This data is presented in Table XVIII and in Fig. 43. The ratio of the differential effective cross section $d\sigma$ of the trail to the differential effective cross section, multiplied by 4π, for scattering by an ideal conducting sphere of radius R_0 is also presented in the table, that is, the value of the ratio

$$\beta_0 = \frac{1}{4\pi}\,\frac{d\sigma}{d\sigma_0}.$$

The value of $d\sigma_0$, just as for $d\sigma$, is taken for backward scattering, that is, when the angle $\psi = \pi$.*

It is seen from these data that the differential effective cross section $d\sigma$ of the principal maximum of the "trail" of a satellite can greatly exceed in the daytime, and in many cases also at night, the effective cross section $d\sigma_0$ of the metallic sphere, and reaches many tens or even hundreds of square meters. However, inasmuch as these values correspond to only one direction of the wave, in estimating the true scattering effect it is necessary to also take the width of the corresponding "lobe" into consideration. The scattering increases in the region of altitudes from 400 to 700 km, and the effective cross section falls off strongly with decreasing wavelength.

We note here that scattering from the "trail" depends, generally speaking, little on the properties and shape of the body itself and on the character of its surface, being determined only by the velocity of the body and its linear dimension R_0. At the same time, scattering from the body itself depends appreciably on its properties. In this respect a smooth metallic sphere is optimal for scattering by a body in the range of frequencies investigated. Other bodies, of similar dimension with different surface character, have considerably smaller values of $d\sigma_0$. We also note that the relative effect of scattering by an inhomogeneous formation, that is, the ratio β_0 increases rapidly with decreasing radius of the sphere R_0.

In fact, when $2\pi R_0/\lambda \ll 1$

$$\frac{d\sigma_0}{do} \sim \frac{R_0^6}{\lambda^4}, \quad \frac{d\sigma}{do} \sim R_0^4 \text{ and } \beta_0 \sim \frac{1}{R_0^2} \to \infty,$$

and when $R_0 \to \infty$

$$\frac{d\sigma_0}{do} \sim R_0^2, \quad \frac{d\sigma}{do} \sim R_0 \text{ and } \beta_0 \sim \frac{1}{R_0} \to 0.$$

§28. Character of the Field of the Scattered Wave at the Point of Observation

Now we consider the overall pattern of the effect of scattering at the point of observation, and we summarize some of the results of the calculations presented above. First let us assume that the body moves so that its velocity vector is close to the direction of the constant magnetic field $\mathbf{H}_0$, and the wave vector $\mathbf{k}$ of the incident electromagnetic wave is normal to $\mathbf{H}_0$ or $\mathbf{V}_0$. The diagram of the scattering has in this case, as we saw, a many-lobed structure so that its principal lobe is directed along the normal to $\mathbf{V}_0$ or $\mathbf{H}_0$, and the side lobes are symmetrically distributed with respect to this normal. In this connection, the angular opening occupied by several lobes is not more than 15 or 20° to the normal to $\mathbf{H}_0$. Then the intensity of the field of the scattered wave decreases monotonically with increase of the angle ϑ_1. Thus, at some point near the earth's surface, the following picture will be observed as the body approaches. First the scattering field increases monotonically, then bursts of intensity occur, corresponding to the positive side maxima (... +2 M, +1 M), to the principal maximum (0), and to the negative maxima (−1 M , −2 M...), after which the field again decreases monotonically. Inasmuch as the effective scattering cross section is rather large only in the principal and in one or two of the side maxima, in practice the field of the scattered wave is rather intense at the point of observation only during a few brief intervals of time δt. Let us estimate the value of δt, when, for example, the body passes at an altitude $h \approx 400$ km. In this case at the principal maximum $d\sigma/do \approx 350$ m^2 for a body of radius $R_0 \approx 2$ m and at a

*For a small conducting sphere ($kR_0 \ll 1$) $d\sigma_0(\psi{=}\pi) = (9R_0^6\omega^4/4c^4)\,do$ (see p. 383 of [33]).

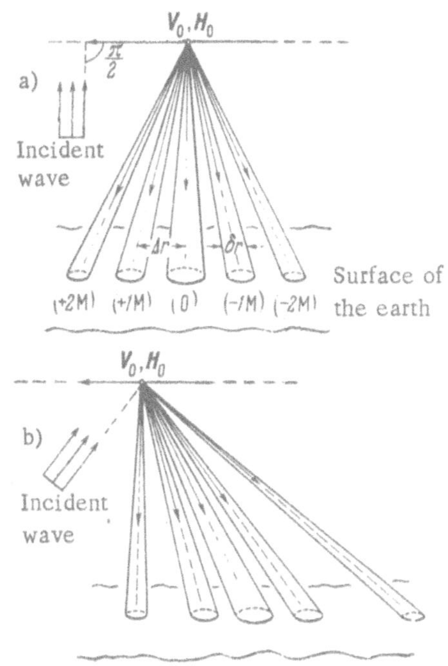

Fig. 44. Schematic diagram of the regions illuminated on the earth's surface by different lobes of the scattered wave: a) $\mathbf{V}_0 \parallel \mathbf{H}_0$; b) $\mathbf{V}_0 \neq \mathbf{H}_0$.

wavelength $\lambda \approx 30$ m, and the width of the principal lobe, if defined according to a decrease of $d\sigma/do$ to the values $(1/4\pi)(d\sigma/do)$, amounts to roughly $\delta\vartheta_1 \approx 0.6°$. Therefore, during the time that it takes the principal lobe to pass the point of observation, that is, during the time $\delta t \sim (h\delta\vartheta_1/V_0 \approx 0.5$ sec, the average intensity of the scattered wave at the point of observation is determined from the value $d\sigma/do \sim 170$ m^2, since $d\sigma/do$ varies in this interval of ϑ_1 approximately according to a linear law. The first two side maxima ($\pm$ M) are still rather intense, since for them $d\sigma_1/do$ is commensurate with $d\sigma_0/do$. The lobes of higher order are already difficult to observe against the general background of the scattering field of the body itself. We see that the effect of scattering at the point of observation appears in the form of bursts of the "received" wave. The duration and relative intensity of the effect will change, of course, for different dimensions, character, and shape of the body's surface, depending on the sensitivity of the indicators.

Figure 44a shows schematically how the surface of the earth is irradiated by the individual lobes of the scattering field when $\mathbf{V}_0 \parallel \mathbf{H}_0$ if the incident wave is normal to $\mathbf{V}_0$, and Fig. 44b is when the incident wave is directed at an angle to $\mathbf{V}_0$ (or $\mathbf{H}_0$). In the latter case the directions of all the lobes are rotated relative to the normal to the magnetic field such that the principal maximum lies in the direction of specular reflection of the wave from the vector $\mathbf{H}_0$.

Now let us assume that the body's motion is not along the magnetic field, but at some not too large angle to it. In this case the scattering diagram has a more complicated form. Its lobes become distorted. However, the general character of the field structure near the earth's surface remains the same. The quantitative changes, however, can be very appreciable. Thus, if the transmission and reception points – the rays OS and OE (see Fig. 31) lie in the same plane with $\mathbf{V}_0$ and $\mathbf{H}_0$ ($\varphi = 0$ or π), then more of the identical lobes of

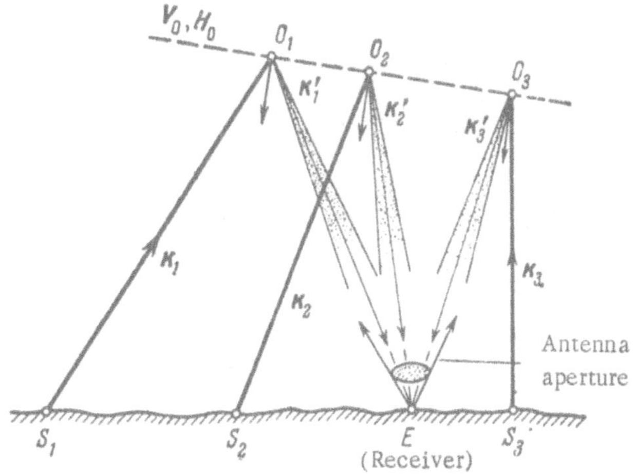

Fig. 45. Schematic diagram of the scattering field for a body irradiated from different points.

TABLE XIX. Characteristics of Intensity Bursts Produced at the Point of Observation by a Wave Scattered by the "Trail" of a Body During the Day ($\alpha = 0$, $\mathbf{V}_0 \parallel \mathbf{H}_0$)

h, km	λ, m	R_0, m	Data on the bursts				
			No. of lobe	β_0	δt, sec	δr, km	Δt, sec
300	30	1	(0)	175	0.6	5	—
			(−1 M)	35	—	—	3.4
		2	(0)	36	1	8	—
			(±1 M)	7	0.8	—	3.4
		3	(0)	14	1.2	9	—
			(±1 M)	2.8	1.0	—	3.4
400	30	1	(0)	293	0.4	3.5	—
			(±1 M)	29	—	—	4.5
		2	(0)	60	0.8	6.5	—
			(±1 M)	6	0.4	3.5	4.5
700	30	1	(0)	29	—	—	—
		2	(0)	6	0.15	1.2	—
300	20	1	(0)	6.8	—	—	—
		2	(0)	1.4	0.25	2	—
400	20	1	(0)	12.7	—	—	—
		2	(0)	2.6	0.25	2	—

smaller intensity are observed (see Fig. 41). On the other hand, if the rays OS and OE lie in a plane perpendicular to the $\mathbf{V}_0\mathbf{H}_0$ plane ($\varphi = \pi/2$ or $\varphi = 3\pi/2$), then the field will be approximately the same as for $\mathbf{V}_0 \parallel \mathbf{H}_0$, and the principal maximum will remain the same in magnitude. Consequently, if the exposure (to electromagnetic radiation) of the body and the reception of the scattered field occur in a plane perpendicular to $\mathbf{V}_0\mathbf{H}_0$, then the effect of scattering at the point of observation does not decrease in comparison with the effect in the case of longitudinal motion of the body.

In conclusion, Table XIX lists, for different altitudes in the ionosphere and for different wavelengths, data about bursts of scattering on the earth's surface during the flight of the body. Here the case when the intensity of the field scattered from the "trail" is larger than the intensity of the field scattered from a metallic sphere of the corresponding size or is commensurate with it ($\beta_0 \geq 1$) is calculated. In the table δr is the width of the illuminated area, $\delta t = \delta r / V_0$ is the duration of the burst at the point of observation, and Δr and Δt are the distances between the centers of the individual areas illuminated by the scattered field and the time intervals between successive bursts at the point of observation.

It follows from Table XIX that the bursts in the intensity of the field of the scattered wave during one passage of the body are being repeated at the point of observation every few seconds, and the duration of a single burst is of the order of or smaller than one second. Naturally, such a theoretical situation occurs only when the body is exposed to a plane wave (traveling) in one direction. If, however, the body is exposed to radiation coming from several points (S_1, S_2, S_3, ...) at different angles, then at one point E near the earth's surface several scattered waves, coming in at different angles, will be observed (Fig. 45) and the time of observation of the scattering effect will increase appreciably.

§29. Effective Scattering Cross Section in the Absence of an External Magnetic Field

In conclusion let us consider the results of calculations of the effective cross section in the case when there is no external constant magnetic field $\mathbf{H}_0$ in the plasma. The corresponding formula is obtained by taking the limit $H_0 \to 0$ in formula (7.20). Of course, one can directly obtain an analogous formula if N_q determined by formula (4.41) is substituted into (7.17). In this connection one can also set $\nu = 0$ in (7.20) since in the absence of a magnetic field the account of collisions has little influence on the effective scattering cross section. As a result $d\sigma$ has the form

$$d\sigma = \left\{\frac{1}{16}\left(\frac{\omega_0}{c}\right)^4 \frac{R_0^4}{q^2}\right\} F(a_0, \chi) \, |G(qR_0, \chi)|^2 \sin^2 \psi_1 \, do, \tag{7.32}$$

where

$$F(a_0, \chi) = \frac{\frac{\pi}{4} a_0^2 \exp(-2a_0^2 \cos^2 \chi) + [a_0 W(a_0 \cos \chi)]^2}{\frac{\pi}{4} a_0^2 \cos^2 \chi \exp(-2a_0^2 \cos^2 \chi) + [1 - a_0 \cos \chi W(a_0 \cos \chi)]^2}. \tag{7.33}$$

Here

$$\left.\begin{aligned} W(x) &= e^{-x^2} \int_0^x e^{t^2} dt, \\ \cos \chi &= \frac{\mathbf{q}\mathbf{V}_0}{qV_0} \end{aligned}\right\} \tag{7.34}$$

and in addition to the notation adopted in (7.20),

$$a_0^2 = \frac{M_i V_0^2}{2\varkappa T}. \tag{7.35}$$

In the ionosphere for $h = 300$, 400, and 700 km we obtain values of a_0 equal to 9.8, 8.6, and 6.2, respectively.

The dependence of the function $F(a_0, \chi)$ on the angle χ is given in Table XX and in Fig. 46 for these values of a_0. Here the value of $F_a(\chi)$ corresponding to some angle χ characterizes the intensity of the scattered wave in the direction $\mathbf{k}'$ for a specified direction $\mathbf{k}$ of the incident wave. Thus, the intensity of the scattered wave varies in the direction of scattering as indicated in Fig. 47. The field of the waves scattered by a moving body represents a smeared out "specular reflection" from the direction $\mathbf{V}_0$ so that the scattering diagram forms in space a double-humped surface of revolution.

TABLE XX. Dependence of the Function $F(a_0, \chi)$ on the Angle χ at Different Heights h in the Ionosphere

χ, deg	h=300km	h=400km	h=700km	χ, deg	h=300km	h=400km	h=700km
20	1.16	1.18	1.20	80	95.70	98.25	50.18
30	1.37	1.39	1.43	81	120.1	108.35	46.09
40	1.77	1.80	1.88	82	135.6	108.74	43.00
50	2.55	2.70	2.79	83	133.4	100.89	39.48
60	4.37	4.80	5.17	84	121.0	90.26	37.06
65	6.37	6.59	8.66	85	106.8	80.86	34.83
70	10.56	11.26	17.70	86	93.4	73.09	33.00
72	12.83	15.05	25.84	87	62.5	67.39	31.75
74	19.08	21.50	37.80	88	79.8	63.77	30.87
76	29.30	34.86	50.04	89	76.6	61.50	30.35
78	50.88	61.2	54.60	90	75.4	60.79	30.18

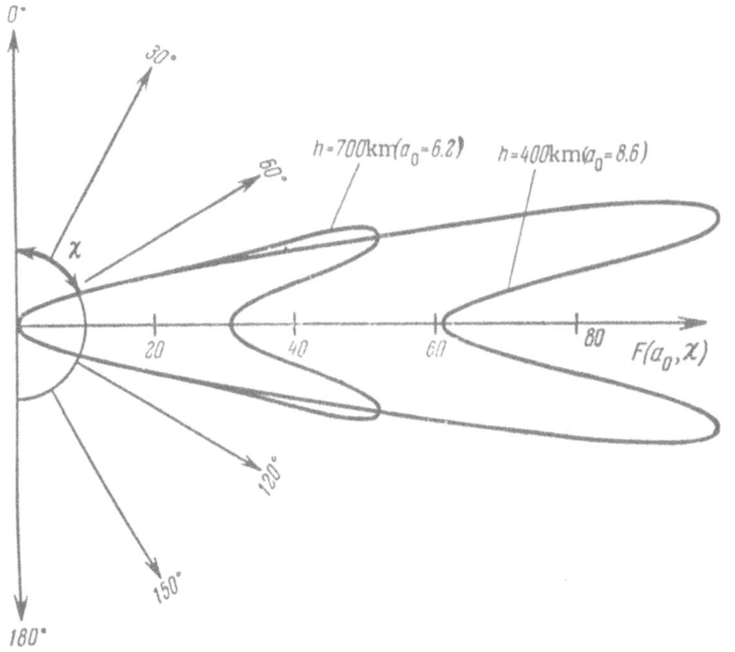

Fig. 46. Dependence of the scattering function $F(a_0, \chi)$ on the angle χ for different values of a_0($h = 300$, 400, 700 km) in the case when $H_0 = 0$.

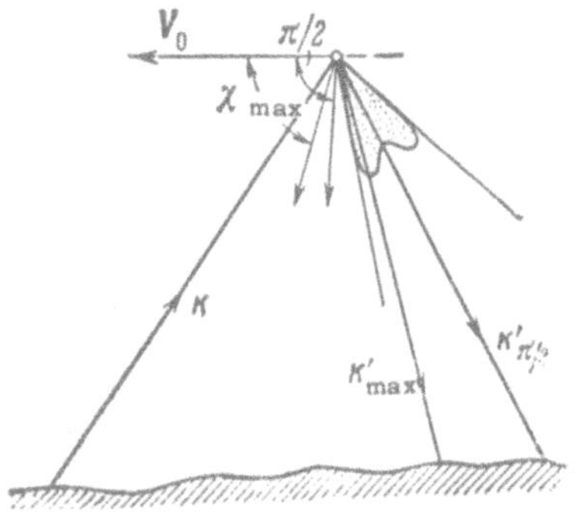

Fig. 47. Schematic diagram of the scattering field in the case when $H_0 = 0$.

TABLE XXI

h, km	300	400	700
$F(a_0, \chi)_{max}$	135.6	108.8	54.6
$F(a_0, \pi/2)$	75.4	60.1	30.2
$(\delta\chi_{max}/2)^\circ$	8	8	12

The axis of revolution of this surface is the velocity $\mathbf{V}_0$. The function $F(a_0, \chi)$ has a minimum exactly at specular reflection, that is, for the vector $\mathbf{k}'$ for which $\mathbf{q} \perp \mathbf{V}_0$. The half-width $\delta\chi/2$ of the effective cross section of this surface, if its is measured from the direction $\chi = \pi/2$ and defined according to the value of χ for which $F(a_0, \chi) = 10^{-1} F(a_0, \chi)_{max}$, varies at different altitudes in the following manner: for $h = 300$, 400, and 700 km $\delta\chi/2$ equals 18, 20, and 30°, respectively.

We see that the width of the lobe increases with increasing altitude. In this connection, with increase of h the values of $F(a_0, \chi)_{max}$, $F(a_0, \pi/2)$, and of the angle $\delta\chi_{max}/2$ between the direction of the maximum F and the direction of $F(a_0, \pi/2)$ vary in accordance with the data presented in Table XXI.

For the indicated values of $F(a_0, \chi)_{max}$ and $F(a_0, \pi/2)$, characterizing the region of maximum scattering only for bodies of very small dimensions, values of the effective cross section $d\sigma/do$ in the direction of maximum scattering are obtained which exceed the corresponding value of $d\sigma_0/do$ for the scattering from a metallic sphere. Thus, for example, when $\lambda = 30$ m and $h = 300$ km for a sphere of radius $R_0 = 0.5$ m, $d\sigma_{max}/do \sim 5 \cdot 10^{-3}$ m^2 and $\beta_0 \sim 7$, i.e., an inhomogeneous formation scatters more strongly than a sphere, and even for $R_0 = 1$ m,$d\sigma_{max}/do \sim 2 \cdot 10^{-2}$ m^2 and $\beta_0 \sim 0.5$, so that the sphere scatters more strongly than the "trail" of the body. We note in conclusion that with decreasing dielectric permittivity ε of the plasma, $d\sigma(\chi) \sim 1/\varepsilon$, as follows from Eq. (7.17). Therefore the scattering must increase strongly if the body is located in a region where ε is close to zero. However, this case requires special consideration, and without a special analysis it is hardly possible to reach any quantitative conclusions with regard to the behavior of $d\sigma$ in the region $\varepsilon \to 0$.

We note that as $\varepsilon \to 0$ one might expect the appearance of an unusual resonance in the scattering, in analogy to the quantum mechanical resonance at a level close to zero [69].

The scattering cross section was calculated in article [77] without account of the influence of electric and external magnetic fields on the motion of the ions. The corresponding integrals were evaluated numerically in [77]. However, the results of [77] are not in agreement with calculations of the scattering cross section obtained with the aid of formula (2.93), which is valid when the electric field is not taken into consideration. We suppose that the reason for this discrepancy is connected with inadequate accuracy of the numerical integrations in article [77].

§30. Scattering by the Trail of a Body Moving in the Lower Ionosphere

In the preceding paragraphs we have evaluated the effective cross section for the scattering of radiowaves by the trail of a body, using expressions for N_q obtained from the kinetic equation. If, however, the body is moving in the lower layers of the ionosphere, where the mean free path of the ions is not very large, then one can evaluate N_q with the aid of the diffusion equations. This has been done in §22. We shall now use the formula obtained in §22 for the effective scattering cross section. First let us evaluate $|N_q|^2$. Multiplying the right side of (5.66) by its complex conjugate, we obtain

$$|N_q|^2 = \frac{\widetilde{I}^2}{(qV_0)^2 + D_a^2(\vartheta_1)\, q^4} = \frac{\widetilde{I}^2}{q^2 V_0^2\left(\cos^2\chi + \frac{q_0^2 D_a^2}{V_0^2}\right)}\ . \tag{7.36}$$

Setting

$$\widetilde{I} = -\pi R_0^2 V_0 N_0$$

and substituting (7.36) into formula (7.17) for $d\sigma$, we find*

$$d\sigma = \frac{R_0^4}{16\varepsilon q^2}\left(\frac{\omega_0}{c}\right)^4 \frac{\sin^2\psi_1}{\cos^2\chi + \frac{q^2 D_a^2}{V_0^2}}\, do, \tag{7.37}$$

where, as above,

$$\cos\chi = \cos\alpha \sin\vartheta_1 + \sin\alpha\cos\vartheta_1\cos\varphi.$$

If the angle ϑ_1 is not too small, so that

$$\sin^2\vartheta_1 \gg \frac{D_\perp^i}{D_\parallel^e},$$

then

$$D_a \approx 2\left(D_\parallel^i \sin^2\vartheta_1 + D_\perp^i \cos^2\vartheta_1\right)$$

and

$$d\sigma = \frac{R_0^4\omega_0^4}{16\, c^4 q^2}\, \frac{\sin^2\psi_1}{\cos^2\chi + \frac{4q^2}{V_0^2}\left(D_\parallel^i \sin^2\vartheta_1 + D_\perp^i \cos^2\vartheta_1\right)^2}\, do. \tag{7.38}$$

---

* We note that for motion in the lower ionosphere, the question may involve the scattering of radiowaves of such low frequency that condition (7.1) is already not satisfied. In this case formula (7.17) is incorrect. However, we shall not list here the very complicated formula for $d\sigma$ when $\omega \sim \omega_H$.

When the body is moving close to the direction of the magnetic field ($\alpha \ll 1$), the maximum value of $d\sigma$ occurs for $\vartheta_1 = 0$ and equals

$$d\sigma_{max} = \frac{R_0^4 \omega_0^4 V_0^2}{64\, c^4 q^4} \frac{\sin^2 \psi_1}{(D_\perp^i)^2} do = \frac{R_0^4}{2^{10} \varepsilon^2} \left(\frac{\omega_0}{\omega}\right)^4 \frac{V_0^2}{(D_\perp^i)^2} \sin^2 \psi_1 \, do. \tag{7.39}$$

for $\sin^2 \vartheta_1 \ll \frac{D_\perp^i}{D_\parallel^e}$,

$$D_a = \frac{2 D_\perp^i D_\perp^e}{D_\perp^i + D_\perp^e}. \tag{7.40}$$

When $\omega_H \gg \nu_{en}$ (see §22), $D_\perp^i \gg D_\perp^e$ and $D_a \approx 2D_\perp^e$, the maximum value $d\sigma_{max}$ under condition (7.40) is given by

$$d\sigma_{max} = \frac{R_0^4 \omega_0^4 V_0^2}{64\, c^4 q^4} \frac{\sin^2 \psi_1}{(D_\perp^e)^2} = \frac{R_0^4}{2^{10} \pi_0 \varepsilon} \left(\frac{\omega_e}{\omega}\right)^4 \frac{V_0^2}{(D_\perp^e)^2} \sin^2 \psi_1 do, \tag{7.41}$$

i.e., it is larger than (7.39) since it contains a small quantity, the square of the coefficients $D_\perp^e$ for transverse diffusion of electrons, in the denominator.

CHAPTER VIII

DISTRIBUTION OF PARTICLES IN A CENTRAL FORCE FIELD

§ 31. Introductory Remarks

Only a rapidly moving body of large size ($R_0 \gg D$, $V_0 \gg v_i$) was considered in the preceding chapters. In this special case the electric charge of the body does not have very great influence on the structure of the disturbed zone because the potential of the field at distances on the order of the Debye radius from the body's surface rapidly goes down to a value of order $\varkappa T/e$. The influence of the potential $\varphi \sim \varkappa T/e$ is small because the energy $MV_0^2/2$ of the ions incident on the body is large in comparison with $e\varphi$. It is understood that for a slowly moving body ($V_0 \lesssim v_i$) and particularly for a body of small dimensions ($R_0 \lesssim D$) these special conditions are now not satisfied and the influence of the body's electric charge becomes very important and even decisive.

In this connection, a field which attracts particles – ions or electrons* – plays an especially important role. In such a field particles having a comparatively small velocity near the attracting body revolve around it, i.e., they move in finite orbits. Such particles accumulate near the body. Because of this the concentration of attracted particles near the body may increase strongly. For example, in the case of the equilibrium distribution, the concentration of particles at the point $\mathbf{r}$ in the vicinity of a body at rest is given by

$$N(r) = N_0 \exp\left[-\frac{U(r)}{\varkappa T}\right], \tag{8.1}$$

where N_0 is the concentration in the undisturbed region far from the body, T is the temperature of the medium, $\varkappa$ is Boltzmann's constant, and $U(\mathbf{r})$ is the potential energy of a particle in the field. Obviously, U vanishes at infinity. It is clear that for sufficiently large values of the quantity $-U(\mathbf{r})/\varkappa T$, the concentration of particles at the point $\mathbf{r}$ is much larger than N_0. In this connection, $N(\mathbf{r})$ is primarily determined by the particles undergoing finite motion, so that the number of particles moving "infinitely," that is, having a positive total energy, $\mathcal{E} = (1/2)Mv^2 + U(\mathbf{r}) > 0$, is negligibly small for $-U(\mathbf{r})/\varkappa T \gg 1$.

However, the attracting body itself violates the equilibrium distribution (8.1) of the particle concentration, if the particles incident on its surface are absorbed (neutralized). Therefore, if the gas is sufficiently rarefied, so that the mean free path l of the particles is large in comparison with the dimensions of the region disturbed by the body, then the character of the particle distribution even in the vicinity of a body at rest may be significantly nonequilibrium. Here the concentration at the point $\mathbf{r}$ may strongly differ both from the equilibrium concentration (8.1) as well as from the concentration of the unperturbed gas N_0. Under these conditions, of course, the velocity distribution of the particles also differs from the equilibrium (Maxwellian) distributions, and it is necessary, as in the case considered above of a rapidly moving body, to first determine the particle distribution function in phase space (in the space of velocities and coordinates), using the kinetic equation for this.

*Generally speaking, a body in a plasma is negatively charged so that the particles attracted to the body are ions, and electrons are repelled. It is possible, however, to specially establish a positive (relative to the plasma) potential φ_0 on the body; then the particles which are attracted will be electrons. In the present chapter, as in the following chapters, we shall not restrict the value of the potential φ_0 of the electric field on the body's surface. Here it is assumed that $H_0 = 0$.

An essential feature which arises in connection with this problem is the presence of two processes for leveling out the disturbances in the particle distribution function produced by the body. One of these is a consequence of the continuous influx of particles from undisturbed regions, that is, from regions very far from the body. The characteristic time for this leveling out process is $t \sim R/v$, where R is the effective size of the disturbed region, v is the average velocity of the particles. The second leveling out process is caused by collisions between particles. The characteristic time for this process is, of course, the time of free flight, $1/\nu$. If collisions are sufficiently infrequent, then $1/\nu \gg R/v$ and leveling out at the expense of particles coming in from infinity plays the major role, which was taken into consideration previously. However, this is only valid for particles undergoing "infinite" (i.e., unbounded) motion. Particles undergoing finite motion, without collisions, cannot be interchanged with particles arriving from infinity since the latter cannot be captured in finite orbits, provided they do not give up part of their energy. Hence it is clear that however infrequent collisions may be, the number and character of the distribution of particles in finite orbits is determined just by them. Moreover, as already indicated above, due to the particles having finite orbits, the concentration near an attracting body may increase strongly. Therefore collisions, even if they are very rare, in essence, play, an important role in those cases when finite (bounded) motion is essential. This circumstance constitutes at the same time the basic difficulty of the problems under consideration, since the solution of the kinetic equation with account of collisions is most complicated.

The electric field in the vicinity of the body is made up of the field created by the body's charge and the field of the space charge which appears in the disturbed zone. If the moving body is a sphere, then the total field possesses axial symmetry around the direction of motion. In the case of a sphere at rest, the field is a central field; the same thing is, of course, valid to a first approximation when the body is moving slowly if ($V_0 \ll v_i$). In addition, the electric field is a central field (it is spherically symmetric) for arbitrary velocity of motion of the body if one can neglect the field of the space charge in comparison with the field due to the body's charge; these conditions are realized, in particular, for the motion of a small charged sphere, whose radius is much smaller than the Debye radius, in a plasma. This case is investigated in [78], where the concentration and flux of unbounded particles in the vicinity of a small moving body are examined. In virtue of what has been said, it is clear that the case of a spherically symmetric field is often realized experimentally. General expressions describing the particle distribution in a rarefied gas located in a spherically symmetric field are derived in the present chapter. In the following chapters, these expressions are used to clarify the character of the plasma disturbances in the vicinity of a body at rest.

The question of the distribution of charged particles and the electric field in the vicinity of a body located in a rarefied plasma are of significant interest not only for problems associated with investigations of the ionosphere and interplanetary media, but also for the theory of probe measurements, which are widely used in laboratory plasma experiments. Therefore, a whole series of articles (see for example, [39-43, 60-62]) have been devoted to their theoretical investigation. The basic goal of these articles is to calculate the flux of particles incident on a charged body in a plasma, since measurements of the flux are the initial steps associated with processing the results of measurements according to the method of Langmuir probes. However, in order to determine the flux of the particles being attracted to the body, it is necessary to know the behavior of the field, and consequently also the distribution of electrons and ions in the vicinity of the body. At the same time, a rigorous solution of this problem encounters substantial difficulties. Therefore the authors of previous articles were compelled to use a series of insufficiently justified approximations.

Thus the widely known article by Langmuir [60, 39] is based on the assumption that a region of space charge, completely screening the field of the body, is formed near a charged body placed in a plasma. Here the potential of the field vanishes on the external boundary of this region, so that the plasma beyond the limits of this region is not disturbed. He also assumed that in the space charge region there are no particles being repelled by the field of the body, and the attracted particles move only in the radial direction. The thermal motion of these particles was not considered, and also particles undergoing bounded motion were neglected. Under such assumptions it is easy to express the concentration of particles being attracted to the space charge region in terms of the potential of the field. The equation obtained for the field was solved by Langmuir and Blodgett [39], and thus was found the field distribution in the space charge region, and the particle flux onto the body's surface was determined.

Later on, however, it was explained (see [40], [61], and [83]) that the electric field penetrates not only into the space charge region, but also into the quasi-neutral plasma for considerable (in comparison with the dimensions of the body) distances. The field distribution in the region of the quasi-neutral plasma was investigated by Bohm, Burhop, and Messy [40]. Here, as before, the thermal motion of the particle being attracted was completely neglected. In the quasi-neutral plasma region, this assumption is a rough approximation, since here the energy of a particle in the field does not exceed its thermal energy. The results of article [40] are more correct under conditions when the temperature of the particle being repelled – electrons (it is assumed that the body is negatively charged) – is much higher than the temperature of the particles attracted – ions. It is precisely for this case that Bohm, Burhop, and Messy obtain an important result, namely, they show that for $T_e \gg T_i$ the flux of ions only depends on the temperature of the electrons.

Articles by Wensel [61], Kagan and Perel' [43], Allen, Boyd, and Raynolds [62], and Bernstein and Rabinowitz [41] are devoted to the further development of the theory in a "heterotemperature" plasma. In particular, in articles [62, 41] the problem was solved numerically without separation of the disturbed region in the vicinity of the body into a quasi-neutral plasma region and a space charge region. In this connection, in the most systematic article by Bernstein and Rabinowitz [41] the problem was formulated within the framework of kinetic theory. To be sure, in order to simplify the calculations of the ion concentration, it was assumed that the ions in the undisturbed plasma do not have a Maxwellian but a monoenergetic velocity distribution, that is, $f_i = C\delta(\mathscr{E} - \mathscr{E}_0)$, where δ is a delta function, $\mathscr{E}_0 = (3/2)kT_i$ is the average energy of the ions. The calculation was carried out here only for $T_e \gg T_i$ ($T_e = 10\,T_i$, $20\,T_i$, and $100\,T_i$), since for $T_e \sim T_i$ there is no basis for the last assumption.

Thus, the problem of the structure of the disturbed zone in the vicinity of a charged body in a plasma remained to a considerable extent unsolved. In the first place, there was no kind of consistent solution of the problem in a single-temperature plasma when $T_e \sim T_i$. In the second place, for the investigation of the space charge region, it was assumed that the particles move only in a radial direction. This assumption turns out to be valid in the case of a large body whose dimension R_0 is much larger than the Debye radius in the plasma. For a small body, however, when $R_0 \ll D$, other conditions are realized. In the third place, in all of the earlier articles the particles undergoing bounded motion in the vicinity of the charged body were not considered in the calculation, and their role may be essential or even decisive. Finally, the stability of the disturbed region in the vicinity of the body was not investigated.

§ 32. Properties of the Motion of Particles in a Central Force Field

First let us turn to an examination of the kinetic equation, which is useful to indicate certain special features of the motion of particles in a central force field (spherically symmetric field) which are important for the following presentation.

A particle of mass M in a central force field $U(\mathbf{r})$ always moves in the one plane $\mathbf{rv}$, where $\mathbf{v}$ is the velocity of the particle and $\mathbf{r}$ is the radius vector drawn from the force center O (see Fig. 48). Its position and velocity in the plane of motion are characterized by the distance r from the center, the angle θ_1, and the two projections of the velocity v_r and v_θ on the direction of the radius vector $\mathbf{r}$ and on the direction orthogonal to it. During the motion of the particle, the following quantities are always conserved: the angular momentum with respect to the center of force

$$\mathscr{M} = Mrv_\theta, \tag{8.2}$$

and the total energy

$$\mathscr{E} = \frac{M}{2}(v_r^2 + v_\theta^2) + U(r) = \frac{Mv_r^2}{2} + \frac{\mathscr{M}^2}{2Mr^2} + U(r). \tag{8.3}$$

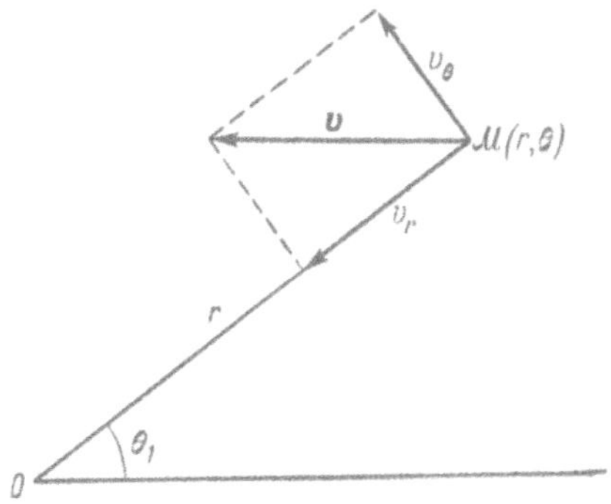

Fig. 48. Schematic diagram of the velocity components of a particle.

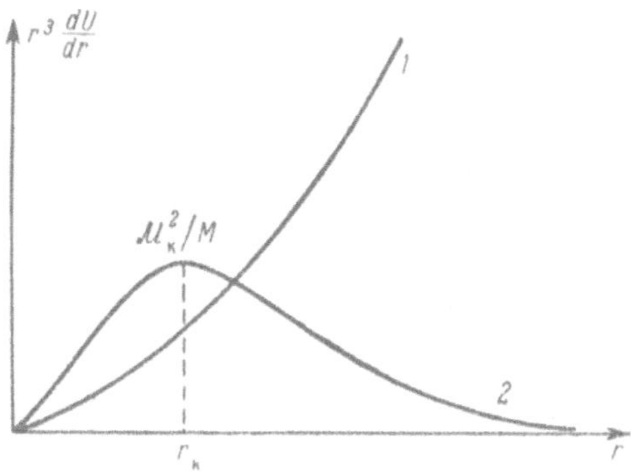

Fig. 49. Illustrating the discussion of the properties of a particle's potential energy U(r).

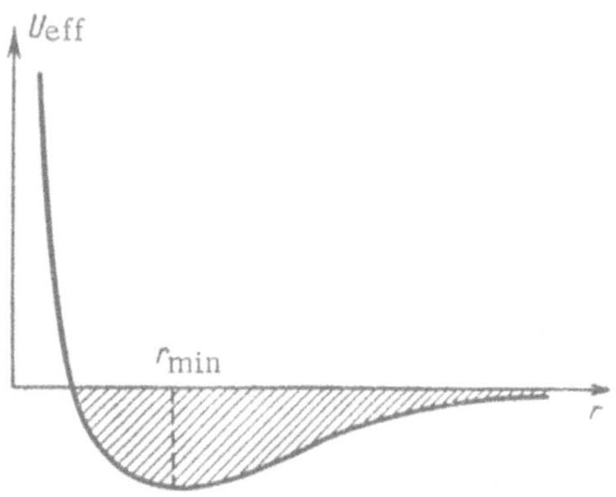

Fig. 50. Form of the effective potential energy $U_{eff}(r)$ for a slowly decreasing field (case 1). The region of bounded motion is crosshatched.

Here M is the mass of the particle and U(r) is the potential energy of the particle in the field. It is natural to assume that $U \to 0$ at infinity. Then, since for an attractive force $F = dU/dr < 0$, an attractive potential is always less than zero, that is, $U(r) < 0$; on the other hand, for a repulsive potential $U(r) > 0$. It is evident from Eq. (8.3) that the character of the changes in the velocity component v_r depends completely on the properties of the effective potential energy,

$$U_{eff}(r) = \frac{\mathcal{M}^2}{2Mr^2} + U(r). \tag{8.4}$$

The basic features of the motion of particles in a central force field therefore become clear if the expression $U_{eff}(r)$ is analyzed. In particular, it is important for us to know how the regions of bounded and unbounded (infinite) motions are separated since, as indicated above, the particle distribution function may have a completely different form in these two regions.

Attractive Potential. An attractive potential has $U(r) < 0$; consequently, the effective potential energy $U_{eff}(r)$ may be both smaller than or larger than zero. Let us consider different characteristic cases of variation of the potential U(r).

Case 1. If U(r) decreases (in absolute value) with increasing r sufficiently slowly so that the decrease of U(r) does not go faster than $1/r^2$, then for any value of the angular momentrum $\mathcal{M}$ the equation

$$\frac{dU_{eff}}{dr} = -\frac{\mathcal{M}^2}{Mr^3} + \frac{dU}{dr}, \quad \text{i.e.,} \quad r^3 \frac{dU}{dr} = \frac{\mathcal{M}^2}{M} \tag{8.5}$$

has one root. In turn, in this case the function $r^3 (dU/dr)$ increases monotonically with increasing r (see curve 1 in Fig. 49), so that Eq. (8.5) always has only one root. The function $U_{eff}(r)$ has, correspondingly, one minimum at the point r_{min}, which is determined by relation (8.5) (see Fig. 50). Particles undergoing bounded motion in the case under consideration have a negative total energy,

$$\mathscr{E} = \frac{Mv_r^2}{2} + U_{eff}(r) < 0. \tag{8.6}$$

The region of bounded motion is crosshatched in Fig. 50. Conversely, the total energy of particles undergoing unbounded motion is always positive. Consequently, in the space of the velocities v_r and v_θ at an arbitrary point r, the regions of bounded and unbounded motions are divided by the curve

$$v_r^2 + v_\theta^2 = -\frac{2}{M} U(r), \tag{8.7}$$

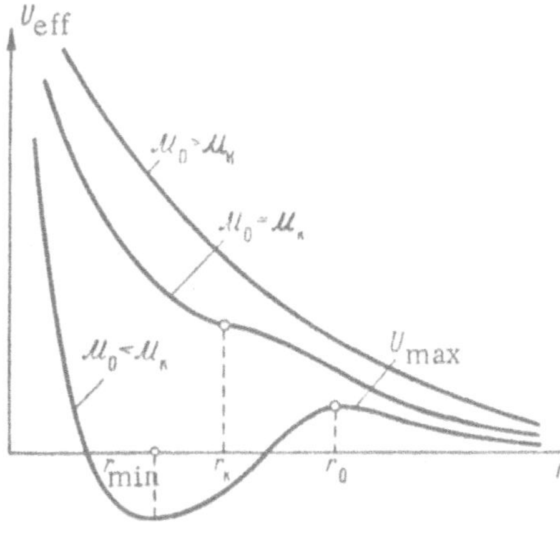

Fig. 51. Form of the effective potential energy $U_{eff}(r)$ (case 2) for different values of the angular momentum.

that is, the "dividing" curve in the space of velocities v_r and v_θ is in this case a circle of radius $[-(2/M)U(r)]^{1/2}$.

Case 2. Let U(r) decrease (in absolute value) with increasing r for small values of r more slowly than $1/r^2$, and for larger values of r – more rapidly than $1/r^2$. In this case the function $r^3(dU/dr)$ is now not a monotonically increasing function of r; this is clear from curve 2 in Fig. 49. Equation (8.5) now has two roots for small values of the angular momentum $\mathcal{M}$ smaller than the critical value $\mathcal{M}_K$ defined below. Accordingly, the effective potential energy $U_{eff}(r)$ has not only a minimum but also a maximum. The dependence of U_{eff} on r is shown in Fig. 51. The minimum of the effective potential energy is located at the point $r = r_{min}$, which is always situated closer to the center than the maximum point r_0. For very small values of the angular momentum $\mathcal{M}$, as is clear from Fig. 49, the point r_{min} tends to zero, but the value of r_0 on the contrary increases without limit. When the value of the angular momentum $\mathcal{M}$ increases, approaching the value $\mathcal{M}_K$, then the distance between the points r_{min} and r_0 on the contrary becomes still smaller. When $\mathcal{M} = \mathcal{M}_K$ both roots merge at the point r_K. The value of r_K is determined from the condition

$$\frac{d}{dr}\left(r^3\frac{dU}{dr}\right)_{r_K} = 0$$

or

$$3\frac{dU}{dr}(r_K) + r_K\frac{d^2U}{dr^2}(r_K) = 0. \qquad (8.8)$$

The distance r_K is uniquely determined by the potential U(r). The point $r = r_{min}$, where the effective potential energy has its minimum, is always located closer to the center than r_K. Conversely, the distance r_0 where U_{eff} has its maximum is always larger than r_K.

The critical value of the angular momentum, $\mathcal{M} = \mathcal{M}_K$, as is clear from (8.5), is given by

$$\mathcal{M}_K = \sqrt{Mr_K^3\frac{dU}{dr}(r_K)}. \qquad (8.9)$$

The value of $\mathcal{M}_K$ also is uniquely determined by the potential U(r). If $\mathcal{M} > \mathcal{M}_K$, then the effective potential energy $U_{eff}(r)$ monotonically increases with decreasing r, as can be seen from Fig. 51. The noted features in the behavior of the effective potential energy significantly manifest themselves in the character of the motion of the particles. In the first place, as is clear from Fig. 51, only particles having angular momentum $\mathcal{M} < \mathcal{M}_K$ can perform bounded motion. In the second place, the total energy of a particle moving in a finite orbit can be both negative as well as positive. It is only necessary that the condition

$$\mathcal{E} = \frac{Mv_r^2}{2} + U_{eff}(r) \leqslant U_{max} \text{ for } r \leqslant r_0, \qquad (8.10)$$

be satisfied, where $U_{max} = U_{eff}(r_0)$ is the value of the effective potential energy at the maximum point. Of course, the value of U_{max} depends on the angular momentum of the particle. A particle performing bounded motion is reflected in the region of a sharp gradient of the potential U(r) even if its total energy $\mathcal{E} > 0$. Similar reflection also occurs for particles performing unbounded motion. It is clear from Fig. 51 that particles coming

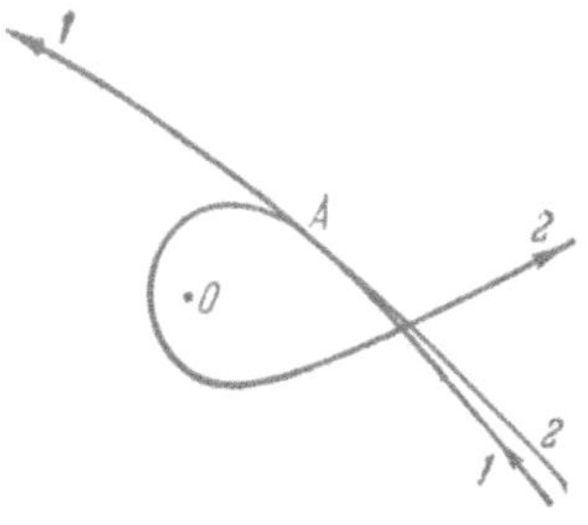

Fig. 52. Trajectories of particles having energies $\mathscr{E} < U_{max}$ (trajectory 1) and $\mathscr{E} > U_{max}$ (trajectory 2) at the point A.

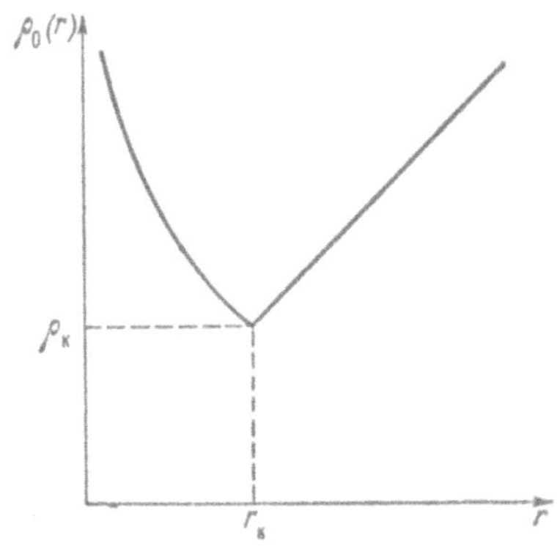

Fig. 53. Dependence of ρ_0 on r in the case of the Debye potential (8.13).

in from infinity and having angular momenta $\mathscr{M} < \mathscr{M}_K$ and energies $\mathscr{E} < U_{max}$ are reflected at a point $r_{tp} \geq r_0 > r_K$. Thus, the point r_{tp} is the turning point of the particle. Naturally, the value of r_{tp} is the minimum distance of the particle's trajectory from the force center. With increasing energy of the particle, when $\mathscr{E} < U_{max}$, the turning point approaches r_0 for a fixed value of the angular momentum $\mathscr{M}$. For $\mathscr{E} \geq U_{max}$ the turning point is transferred by a jump into the region of small values of $r < r_{min}$. Here the trajectory of the particle changes completely, as is evident from Fig. 52, where the trajectories of two particles, having the same angular momenta $\mathscr{M}$, but different values of the energy, are shown: $\mathscr{E} < U_{max}$ (trajectory 1) and $\mathscr{E} > U_{max}$ (trajectory 2). Thus, when $\mathscr{E} = U_{max}$ the character of the trajectory and the position of the turning point change very rapidly for particles undergoing unbounded motion and having angular momenta $\mathscr{M} < \mathscr{M}_K$. With increasing value of the angular momentum $\mathscr{M}$ and consequently with decreasing distance between the two roots, r_{min} and r_0, of Eq. (8.5), the magnitude of the jump of the turning point decreases. For $\mathscr{M} \gg \mathscr{M}_K$, when both roots combine, the position of the turning point no longer experiences any jump but changes continuously with change of the energy or angular momentum of the particle.

Now let us establish the shape of the curve in the space of velocities v_r and v_θ which divides, at a specified point r, the regions of bounded and unbounded motions. As is clear from Eq. (8.10), particles whose total energy

$$\mathscr{E} = \frac{Mv_r^2}{2} + \frac{Mv_\theta^2}{2} + U(r) \leqslant U_{max}. \tag{8.10a}$$

perform bounded motion. Here U_{max} is the value of the effective potential energy U_{eff} at the point r_0, that is,

$$U_{max} = U(r_0) + \frac{\mathscr{M}^2}{2Mr_0^2} = U(r_0) + \frac{Mv_\theta^2}{2}\frac{r^2}{r_0^2}, \tag{8.4a}$$

$$r_0^3 \frac{dU}{dr_0} = \frac{\mathscr{M}^2}{M} = Mv_\theta^2 r^2, \tag{8.5a}$$

where r_0 is the larger of the two roots of the latter equation. The maximum point r_0 and the value of U_{max} itself vary depending on the angular momentum of the particle. Therefore, the curve dividing the regions of bounded and unbounded motions in the velocity space of v_r and v_θ for a given value of the point r also depends on the angular momentum of the particle, and is the parametric solution of Eqs. (8.10a), (8.4a), and (8.5a) with the parameter r_0, that is, it is given by the expressions

$$v_r^2 = -\frac{2}{M}[U(r) - U(r_0)] - \frac{r_0}{M}\frac{dU}{dr}(r_0)\left(\frac{r_0^2}{r^2} - 1\right), \tag{8.11a}$$

$$v_\theta^2 = \frac{r_0^3}{Mr^2}\frac{dU}{dr}(r_0). \tag{8.11b}$$

The parameter r_0 in Eqs. (8.11) varies from a certain minimum value, which we shall denote by ρ_0, to infinity. Since a decrease of r_0 corresponds to a decrease of the velocity v_r^2, it is clear that the minimum value of r_0 occurs when $v_r^2 = 0$. Therefore, ρ_0 is the maximum root of Eq. (8.11a) when $v_r^2 = 0$, that is

$$U(\varrho_0) - U(r) - \frac{\varrho_0}{2}\frac{dU}{dr}(\varrho_0)\left(\frac{\varrho_0^2}{r^2} - 1\right) = 0. \tag{8.12}$$

Hence it is clear that ρ_0 depends on r. In particular, $\rho_0 = r$ is always a root of Eq. (8.12). For $r > r_K$ this root is unique. On the other hand, if $r < r_K$ then Eq. (8.12) has another root $\rho_0(r)$ which is always larger than r_K (since ρ_0 is one of the values of r_0, but r_0 is the maximum root of Eq. (8.5) and r_0 is always greater than or equal to r_K). In particular, for small values of r ($r \to 0$), Eq. (8.12) takes the form

$$-r^2 U(r) = \frac{\varrho_0^3}{2}\frac{dU}{dr}(\varrho_0). \tag{8.12a}$$

Inasmuch as the potential energy increases (in absolute value) more slowly than $1/r^2$ as $r \to 0$, then $r^2U(r) \to 0$. On the other hand, for larger values of r the potential falls off more rapidly than $1/r^2$, and therefore the right side of Eq. (8.12a) tends to zero as $\rho_0 \to \infty$. Thus, the function $\rho_0(r) \to \infty$ as $r \to 0$. For a Debye potential, $U(r) = -U_0(D/r)\exp(-r/D)$, the dependence of ρ_0 on r is shown in Fig. 53.

It is clear from (8.11) that the curve $v_r(v_\theta)$, separating the regions of bounded and unbounded motions, at an arbitrary point r now essentially depends not only on the value of U at the point r, as it would in case 1 considered above of a slowly decreasing potential U(r), but it also depends on the overall behavior of the potential U from infinity to the point $\rho_0(r)$. In this connection, $v_\theta = 0$ and $v_r = [-(2/M)U(r)]^{1/2}$ as $r \to \infty$. With increase of v_θ, the function $v_r(v_\theta)$ decreases; however, the curve $v_r(v_\theta)$ goes higher than the dividing curve (8.7), $v_r = [-(2/M)U(r) - v_\theta^2]^{1/2}$ (see the dot-dashed line in Fig. 54); the value of v_r tends to zero for $v_\theta = v_{\theta 0} = (\rho_0^3/Mr^2)(dU/dr)(\rho_0)$.

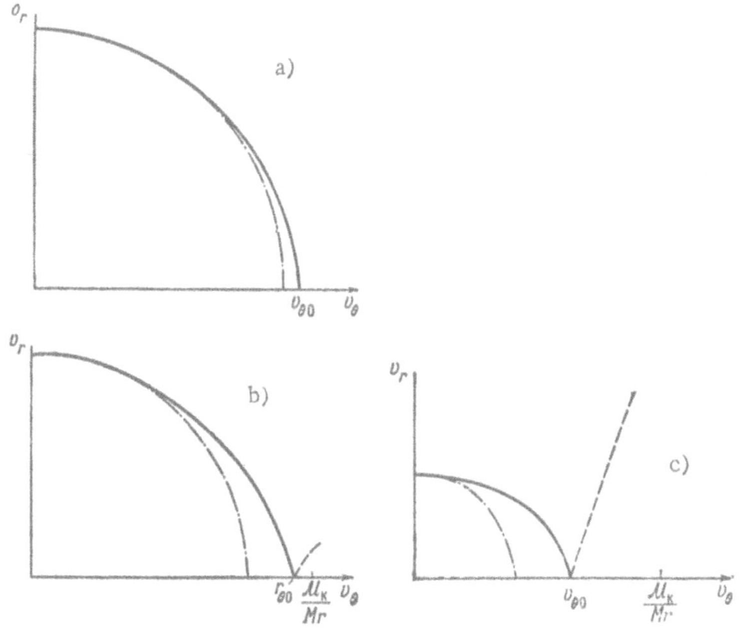

Fig. 54. Boundary of the regions of bounded and unbounded motions for different values of r for the Debye potential (8.13): a) r = D; b) r = 2D; c) r = 4D. The dot-dashed line indicates the dividing line (circle) in the case of a slowly decreasing potential.

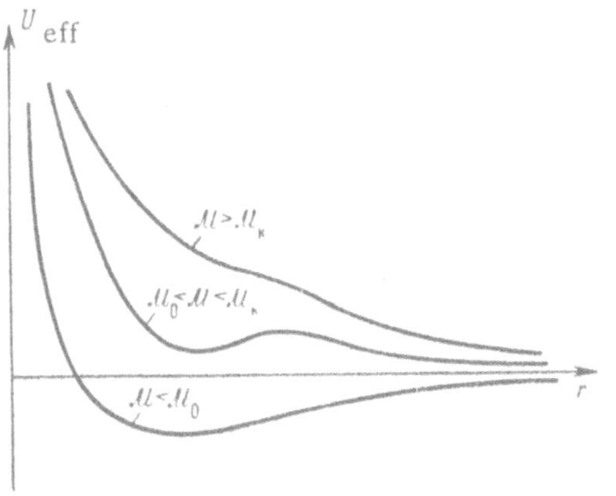

Fig. 55. Form of $U_{eff}(r)$ (case 3) for different values of the angular momentum $\mathcal{M}$.

We note that for $r \leq r_K$ Eqs. (8.11a) and (8.11b) have a solution only when $v_\theta \leq v_{\theta 0}$. If, however, $r > r_K$, then a solution exists even when $v_\theta > v_{\theta 0}$. To be sure, this solution is obtained now for values of the parameter r_0 smaller than $\rho_0(r) = r$, and consequently the corresponding part of the curve $v_r(v_\theta)$ indicated by the dashed lines in Figs. 54b and c now does not separate the regions of bounded and unbounded motions, since bounded motion is impossible for $v_\theta > v_{\theta 0}$. However, the particles performing unbounded motion here are not equivalent: some of them, as we saw above, have turning points in the region of large values $r > r_0 \geq r_K$, and the remainder have turning points in the region of small values $r < r_{min} \leq r_K$. The indicated dashed curve exactly separates the two groups of particles.

As an example, let us consider the Debye potential

$$U(r) = -U_0 \frac{D}{r} \exp\left(-\frac{r}{D}\right), \tag{8.13}$$

which falls off with increasing r in just the right fashion: for small values of r it decreases more slowly than $1/r^2$, and for larger values of r it decreases more rapidly than $1/r^2$. Solving Eq. (8.8), we find that in this case

$$r_K = \frac{1+\sqrt{5}}{2} D \simeq 1.61D$$

and

$$\mathcal{M}_K = \left(\frac{9}{4}+\sqrt{5}\right)^{\frac{1}{2}} \exp\left(-\frac{1+\sqrt{5}}{4}\right) D\sqrt{MU_0} \simeq 0{,}94 D\sqrt{MU_0}.$$

The parametric solutions of Eqs. (8.11a) and (8.11b) here assumes the form

$$\frac{Mv_r^2}{U_0} = 2\frac{D}{r} e^{-\frac{r}{D}} - \frac{D^2}{r^2} e^{-x}\left[x(x+1) - \frac{r^2}{D^2}\frac{x-1}{x}\right], \quad \frac{Mv_\theta^2}{U_0} = \frac{D^2}{r^2} x(x+1) e^{-x},$$

where the parameter x varies from $\rho_0(r)$ to ∞. The shape of the curve $\rho_0(r)$ is given in Fig. 53, and the shape of the dividing curves for the values $r = D$ (i.e., $r < r_K$), $r = 2D$, and $r = 4D$ (i.e., for $r > r_K$) are shown in Fig. 54.

Case 3. Let U(r) decrease for large values of r ($r \to \infty$) exactly like $1/r^2$, and for small values of r it falls off initially more slowly and then more rapidly than $1/r^2$. Just this case is of most interest for the questions considered by us in this book. In fact, as was shown above in Chapter IV (also see Chapter IX), at very large distances from the body, when $r \gg R_0$, $r \gg D$, the electric field always falls off exactly like $1/r^2$. At distances comparable with the Debye radius, the field falls off exponentially abruptly, that is, more rapidly than $1/r^2$, because of Debye screening. Finally, if the dimension of the body is small in comparison with the Debye radius, screening is not important in its immediate vicinity ($r < D$), and the field fall falls off according to Coulomb's law, that is, more slowly than $1/r^2$.

Although this case is close to the case considered above, there is a certain difference between them. In fact, from the behavior of the effective potential energy $U_{eff}(r)$, shown for various values of the angular momentum $\mathcal{M}$ in Fig. 55, it is clear that the function $U_{eff}(r)$ is absent now not only for values of $\mathcal{M} > \mathcal{M}_K$, but also for values of $\mathcal{M} < \mathcal{M}_0$, where $\mathcal{M}_0$ is some chosen value of the angular momentum. It is easy to determine, considering that for large values of r the potential energy U(r) falls off exactly as $1/r^2$. In fact, $U(r) = -(C/r^2) + U_1(r)$, where $U_1(r)$ is a function which decreases in absolute value with increasing r faster than $1/r^2$. Substituting this expression into (8.4), we have

$$U_{\text{eff}}(r) = \left(\frac{\mathcal{M}^2}{2M} - C\right)\frac{1}{r^2} + U_1(r).$$

From here it is clear that a maximum or minimum of the effective potential energy is possible only when

$$\mathcal{M} > \mathcal{M}_0 = \sqrt{2MC}.$$

The regions of bounded and unbounded motions for values of the angular momentum $\mathcal{M} < \mathcal{M}_0$, as in case 1, are separated by the curve $\mathcal{E} = 0$, that is,

$$v_r^2 + v_\theta^2 = -\frac{2}{M}U(r). \tag{8.7}$$

For $\mathcal{M} > \mathcal{M}_0$ the boundary of the regions of bounded and unbounded motions is determined, as in case 2, by conditions (8.11). If the field falls off more rapidly than $1/r^2$ for large values of r, then $C = 0$, $\mathcal{M}_0 = 0$, and case 3 considered here coincides with case 2. If the field decreases exactly like $1/r^2$, then

$$\mathcal{M}_0 = \sqrt{2MC} = \sqrt{-2Mr^2U(r)},$$

that is, the boundary of the regions of bounded and unbounded motions is determined by condition (8.7), the same as in case 1.

Repulsive Potential. In this case U(r) is a monotonic positive function ($U(r) \geq 0$). Consequently, $U_{\text{eff}}(r)$ is also a positive monotonically decreasing function of r, and the motion of particles in such a field is always unbounded.

§ 33. Solution of the Kinetic Equation

a. The Kinetic Equation in the Case of a Central Force Field. In a central force field U(r), the distribution function naturally depends on r, θ_1, and the velocities v_r and v_θ shown in Fig. 48. It is important, however, that this system of coordinates is not an inertial system – the velocities v_r and v_θ change in time because of the rotation of the coordinate system affiliated with motion of a particle along its trajectory. With account of this observation, the kinetic equation has the form

$$\frac{\partial f}{\partial t} + v_r\frac{\partial f}{\partial r} + \frac{v_\theta}{r}\frac{\partial f}{\partial \theta_1} + \left(-\frac{1}{M}\frac{dU}{dr} + \left[\frac{dv_r}{dt}\right]\right)\frac{\partial f}{\partial v_r} + \left[\frac{dv_\theta}{dt}\right]\frac{\partial f}{\partial v_\theta} + Y = 0, \tag{8.14}$$

where Y is the collision integral (see §3), and the terms (dv_r/dt) and (dv_θ/dt) describe the change of the velocities v_r and v_θ caused by rotation of the coordinate system.

In order to determine (dv_r/dt) and (dv_θ/dt), let us examine how the velocities v_r and v_θ are changed if a particle moves along the trajectory from the point $\mathbf{r}$ to the point $\mathbf{r} + d\mathbf{r}$ during the time interval dt (see Fig. 56a). Here the coordinate system is rotated through the angle $d\theta_1$, where $d\theta_1/dt = v_\theta/r$. During rotation of the coordinate system through an angle $d\theta_1$, the projections of the velocity $\mathbf{v}$ on the directions of v_r and v_θ (Fig. 56b) are also changed. As a result,

$$v_r = v_r' + v_\theta' \, d\theta_1 + O(d\theta_1^2),$$
$$v_\theta = v_\theta' - v_r' \, d\theta_1 + O(d\theta_1^2).$$

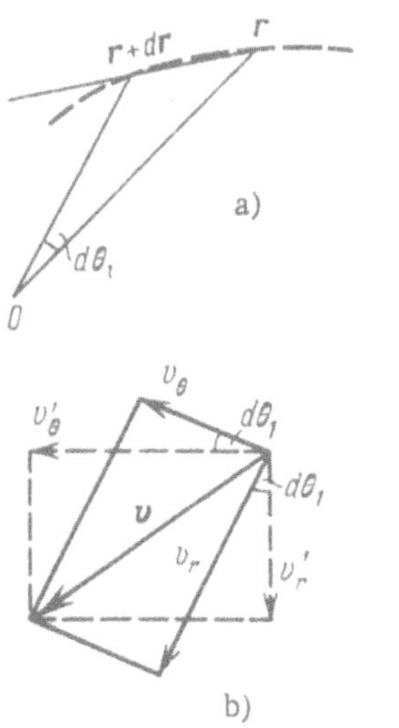

Fig. 56. Change of the velocities v_r and v_θ during motion of the particle along its tracjectory.

as is evident from Fig. 56b. Considering now that $v_r = v_r(t+dt)$, $v_\theta = v_\theta(t+dt)$, but $v'_r = v_r(t)$ and $v'_\theta = v_\theta(t)$, we find

$$\left[\frac{dv_r}{dt}\right] = \frac{v_r(t+dt) - v_r(t)}{dt} = v_\theta \cdot \frac{d\theta_1}{dt} = \frac{v_\theta^2}{r},$$

$$\left[\frac{dv_\theta}{dt}\right] = \frac{v_\theta(t+dt) - v_\theta(t)}{dt} = -\frac{v_r v_\theta}{r}.$$

Thus, the kinetic equation in a central force field has the following form in terms of the variables r, θ_1, v_r, and v_θ:

$$\frac{\partial f}{\partial t} + \frac{v_\theta}{r}\frac{\partial f}{\partial \theta_1} + v_r\frac{\partial f}{\partial r} + \left(\frac{v_\theta^2}{r} - \frac{1}{M}\frac{dU}{dr}\right)\frac{\partial f}{\partial v_r} - \frac{v_\theta v_r}{r}\frac{\partial f}{\partial v_\theta} + Y = 0. \tag{8.15}$$

b. Steady-State Solution of the Kinetic Equation in the Absence of Collisions. The stationary (steady-state) distribution function in the absence of collisions satisfies the equation

$$v_r\frac{\partial f}{\partial r} + \frac{v_\theta}{r}\frac{\partial f}{\partial \theta_1} + \left(\frac{v_\theta^2}{r} - \frac{1}{M}\frac{dU}{dr}\right)\frac{\partial f}{\partial v_r} - \frac{v_\theta v_r}{r}\frac{\partial f}{\partial v_\theta} = 0. \tag{8.16}$$

Equation (8.16) is a linear equation and can be solved by the method of characteristics. In other words, the solution of Eq. (8.16) is an arbitrary function of integrals of the characteristic equations

$$\frac{dr}{v_r} = \frac{dv_r}{\frac{v_\theta^2}{r} - \frac{1}{M}\frac{dU}{dr}}, \tag{8.17}$$

$$\frac{dr}{v_r} = -\frac{r\,dv_\theta}{v_\theta v_r}, \tag{8.18}$$

$$\frac{dr}{v_r} = \frac{r\,d\theta_1}{v_\theta}. \tag{8.19}$$

Equation (8.18) can be integrated at once, whence

$$r = \frac{C_1}{v_\theta} = \frac{\mathcal{M}}{Mv_\theta}; \quad Mv_\theta r = \mathcal{M}. \tag{8.20}$$

is obtained. Substituting $v_\theta = \mathcal{M}/Mr$ into (8.17) and integrating it, we find

$$\frac{M}{2}v_r^2 + \frac{\mathcal{M}^2}{2Mr^2} + U(r) = C_2 = \mathcal{E}. \tag{8.21}$$

Now expressing the velocity v_r in terms of $\mathcal{E}$ and $\mathcal{M}$ one can also integrate the last equation (8.19), from where it follows that

$$\theta_1 - \frac{\mathcal{M}}{(2M)^{1/2}}\int^r \frac{dr}{r^2\sqrt{\mathcal{E} - U(r) - \frac{\mathcal{M}^2}{2Mr^2}}} = \text{const} = \theta_{10}. \tag{8.22}$$

Thus, the characteristic surfaces for Eq. (8.16) are integrals of motion of the particle, i.e., the energy and angular momentum of the particle, and also the constant in (8.22), equal to θ_{10}.

As a result we find that the solution of Eq. (8.16) is an arbitrary function of energy, angular momentum, and the angle θ_{10}, that is,

$$f=f\left[\frac{M}{2}(v_r^2+v_\theta^2)+U(r);\quad rv_\theta;\ \theta_1-\frac{\mathcal{M}}{\sqrt{2M}}\int^r\frac{dr}{r^2\sqrt{\mathcal{E}-U(r)-\frac{\mathcal{M}^2}{2Mr^2}}}\right]. \tag{8.23}$$

The exact form of the function $f(\mathcal{E},\ \mathcal{M},\ \theta_{10})$ is determined by the boundary conditions or the initial conditions of the problem. It is significantly different for particles undergoing unbounded and bounded motions.

For particles undergoing unbounded motion, it is necessary to impose a boundary condition at infinity. At the same time, at infinity the field U(r) vanishes and the motion of the particles is undisturbed; the distribution function here is Maxwellian. For example, at infinity let the gas be at rest relative to the force center. In this case

$$f_{r\to\infty}=N_0\left(\frac{M}{2\pi\varkappa T}\right)^{3/2}\exp\left[-\frac{M(v_r^2+v_\theta^2)}{2\varkappa T}\right]=N_0\left(\frac{M}{2\pi\varkappa T}\right)^{3/2}\exp\left(-\frac{\mathcal{E}}{\varkappa T}\right), \tag{8.24}$$

where N_0 is the concentration of particles in the undisturbed gas. Consequently, in the entire region of unbounded motion

$$f_{\text{unb}}(v_r,\ v_\theta,\ r)=N_0\left(\frac{M}{2\pi\varkappa T}\right)^{3/2}\exp\left(-\frac{\mathcal{E}}{\varkappa T}\right)=N_0\left(\frac{M}{2\pi\varkappa T}\right)^{3/2}\exp\left[-\frac{M(v_r^2+v_\theta^2)}{2\varkappa T}-\frac{U(r)}{\varkappa T}\right]. \tag{8.25}$$

The distribution (8.25) is established for any initial distribution of particles in a disturbed region of finite dimensions because of their intermixing with other particles arriving from undisturbed regions of space. During the course of time, the initial disturbance of the concentration of unbounded particles falls off here in proportion to $(R/vt)^3 \sim 1/t^3$. An expression for the particle distribution function in the case when the body is moving and the field is a central force field is obtained in article [78].

Particles undergoing bounded motion in the absence of collisions are never associated with the particles undergoing unbounded motion, and therefore they are never associated with conditions at infinity. Their distribution is determined by the initial conditions. One can examine, for example, such a formulation of the physical problem. At the initial moment of time, let the disturbing field U = 0 and the particles be distributed according to Maxwell's law. Now let us assume that a field U(r) is slowly turned on, so that $(1/U)\ dU/dt \ll v/R$, where v is the average velocity of the particles and R is the size of the disturbed region. Then equilibrium will always be reached in the region of positive energies and the distribution will be a Maxwell-Boltzmann distribution. In this connection, particles with negative energies do not appear. Therefore, in the considered case

$$f\ \mathrm{b}\ (v_r,\ v_\theta,\ r)=\begin{cases}N_0\left(\frac{M}{2\pi\varkappa T}\right)^{3/2}\exp\left[-\frac{M(v_r^2+v_\theta^2)}{2\varkappa T}-\frac{U(r)}{\varkappa T}\right], & \mathcal{E}\geqslant 0,\\ 0, & \mathcal{E}<0.\end{cases} \tag{8.26}$$

We note that the boundary separating the regions of bounded and unbounded motions only depends on the energy and angular momentum of the particles so that for any arbitrary function $f_b(\mathcal{E},\mathcal{M})$ the total distribution function satisfies Eq. (8.16).

c. Influence of an Absorbing Surface. It would be possible to take the presence of an absorbing surface into consideration by introducing a special term [see (1.11) and (1.22)] into the right side of the kinetic equation. However, here we shall use the boundary conditions written in another form. We consider that at a given point r those particles whose trajectories went through the absorbing surface S, that is, through a sphere of radius R_0, are absent. Then, from the law of conservation of energy [Eq. (8.3)] it follows that

$$\frac{Mv_r^2}{2}+\frac{\mathcal{M}^2}{2Mr^2}+U(r)=\frac{Mv_{r0}^2}{2}+\frac{\mathcal{M}^2}{2MR_0^2}+U(R_0). \tag{8.3a}$$

Since the particles passing through the surface are absent, then $f(v_r, v_\theta, r, \theta_1)=0$ when $v_{r0}\geq 0$. If Eq. (8.3a) is taken into consideration, then one can rewrite this condition in the form

$$f_{\text{unb}}(v_r,\ v_\theta,\ r,\ \theta_1)=0, \tag{8.27}$$

provided

$$v_r^2+v_\theta^2\left(1-\frac{r^2}{R_0^2}\right)+\frac{2}{M}\left[U(r)-U(R_0)\right]\geqslant 0, \tag{8.28a}$$

$$v_r\geqslant 0. \tag{8.28b}$$

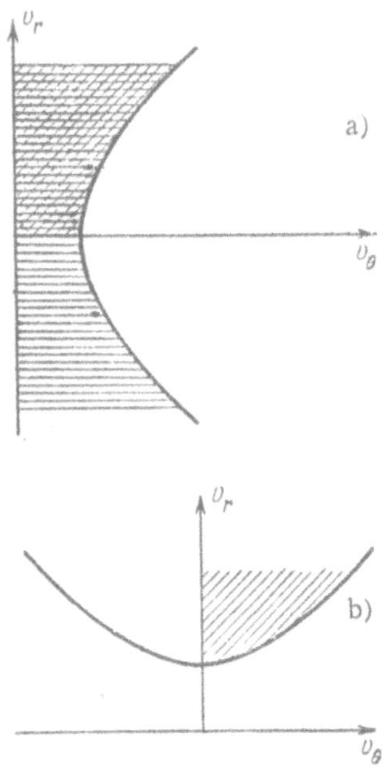

Fig. 57. Regions in which the distribution function vanishes because of the absorption of particles on the surface of the body: a) particles which are attracted; b) particles which are repelled.

Equation (8.28a) determines the hyperbola which is shown in Fig. 57a for an attractive potential, and in Fig. 57b for a repulsive potential.

Those regions in which the distribution function vanishes because the particles are absorbed by the body [conditions (8.28a) and (8.28b)] are crosshatched with slanting lines in the figure.

Particles undergoing bounded motion are not absorbed only when their orbits do not pass through the body's surface. Therefore, $f_b=0$ for $v_r<0$, only when condition (8.28a) is satisfied. The region where the function f_b vanishes is indicated on Fig. 57a by horizontal crosshatched lines.

It is clear that the size of the region in which bounded motion is possible decreases with increasing radius of the absorbing surface. When $R_0\geq r_K$, where r_K is a characteristic dimension determined by formula (8.8), bounded motion is impossible in general, since for bounded motion the minimum distance of the particle from the body's center is always less than r_K. Consequently, r_K is an important characteristic of bounded motion. Therefore, one can regard a body as large if $R_0\gg r_K$, and small when $R_0\ll r_K$. In particular, in a plasma when $\varphi\sim\varkappa T/e$, $r_K\sim D$ (see §34 below), so that the body is large if $R_0\gg D$ and small if $R_0\ll D$.

d. Consideration of the Influence of Collisions. Since collisions are infrequent, in the region of unbounded motion account of collisions leads to small correction terms of order R/l, where l is the mean free path of the particles. Consequently, one can neglect the influence of collisions on the distribution function of unbounded particles for sufficiently large values of l. A substantially different picture is observed in the region of bounded motion. Here, in the approximation in which collisions are not taken into consideration, we only establish that the distribution function is some function of the integrals of motion of the particle.

$$f_b=f_b\ (\mathcal{E},\ \mathcal{M}). \tag{8.29}$$

One can determine the exact form of the function $f_b(\mathcal{E}, \mathcal{M})$ only by solving the steady-state equation (8.15) with account of the collision integral. To be sure, if the particles are not absorbed by the body's surface, then the equilibrium Maxwell-Boltzmann distribution is always established in the region of bounded motion. However, in the most interesting case of an absorbing surface, the distribution is substantially nonequilibrium. Under these conditions there is a steady-state flux of particles, determined by the collisions, from the region of unbounded motion into the region of bounded motion, and from there onto the surface of the body. The distribution function f_b is the solution of the steady-state equation (8.15) in the presence of that flux. In this connection, inasmuch as the distribution function (8.29) becomes zero without the collision part of Eq. (8.15), it reduces to the form

$$\overline{Y}\,[f_b] = 0, \tag{8.30}$$

where $\overline{Y}$ is the averaged collision integral, written down in energy and angular momentum space. In fact, in the case we are considering the collisions are infrequent so that from collision to collision some intermixing between different coordinate points always occurs because of the motion of the particles; this intermixing must be taken into consideration by the indicated averaging. Such an averaging in the usual collision integral is carried out in practice over a time interval t such that $1/\nu \gg t \gg 1/\omega$, where $1/\nu$ is the time of free flight, and ω is the frequency of revolution of a particle performing bounded motion. One can find a general proof of the feasibility of such an averaging and the explicit form of the averaged operator Y for a number of cases in, for example, [36, 37]. Of course, the form of the operator Y after averaging depends on the potential U(r) and on the nature of the particle collisions. The corresponding calculations for the most simple case are presented below in §35.

§ 34. Concentration of Particles in the Vicinity of a Body at Rest

The concentration of particles in the vicinity of a body at rest,

$$N(r) = 2\pi \int_{\text{unb}} f_{\text{unb}}(v_r, v_\theta, r)\, v_\theta\, dv_\theta\, dv_r + 2\pi \int_{\text{b}} f_{\text{b}}(v_r, v_\theta, r)\, v_\theta\, dv_\theta\, dv_r$$

is naturally determined by two integrals over the regions, respectively, of bounded and unbounded motions of the particles.

As already shown above, the boundary separating the regions of bounded and unbounded motions is determined by conditions (8.7) and (8.11). In this section we shall consider the concentration of particles undergoing unbounded motion – N_{unb}. We note that the boundary separating the regions of unbounded and bounded motions for particles being attracted to the body substantially depends on the nature of the variation of the potential of the field. (We shall denote the concentration of attracted particles undergoing unbounded motion by N^+_{unb}.) The expressions for N^+_{unb} vary accordingly. Therefore, it is necessary to carry out calculations of the concentration of attracted particles separately for each of the three characteristic cases considered above of variation of the attractive potential U(r).

Case 1. U(r) falls off (in absolute value) with increasing r always more slowly than $1/r^2$.

First let us assume that the size of the absorbing body is infinitesimal, that is, let us consider the field of an attractive center. The boundary of the region of unbounded motion is determined by condition (8.7). Integrating the distribution function (8.25) in this region, we have

$$N^+_{\text{unb}}(r) = \frac{N_0}{\sqrt{2\pi}}\left(\frac{M}{\varkappa T}\right)^{3/2} \iint_{\text{unb}} \exp\left[-\frac{M(v_r^2+v_\theta^2)}{2\varkappa T} - \frac{U(r)}{\varkappa T}\right] v_\theta\, dv_\theta\, dv_r$$

$$= \frac{N_0}{\sqrt{2\pi}}\left(\frac{M}{\varkappa T}\right)^{3/2} \exp\left[-\frac{U(r)}{\varkappa T}\right] \iint\limits_{\substack{v_r^2+v_\theta^2 \geq -\frac{2}{M}U(r);\\ v_\theta \geq 0}} \exp\left[-\frac{M(v_r^2+v_\theta^2)}{2\varkappa T}\right] v_\theta\, dv_\theta\, dv_r$$

$$= \sqrt{\frac{2}{\pi}}\, N_0 \left(\frac{M}{\varkappa T}\right)^{3/2} \exp\left[-\frac{U(r)}{\varkappa T}\right] \left\{ \int_0^{\sqrt{-\frac{2}{M}U(r)}} \exp\left(-\frac{Mv_r^2}{2\varkappa T}\right) dv_r \int_{\sqrt{-\frac{2}{M}U(r)-v_r^2}}^{\infty} \right. \times$$

$$\times \exp\left(-\frac{Mv_\theta^2}{2\varkappa T}\right) v_\theta\, dv_\theta + \int\limits_{\sqrt{-\frac{2}{M}U(r)}}^{\infty} \exp\left(-\frac{Mv_r^2}{2\varkappa T}\right) dv_r \int\limits_0^{\infty} \exp\left(-\frac{Mv_\theta^2}{2\varkappa T}\right) v_\theta\, dv_\theta \Bigg\},$$

or after simple calculations

$$N_{\text{unb}}^{+}(r) = N_0 \left\{ \frac{2}{\sqrt{\pi}} \sqrt{-\frac{U(r)}{\varkappa T}} + \exp\left[-\frac{U(r)}{\varkappa T}\right]\left[1 - \Phi\left(\sqrt{-\frac{U(r)}{\varkappa T}}\right)\right]\right\}, \tag{8.31}$$

where $\Phi(x)$ is the probability integral.

It is important to note that in the case considered, the concentration of particles at the point r is completely determined by the value of the potential U at that same point r. For large values of $-U(r)/\varkappa T$ the concentration of particles increases like $(-U(r)/\varkappa T)^{1/2}$, and for small values of $-U(r)/\varkappa T$ the value of N(r) is close to N_0; thus, if $-U(r)/\varkappa T \ll 1$, then

$$N_{\text{unb}}^{+}(r) = N_0 \left[1 - \frac{U(r)}{\varkappa T} + \ldots\right], \tag{8.31a}$$

and for $-U(r)/\varkappa T \gg 1$

$$N_{\text{unb}}^{+}(r) = \frac{2}{\sqrt{\pi}} N_0 \sqrt{-\frac{U(r)}{\varkappa T}} \left[1 - \frac{\varkappa T}{2U(r)} - \frac{(\varkappa T)^2}{4U^2(r)} - \ldots\right]. \tag{8.31b}$$

If we have not simply a force center, but an absorbing body, then for the calculation of averages it is necessary to take into consideration the cutoff of the distribution function because of absorption of particles, which was mentioned in the preceding conditions (8.7), (8.28a), and (8.28b) (see Fig. 57), that is, for a sphere of radius R_0

$$v_r^2 + v_\theta^2 \geqslant -\frac{2}{M} U(r) \tag{8.7}$$

and

$$v_r^2 + v_\theta^2\left(1 - \frac{r^2}{R_0^2}\right) \leqslant -\frac{2}{M}\left[U(r) - U(R_0)\right] \text{ for } v_r \geqslant 0. \tag{8.28}$$

Now integrating the distribution function (8.25), we find that

$$N_{\text{unb}}^{+}(r) = \frac{N_0}{\sqrt{2\pi}}\left(\frac{M}{\varkappa T}\right)^{3/2} \iint \exp\left[-\frac{M(v_r^2+v_\theta^2)}{2\varkappa T} - \frac{U(r)}{\varkappa T}\right] v_\theta\, dv_\theta\, dv_r = \frac{N_0}{\sqrt{2\pi}}\left(\frac{M}{\varkappa T}\right)^{3/2} \exp\left[-\frac{U(r)}{\varkappa T}\right]$$

$$\times \Bigg\{ \iint\limits_{\substack{v_r \leqslant 0 \\ v_r^2+v_\theta^2 \geqslant -\frac{2}{M}U(r)}} \exp\left[-\frac{M(v_r^2+v_\theta^2)}{2\varkappa T}\right] v_\theta\, dv_\theta\, dv_r + \iint\limits_{\substack{v_r > 0 \\ v_r^2+v_\theta^2 \geqslant -\frac{2}{M}U(r) \\ v_r^2+v_\theta^2\left(1-\frac{r^2}{R_0^2}\right) \leqslant -\frac{2}{M}[U(r)-U(R_0)]}} \exp\left[-\frac{M(v_r^2+v_\theta^2)}{2\varkappa T}\right] v_\theta\, dv_\theta\, dv_r \Bigg\}.$$

In the first term, the integration is performed in the same way as above for the derivation of formula (8.31), so that

$$N_{\text{unb}}^{+}(r) = N_0 \left\{ \frac{1}{\sqrt{\pi}} \sqrt{-\frac{U(r)}{\varkappa T}} + \frac{1}{2} \exp\left[-\frac{U(r)}{\varkappa T}\right] \times \right.$$

$$\left.\left[1 - \Phi\left(\sqrt{-\frac{U(r)}{\varkappa T}}\right)\right]\right\} + \frac{N_0}{\sqrt{2\pi}}\left(\frac{M}{\varkappa T}\right)^{3/2} \exp\left[-\frac{U(r)}{\varkappa T}\right] \times$$

$$\times\left\{\int\limits_0^{\sqrt{-\frac{2}{M}[U(r)-U(R_0)]}}\exp\left(-\frac{Mv_r^2}{2\varkappa T}\right)dv_r\int\limits_{\sqrt{-\frac{2}{M}U(r)-v_r^2}}^{\infty}\exp\left(-\frac{Mv_\theta^2}{2\varkappa T}\right)v_\theta\,dv_\theta\right.$$

$$\left.+\int\limits_{\sqrt{-\frac{2}{M}\left[U(r)-\frac{R_0^2}{r^2}U(R_0)\right]}}^{\infty}\exp\left(-\frac{Mv_r^2}{2\varkappa T}\right)dv_r\int\limits_{\sqrt{\frac{v_r^2+\frac{2}{M}[U(r)-U(R_0)]}{\frac{r^2}{R_0^2}-1}}}^{\infty}\exp\left(-\frac{Mv_\theta^2}{2\varkappa T}\right)v_\theta\,dv_\theta\right\}.$$

All of the integrals are easily evaluated. Finally we have

$$N^+_{\text{unb}}(r)=N_0\left\{\frac{1}{\sqrt{\pi}}\sqrt{-\frac{U(r)}{\varkappa T}}\left(1+\sqrt{1-\frac{R_0^2U(R_0)}{r^2U(r)}}\right)+\frac{1}{2}\exp\left[-\frac{U(r)}{\varkappa T}\right]\left[1-\Phi\left(\sqrt{-\frac{U(r)}{\varkappa T}}\right)\right]\right.$$

$$\left.+\frac{1}{2}\sqrt{1-\frac{R_0^2}{r^2}}\exp\left[-\frac{r^2U(r)-R_0^2U(R_0)}{(r^2-R_0^2)\varkappa T}\right]\left[1-\Phi\left(\sqrt{-\frac{r^2U(r)-R_0^2U(R_0)}{(r^2-R_0^2)\varkappa T}}\right)\right]\right\}. \quad (8.32)$$

It is clear from Eq. (8.32) that the concentration of particles at the point r in the presence of an absorbing surface now depends not only on the value of the potential U at that same point r, but also on the value of U on the surface of the sphere, and also on the ratio r/R_0. With increasing values of the ratio r/R_0, when $r/R_0\to\infty$, the role of absorption becomes insignificant, and formula (8.32) coincides with (8.31):

$$N^+_{\substack{\text{unb}\\ r/R_0\to\infty}}(r)=N_0\left\{\frac{2}{\sqrt{\pi}}\sqrt{-\frac{U(r)}{\varkappa T}}+\exp\left[-\frac{U(r)}{\varkappa T}\right]\left[1-\Phi\left(\sqrt{-\frac{U(r)}{\varkappa T}}\right)\right]\right\}.$$

Conversely, when $r\to R_0$

$$N^+_{\substack{\text{unb}\\ r/R_0\to 1}}(r)=\frac{N_0}{2}\left\{\frac{2}{\sqrt{\pi}}\sqrt{-\frac{U(r)}{\varkappa T}}+\exp\left[-\frac{U(r)}{\varkappa T}\right]\left[1-\Phi\left(\sqrt{-\frac{U(r)}{\varkappa T}}\right)\right]\right\},$$

that is, the concentration of particles near the surface is decreased by a factor of two, because of absorption.

Case 2. The potential U falls off (in absolute value) with increasing r more slowly than $1/r^2$ for small r, and faster than $1/r^2$ for large values of r. Again we first consider the case of the field of an attractive center. Here the boundary of the region of unbounded motion is now determined by conditions (8.11) (see Fig. 54). Integrating the distribution function (8.25) in this region, we find that the concentration of particles is given by

$$N^+_{\text{unb}}(r)=\frac{N_0}{\sqrt{2\pi}}\left(\frac{M}{\varkappa T}\right)^{3/2}\iint\exp\left[-\frac{M(v_r^2+v_\theta^2)}{2\varkappa T}-\frac{U(r)}{\varkappa T}\right]v_\theta\,dv_\theta\,dv_r=\frac{2N_0}{\sqrt{2\pi}}\left(\frac{M}{\varkappa T}\right)^{3/2}\exp\left[-\frac{U(r)}{\varkappa T}\right]$$

$$\times\left\{\int\limits_{\sqrt{-\frac{2}{M}U(r)}}^{\infty}\exp\left(-\frac{Mv_r^2}{2\varkappa T}\right)dv_r\int\limits_0^{\infty}\exp\left(-\frac{Mv_\theta^2}{2\varkappa T}\right)v_\theta\,dv_\theta\right.$$

$$\left.+\int\limits_0^{\sqrt{-\frac{2}{M}U(r)}}\exp\left(-\frac{Mv_r^2}{2\varkappa T}\right)dv_r\int\limits_{v_\theta(v_r)}^{\infty}\exp\left(-\frac{Mv_\theta^2}{2\varkappa T}\right)v_\theta\,dv_\theta\right\}$$

$$=N_0\exp\left[-\frac{U(r)}{\varkappa T}\right]\left\{1-\Phi\left(\sqrt{-\frac{U(r)}{\varkappa T}}\right)+\sqrt{\frac{2M}{\pi\varkappa T}}\int\limits_0^{\sqrt{-\frac{2}{M}U(r)}}\exp\left[-\frac{M(v_r^2+v_\theta^2(v_r))}{2\varkappa T}\right]dv_r\right\}, \quad (8.33)$$

where $v_\theta(v_r)$ is determined by Eqs. (8.11). In the last integral one can change over to a new variable r_0, related to $v_\theta(v_r)$ by relationship (8.11). As a result the implicit function $v_\theta^2(v_r)$ vanishes. In fact, from Eq. (8.11a) we find that

$$\frac{dv_r}{dr_0}=\frac{\left(1-\frac{r_0^2}{r^2}\right)\left(3\frac{dU}{dr_0}+r_0\frac{d^2U}{dr_0^2}\right)}{2M^{1/2}\sqrt{2\left[U(r_0)-U(r)\right]-r_0\frac{dU(r_0)}{dr_0}\left(\frac{r_0^2}{r^2}-1\right)}},$$

where $r_0\to\rho_0(r)$ as $v_r\to 0$, and $r_0\to\infty$ as $v_r\to[-(2/M)U(r)]^{1/2}$ (see §32). Therefore

$$\left(\frac{2M}{\pi\varkappa T}\right)^{1/2}\int\limits_0^{\sqrt{-\frac{2}{M}U(r)}}\exp\left[-\frac{M\left(v_r^2+v_\theta^2(v_r)\right)}{2\varkappa T}\right]dv_r$$

$$=\frac{1}{2\sqrt{\pi\varkappa T}}\int\limits_{\rho_0(r)}^{\infty}\frac{\left(1-\frac{r_0^2}{r^2}\right)\left(3\frac{dU}{dr_0}+r_0\frac{d^2U}{dr_0^2}\right)}{\left[U(r_0)-U(r)-\frac{r_0}{2}\frac{dU}{dr_0}\left(\frac{r_0^2}{r^2}-1\right)\right]^{1/2}}$$

$$\times\exp\left\{-\frac{M}{2\varkappa T}\left[\frac{2}{M}U(r_0)-\frac{2}{M}U(r)-\frac{r_0}{M}\frac{dU}{dr_0}\left(\frac{r_0^2}{r^2}-1\right)+\frac{r_0^3}{Mr^2}\frac{dU}{dr_0}\right]\right\}dr_0$$

$$=\frac{1}{2\sqrt{\pi\varkappa T}}\exp\left[\frac{U(r)}{\varkappa T}\right]\int\limits_{\rho_0(r)}^{\infty}\frac{\left(1-\frac{r_0^2}{r^2}\right)\left(3\frac{dU}{dr_0}+r_0\frac{d^2U}{dr_0^2}\right)}{\left[U(r_0)-U(r)-\frac{r_0}{2}\frac{dU}{dr_0}\left(\frac{r_0^2}{r^2}-1\right)\right]^{1/2}}\exp\left[-\frac{U(r_0)}{\varkappa T}-\frac{r_0}{2\varkappa T}\frac{dU}{dr_0}\right]dr_0.$$

Using the latter transformation, we finally have

$$N_{\text{unb}}^{+}(r)=N_0\left\{\exp\left[-\frac{U(r)}{\varkappa T}\right]\left[1-\Phi\left(\sqrt{-\frac{U(r)}{\varkappa T}}\right)\right]+\right.$$

$$\left.+\frac{1}{2\sqrt{\pi\varkappa T}}\int\limits_{\rho_0(r)}^{\infty}\frac{\left(1-\frac{r_0^2}{r^2}\right)\left(3\frac{dU}{dr_0}+r_0\frac{d^2U}{dr_0^2}\right)}{\left[U(r_0)-U(r)-\frac{r_0}{2}\frac{dU}{dr_0}\left(\frac{r_0^2}{r^2}-1\right)\right]^{1/2}}\exp\left[-\frac{U(r_0)}{\varkappa T}-\frac{r_0}{2\varkappa T}\frac{dU}{dr_0}\right]dr_0\right\}, \tag{8.34}$$

where $\rho_0(r)$ is the maximum root of Eq. (8.12).

As indicated above (see §32 and Fig. 53), the root $\rho_0(r)$ is always larger than r_K. Therefore, in the entire region of variation of the parameter r_0 in the integral term in formula (8.34), the potential $U(r_0)$ falls off with increasing r_0 more rapidly than $1/r_0^2$, that is, $-U(r_0)-(1/2)r_0(dU/dr_0)$ is always less than zero, and the exponential term is smaller than unity. If the relation

$$\frac{U(r_0)}{\varkappa T}+\frac{r_0}{2\varkappa T}\frac{dU}{dr_0}\ll 1, \tag{8.35}$$

is satisfied over the entire range of variation of the parameter r_0, then expanding the exponent in expression (8.34), we find that

$$N_{\text{unb}}^{+}(r)=N_0\left\{\exp\left[-\frac{U(r)}{\varkappa T}\right]\left[1-\Phi\left(\sqrt{-\frac{U(r)}{\varkappa T}}\right)\right]\right.$$

$$+\frac{1}{2\sqrt{\pi\varkappa T}}\int\limits_{\rho_0(r)}^{\infty}\frac{\left(1-\frac{r_0^2}{r^2}\right)\left(3\frac{dU}{dr_0}+r_0\frac{d^2U}{dr_0^2}\right)}{\left[U(r_0)-U(r)-\frac{r_0}{2}\frac{dU}{dr_0}\left(\frac{r_0^2}{r^2}-1\right)\right]^{1/2}}\times$$

$$\times\left[1-\frac{U(r_0)+\frac{r_0}{2}\frac{dU}{dr_0}}{\varkappa T}+\frac{\left(U(r_0)+\frac{r_0}{2}\frac{dU}{dr_0}\right)^2}{2(\varkappa T)^2}-\ldots\right]dr_0\Big\}$$

$$=N_0\left\{\left[1-\Phi\left(\sqrt{-\frac{U(r)}{\varkappa T}}\right)\right]\exp\left[-\frac{U(r)}{\varkappa T}\right]+\frac{2}{\sqrt{\pi\varkappa T}}\left[\sqrt{U(r_0)-U(r)-\frac{r_0dU}{2dr_0}\left(\frac{r_0^2}{r^2}-1\right)}\right]_{\rho_0(r)}^{\infty}\right.$$

$$\left.-\frac{2}{\sqrt{\pi}}\sqrt{-\frac{U(r)}{\varkappa T}}\left[\frac{A_1(r)}{\varkappa T}-\frac{B_1(r)}{(\varkappa T)^2}+\ldots\right]\right\}$$

$$=N_0\left\{\left[1-\Phi\left(\sqrt{-\frac{U(r)}{\varkappa T}}\right)\right]\exp\left[-\frac{U(r)}{\varkappa T}\right]+\frac{2}{\sqrt{\pi}}\sqrt{-\frac{U(r)}{\varkappa T}}\left[1-\frac{A_1(r)}{\varkappa T}+\frac{B_1(r)}{(\varkappa T)^2}-\ldots\right]\right\} \quad (8.36)$$

where

$$A_1(r)=\frac{1}{4}\int_{\rho_0(r)}^{\infty}\frac{\left(1-\frac{r_0^2}{r^2}\right)\left(3\frac{dU}{dr_0}+r_0\frac{d^2U}{dr_0^2}\right)\left(U(r_0)+\frac{r_0}{2}\frac{dU}{dr_0}\right)}{\sqrt{U^2(r)-U(r)U(r_0)+\frac{r_0}{2}U(r)\frac{dU}{dr_0}\left(\frac{r_0^2}{r^2}-1\right)}}dr_0,$$

$$B_1(r)=\frac{1}{8}\int_{\rho_0(r)}^{\infty}\frac{\left(1-\frac{r_0^2}{r^2}\right)\left(3\frac{dU}{dr_0}+r_0\frac{d^2U}{dr_0^2}\right)\left(U(r_0)+\frac{r_0}{2}\frac{dU}{dr_0}\right)^2}{\sqrt{U^2(r)-U(r)U(r_0)+\frac{r_0}{2}U(r)\frac{dU}{dr_0}\left(\frac{r_0^2}{r^2}-1\right)}}dr_0.$$

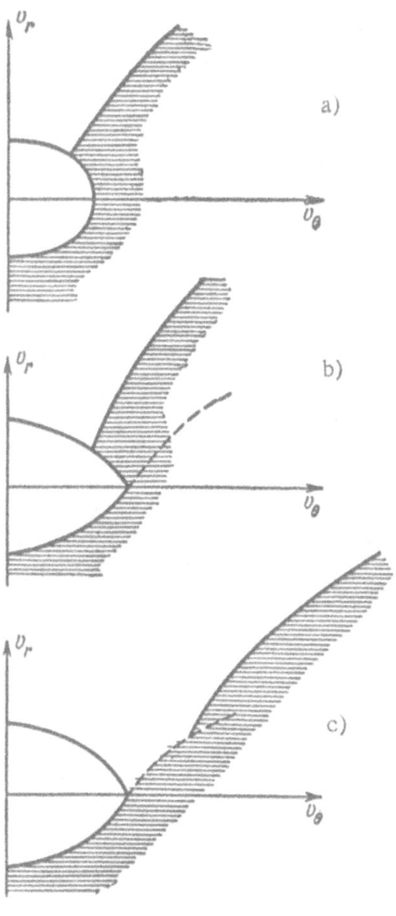

Fig. 58. Domain of integration in velocity space for different values of r.

For small values of r, that is, when $r \ll r_K$ and $\rho_0(r)\to\infty$ [see §32, Fig. 53, and formula (8.12a)], condition (8.35) is always satisfied. In this case, if the correction terms $A_1(r)$ and $B_1(r)$ are neglected in formula (8.36), then it coincides with formula (8.31), that is

$$N^+_{\text{unb}}(r)=N_1(r)=N_0\left\{\frac{2}{\sqrt{\pi}}\sqrt{-\frac{U(r)}{\varkappa T}}\right.$$

$$\left.+\left[1-\Phi\left(\sqrt{-\frac{U(r)}{\varkappa T}}\right)\right]\exp\left[-\frac{U(r)}{\varkappa T}\right]\right\}. \quad (8.36a)$$

This answer is understandable, inasmuch as the potential U(r) decreases more rapidly than $1/r^2$, only beginning with the point r_K. Therefore, if $r \ll r_K$, then the presence of a region of rapid decrease of the potential does not manifest itself in the concentration of particles at the point r.

Since in fact the exponential term is always less than unity, so at any arbitrary distance r the concentration of particles $N^+_{\text{unb}}(r)$ is smaller than $N_1(r)$. This is explained by the fact that in formula (8.34) the presence of a region of sharp decrease of the potential U(r) is taken into consideration, as a consequence of which part of the particles coming from infinity do not travel completely up to the point r_K, but are "reflected" at larger distances from the center (see §32). It is important to emphasize that, in contrast to the above considered case of a slowly decreasing potential U(r), the value of the concentration of particles at the point r now significantly depends on the overall behavior of the potential U(r), and not only on the value of U at the point r.

In the presence of an absorbing body, for calculation of the concentration it is now necessary, as above, to consider the cutoff of the particle distribution function caused by their absorption on the surface of the body – a sphere of radius R_0. The region of integration in the space of velocities v_r and v_θ is then determined by Eqs. (8.11) and (8.28). It is shown in Fig. 58 (the crosshatched region) for different distances r from the center of the body. Here the character of the region of integration is different depending on the relation between the distance r, the size of the body R_0, and the distance r_K, which characterizes the variation of the potential field U(r) (see §32). For $r < r_K$ the potential U(r) decreases (in absolute value) with increasing r more slowly than $1/r^2$, and for $r > r_K$ it decreases more rapidly than $1/r^2$. The region of integration for distances $r < r_K$ is shown in Fig. 58a and for $r > r_K$ it is shown in Figs. 58b and c. Here in the case illustrated in Fig. 58b, the radius R_0 of the body is smaller than $R_{0K}(r)$, where R_{0K} is defined by the relation

$$-U(R_{0K})+U(r)+\frac{r}{2}\frac{dU}{dr}\left(1-\frac{r^2}{R_{0K}^2}\right)=0,$$

which follows from condition (8.11a) for $v_r = 0$. In the case represented in Fig. 58c, however, on the contrary $R_0 > R_{0K}(r)$. As is clear from the last figure, in this case not only the curve dividing the regions of unbounded and bounded motions (the continuous lines) in important, but also the curve in the region of unbounded motion (the dashed line in Figs. 58b and c), separating the particles whose turning points are located in regions close to and far away from the body's center (see §32 and Figs. 52, 54b, and 54c), is essential. Integrating the distribution function (8.25) in the same way as before, we obtain

$$N_{\text{unb}}^{+}(r)=\frac{N_0}{2}\left\{\left[1-\Phi\left(\sqrt{-\frac{U(r)}{\varkappa T}}\right)\right]\exp\left[-\frac{U(r)}{\varkappa T}\right]\right.$$

$$+\sqrt{1-\frac{R_0^2}{r^2}}\left[1-\Phi\left(\sqrt{\frac{\varrho_0^3(R_0)}{2R_0^2\varkappa T}\frac{dU}{dr}(\varrho_0(R_0))-\frac{U(r)-U(R_0)}{\varkappa T\left(1-\frac{R_0^2}{r^2}\right)}}\right)\right]\exp\left[\frac{R_0^2U(R_0)-r^2U(r)}{(r^2-R_0^2)\varkappa T}\right]$$

$$+\frac{1}{2\sqrt{\pi\varkappa T}}\int_{\rho_0(r)}^{\infty}\frac{\left(1-\frac{r_0^2}{r^2}\right)\left(3\frac{dU}{dr_0}+r_0\frac{d^2U}{dr_0^2}\right)}{\left[U(r_0)-U(r)-\frac{r_0}{2}\frac{dU}{dr_0}\left(\frac{r_0^2}{r^2}-1\right)\right]^{1/2}}\exp\left[-\frac{U(r_0)}{\varkappa T}-\frac{r_0}{2\varkappa T}\frac{dU}{dr_0}\right]dr_0+\frac{1}{2\sqrt{\pi\varkappa T}}$$

$$\left.\times\int_{\rho_0(r)}^{\rho(R_0)}\frac{\left(1-\frac{r_0^2}{r^2}\right)\left(3\frac{dU}{dr_0}+r_0\frac{d^2U}{dr_0^2}\right)}{\left[U(r_0)-U(r)-\frac{r_0}{2}\frac{dU}{dr_0}\left(\frac{r_0^2}{r^2}-1\right)\right]^{1/2}}\exp\left[-\frac{U(r_0)}{\varkappa T}-\frac{r_0}{2\varkappa T}\frac{dU}{dr_0}\right]dr_0\right\}. \tag{8.37}$$

The value of $N_{\text{unb}}^{+}(r)$ determined by formula (8.37) is always smaller than the corresponding value of $N_{\text{unb}}^{+}(r)$ in the absence of an absorbing surface, determined by formula (8.34). For $R_0/r \to 0$, when the radius of the surface is small, formulas (8.37) and (8.34) become identical; if $r \to R_0$, then because of the presence of an absorbing surface, the concentration of particles decreases by a factor of two. If the body's radius $R_0 \ll r_K$, then (8.37) naturally goes over into (8.32). On the other hand, if the potential U(r) falls off abruptly near the body's surface, then in the region where $-U(r)/\varkappa T \ll 1$,

$$N_{\text{unb}}^{+}(r)=\frac{N_0}{2}\left(1+\sqrt{1-\frac{R_0^2}{r^2}}\right).$$

Case 3. The potential U(r) falls off (in absolute value) with increasing r initially more slowly than $1/r^2$, then more rapidly than $1/r^2$, and for large values of r it falls off exactly like $1/r^2$. In other words, as $r \to \infty$ let

$$U(r)=-\frac{C}{r^2}=-C_0\frac{\varkappa TR_0^2}{r^2},$$

where C_0 is a dimensionless constant determining the character of the variation of the potential U(r) for large values of r.

In these cases, for small values of the angular momentum $\mathcal{M} < \mathcal{M}_0 = (2MC)^{1/2}$ the boundary of the region of unbounded motion is determined by condition (8.7), and for $\mathcal{M} > \mathcal{M}_0$, it is determined by condition (8.11). Integrating, as usual, the distribution function over the region of unbounded motion in the field of an attractive (force) center, we find

$$N^+_{\text{unb}}(r) = N_0\left\{\left[1-\Phi\left(\sqrt{-\frac{U(r)}{\varkappa T}}\right)\right]\exp\left[-\frac{U(r)}{\varkappa T}\right] + \frac{2}{\sqrt{\pi}}\left(\sqrt{-\frac{U(r)}{\varkappa T}} - \sqrt{-\frac{U(r)}{\varkappa T} - C_0\frac{R_0^2}{r^2}}\right) + \right.$$

$$\left. + \frac{1}{2\sqrt{\pi\varkappa T}}\int\limits_{\rho_0(r)}^{\infty}\frac{\left(1-\frac{r_0^2}{r^2}\right)\left(3\frac{dU}{dr_0}+r_0\frac{d^2U}{dr_0^2}\right)}{\left[U(r_0)-U(r)-\frac{r_0}{2}\frac{dU}{dr_0}\left(\frac{r_0^2}{r^2}-1\right)\right]^{1/2}}\exp\left[-\frac{U(r_0)}{\varkappa T}-\frac{r_0}{2\varkappa T}\frac{dU}{dr_0}\right]dr_0\right\}. \tag{8.38}$$

Formula (8.38) differs from (8.34) by the fact that an additional term appears because of integration with respect to the angular momentum $\mathcal{M} < \mathcal{M}_0$. If the potential of the field falls off faster than $1/r^2$ at large distances r, then $C_0 = 0$ and formula (8.38) goes over into (8.34). If, however, U(r) always decreases like $1/r^2$, then $-[U(r)/\varkappa T] - C_0(R_0^2/r^2) \equiv 0$ and, in addition, the integral term is equal to zero, so that formula (8.38) goes over in this case into (8.31).

With account of the absorbing surface, the region of integration is given by condition (8.28), and also by conditions (8.7) for $v_\theta < (2C_0\varkappa TR_0^2/Mr^2)^{1/2}$ and by condition (8.11) for $v_\theta > (2C_0\varkappa TR_0^2/Mr^2)^{1/2}$. As a result of the integration, an expression for $N^+_{\text{unb}}(r)$ is obtained which differs from (8.37) only by the additional term

$$\Delta N^+_{\text{unb}} = \frac{N_0}{2}\left\{\frac{2}{\sqrt{\pi}}\left(\sqrt{-\frac{U(r)}{\varkappa T}} - \sqrt{-\frac{U(r)}{\varkappa T} - C_0\frac{R_0^2}{r^2}}\right)\right\} \tag{8.37a}$$

for $R_0 > R_{0K}(r)$ and

$$\Delta N^+_{\text{unb}} = \frac{N_0}{2}\left\{\frac{2}{\sqrt{\pi}}\left(\sqrt{-\frac{U(r)}{\varkappa T}} + \sqrt{-\frac{U(r)}{\varkappa T}+\frac{R_0^2U(R_0)}{r^2\varkappa T}} - 2\sqrt{-\frac{U(r)}{\varkappa T} - C_0\frac{R_0^2}{r^2}}\right) + \sqrt{1-\frac{R_0^2}{r^2}}\times\right.$$

$$\left.\times\left[\Phi\left(\sqrt{\frac{\varrho_0^3(R_0)}{2R_0^2\varkappa T}\frac{dU}{dr}(\varrho_0(R_0)) - \frac{U(r)-U(R_0)}{\varkappa T\left(1-\frac{R_0^2}{r^2}\right)}}\right) - \Phi\left(\sqrt{\frac{R_0^2U(R_0)-r^2U(r)}{(r^2-R_0^2)\varkappa T}}\right)\right]\exp\left[\frac{R_0^2U(R_0)-r^2U(r)}{(r^2-R_0^2)\varkappa T}\right]\right\} \tag{8.37b}$$

for $R_0 < R_0K(r)$. For example, in the case when $R_0 > R_0K(r)$

$$N^+_{\text{unb}}(r) = \frac{N_0}{2}\left\{\left[1-\Phi\left(\sqrt{-\frac{U(r)}{\varkappa T}}\right)\right]\exp\left[-\frac{U(r)}{\varkappa T}\right] + \right.$$

$$+\frac{2}{\sqrt{\pi}}\left(\sqrt{-\frac{U(r)}{\varkappa T}} - \sqrt{-\frac{U(r)}{\varkappa T} - C_0\frac{R_0^2}{r^2}}\right) + \sqrt{1-\frac{R_0^2}{r^2}}$$

$$\left[1-\Phi\left(\sqrt{\frac{\varrho^3(R_0)}{2R_0^2\varkappa T}\frac{dU}{dr}(\varrho_0(R_0)) - \frac{U(r)-U(R_0)}{\varkappa T\left(1-\frac{R_0^2}{r^2}\right)}}\right)\right]\exp\left[\frac{R_0^2U(R_0)-r^2U(r)}{(r^2-R_0^2)\varkappa T}\right]$$

$$+\frac{1}{2\sqrt{\pi\varkappa T}}\int\limits_{\rho_0(r)}^{\infty}\frac{\left(1-\frac{r_0^2}{r^2}\right)\left(3\frac{dU}{dr_0}+r_0\frac{d^2U}{dr_0^2}\right)}{\left[U(r_0)-U(r)-\frac{r_0}{2}\frac{dU}{dr_0}\left(\frac{r_0^2}{r^2}-1\right)\right]^{1/2}}\exp\left[-\frac{U(r_0)}{\varkappa T}-\frac{r_0}{2\varkappa T}\frac{dU}{dr_0}\right]dr_0 +$$

$$\left. +\frac{1}{2\sqrt{\pi\varkappa T}}\int\limits_{\rho_0(r)}^{\rho_0(R_0)}\frac{\left(1-\frac{r_0^2}{r^2}\right)\left(3\frac{dU}{dr_0}+r_0\frac{d^2U}{dr_0^2}\right)}{\left[U(r_0)-U(r)-\frac{r_0}{2}\frac{dU}{dr_0}\left(\frac{r_0^2}{r^2}-1\right)\right]^{1/2}}\exp\left[-\frac{U(r_0)}{\varkappa T}-\frac{r_0}{2\varkappa T}\frac{dU}{dr_0}\right]dr_0\right\}. \tag{8.38}$$

Thus, we see that depending on the nature of the variation of the potential U(r) in the vicinity of a body at rest (cases 1-3), the form of the equations determining the concentration of particles changes substantially. In this connection, it is natural that the formulas become more complicated in the presence of a body with an absorbing surface. It is important to note here that in all cases the concentration of particles depends significantly on the potential of the field, and can be determined only if the function U(r) is known. In simple cases the formulas for the concentration of particles contain tabulated functions, and in more complicated cases they also contain one-dimensional integrals requiring numerical integration.

Up to now we have considered the behavior of particles attracted to the body. Now let us go on to the case of particles in a repulsive field. In this case the potential $U(r) > 0$ and the motion of the particles is always unbounded, as indicated above. Therefore the Maxwell-Boltzmann distribution is valid in the field of a repulsive center of force, so that the particle concentration $N^-(r)$ is given by

$$N^-(r) = N_0 \exp\left[-\frac{U(r)}{\varkappa T}\right]. \tag{8.40}$$

Upon account of the absorption of particles on the surface of a sphere of radius R_0, the domain of integration is limited by conditions (8.28)

$$v_r^2 + v_\theta^2\left(1 - \frac{r^2}{R_0^2}\right) \geqslant \frac{2}{M}[U(R_0) - U(r)] \quad \text{for} \quad v_r > 0 \tag{8.28}$$

(also see Fig. 57b). In this case, the concentration of particles is given by

$$N^-(r) = \frac{N_0}{\sqrt{2\pi}}\left(\frac{M}{\varkappa T}\right)^{3/2} \iint \exp\left[-\frac{M(v_r^2+v_\theta^2)}{2\varkappa T} - \frac{U(r)}{\varkappa T}\right] v_\theta \, dv_\theta \, dv_r = \frac{N_0}{\sqrt{2\pi}}\left(\frac{M}{\varkappa T}\right)^{3/2}$$

$$\times \left\{ \int\limits_{-\infty}^{\sqrt{\frac{2}{M}[U(R_0)-U(r)]}} \exp\left(-\frac{Mv_r^2}{2\varkappa T}\right) dv_r \int\limits_0^\infty v_\theta \exp\left(-\frac{Mv_\theta^2}{2\varkappa T}\right) dv_\theta + \int\limits_{\sqrt{\frac{2}{M}[U(R_0)-U(r)]}}^{\infty} \exp\left(-\frac{Mv_r^2}{2\varkappa T}\right) dv_r \right.$$

$$\left. \times \int\limits_{\sqrt{\frac{\frac{2}{M}[U(r)-U(R_0)]+v_r^2}{\frac{r^2}{R_0^2}-1}}}^{\infty} v_\theta \exp\left(-\frac{Mv_\theta^2}{2\varkappa T}\right) dv_\theta \right\} \exp\left[-\frac{U(r)}{\varkappa T}\right].$$

Integrating, we obtain

$$N^-(r) = \frac{N_0}{2}\left\{1 + \Phi\left(\sqrt{\frac{U(R_0)-U(r)}{\varkappa T}}\right) + \sqrt{1 - \frac{R_0^2}{r^2}}\left[1 - \Phi\left(\sqrt{\frac{r^2U(R_0)-r^2U(r)}{(r^2-R_0^2)\varkappa T}}\right)\right]\right.$$

$$\left. \times \exp\left[\frac{R_0^2U(R_0)-R_0^2U(r)}{(r^2-R_0^2)\varkappa T}\right]\right\} \exp\left[-\frac{U(r)}{\varkappa T}\right]. \tag{8.41}$$

As $r \to R_0$, $U(r) \to U(R_0)$ and $N^-(r) \to (N_0/2)\exp[-U(r)/\varkappa T]$; for $r/R_0 \to \infty$, $U(r) \ll U(R_0)$ and $N^-(r) \to N_0 \exp[-U(r)/\varkappa T]$.

§ 35. Distribution of the Particles Performing Bounded Motion

It has already been indicated above that the character of the distribution of particles in bounded orbits in the vicinity of an absorbing body is determined by the collisions of the particles. The distribution function here satisfies equation (8.30)

$$\overline{Y}(f_b) = 0, \tag{8.30}$$

where $\overline{Y}$ is the averaged collision integral. Its form, of course, depends on the nature of the collisions of the particles.

a. The Averaged Collision Integral. Let us dwell on the simplest case, when the particles being attracted to the center collide with particles on which the field does not act. Such a situation is realized in a weakly ionized plasma, where the charged particles collide with neutral particles whose distribution does not depend on the field. Equation (8.30) will then be linear. In this connection, if we consider an attractive force center, that is, an infinitely small body whose surface absorbs particles, and we assume that it is positively charged so that the attracted particles are electrons, the the form of the collision integral is simplified considerably.

Because of the smallness of the electron's mass, during each elastic collision of an electron with the neutral particles, its momentum varies strongly and its energy only changes insignificantly. The average relative change of the energy associated with a single collision is m/M times smaller than the relative change of the electron's momentum. This leads to the result that the equilibrium distribution of the electrons with respect to energy is established much more slowly than with respect to other variables. Consequently, the non-equilibrium distribution function basically depends on only the energy of the electron.

During each collision, the electron gives up only a small part of its energy. This enables one to represent the collision integral in energy space in differential form as in (1.35). The kinetic equation itself is here an equation of the Fokker-Planck type, that is, a diffusion equation in energy space

$$\frac{\partial f}{\partial t}=-A(\mathcal{E})\frac{\partial}{\partial \mathcal{E}}[I(\mathcal{E})], \tag{8.42}$$

where $I(\mathcal{E})$ is the particle flux in energy space, and A is a normalization coefficient which also depends on the energy $\mathcal{E}$. The flux must vanish in the case of the equilibrium Boltzmann distribution. Therefore,

$$I(\mathcal{E})=-\frac{D(\mathcal{E})}{2A(\mathcal{E})}\left(\frac{\partial f}{\partial \mathcal{E}}+\frac{1}{\varkappa T}f\right) \tag{8.43}$$

(see, for example, [23]) where $D(\mathcal{E})$ is also some coefficient whose significance will be explained below.

Thus, in the case considered, the distribution function of the particles in bounded orbits only depends on their energy and is determined by Eq. (8.30), which is now written in the form

$$\overline{Y}(f)=A(\mathcal{E})\frac{\partial}{\partial \mathcal{E}}\left\{\frac{D(\mathcal{E})}{2A(\mathcal{E})}\left[\frac{\partial f}{\partial \mathcal{E}}+\frac{1}{\varkappa T}f\right]\right\}=0. \tag{8.44}$$

It is necessary for us to determine only the unknown functions $A(\mathcal{E})$ and $D(\mathcal{E})$.

An equation of the Fokker-Planck type, as indicated in §3, is applicable not only in the case of a weakly ionized plasma in order to describe the collisions of electrons with plasma in order to describe the collisions of electrons with neutral particles, but it is also applicable under a much wider range of conditions, for example, in a strongly ionized plasma for Coulomb interactions between the particles. Here the distribution function of the bounded particles depends, generally speaking, now not only on the energy but also on the angular momentum of the particle. Therefore, in the general case when the Fokker-Planck approximation is applicable for the collision integral, Eq. (8.44) can be represented in the form

$$\overline{Y}(f)=A(\mathcal{E},\ \mathcal{M})\,\mathrm{div}_{\mathcal{E},\mathcal{M}}I(\mathcal{E},\ \mathcal{M}), \tag{8.45}$$

where $A(\mathcal{E}, \mathcal{M})$ is a normalization coefficient as before, and $I(\mathcal{E}, \mathcal{M})$ is the particle flux in the space of energy and angular momentum.

b. Determination of the Normalization Coefficient. The normalization coefficient $A(\mathcal{E})$ in Eq. (8.44) is the reciprocal volume in configuration space ($\mathbf{r}$, $\mathbf{v}$) for a given value of the energy $\mathcal{E}$. In fact, let N be the total number of bounded particles whose energies vary between the values $\mathcal{E}_1$ and $\mathcal{E}_2$, that is,

$$N = \int\limits_{\mathcal{E}_1 \leqslant \mathcal{E} \leqslant \mathcal{E}_2} f\, d\Gamma = \int\limits_{\mathcal{E}_1 \leqslant \mathcal{E} \leqslant \mathcal{E}_2} f\, dv_x\, dv_y\, dv_z\, dx\, dy\, dz.$$

By virtue of the law of conservation of the number of particles, the change of N per unit time is equal to the difference of particle fluxes in configuration space $\mathbf{r}$, $\mathbf{v}$ through the surfaces of constant energy $\mathcal{E}_1$ and $\mathcal{E}_2$, that is

$$\frac{dN}{dt} = I(\mathcal{E}_1) - I(\mathcal{E}_2).$$

On the other hand, integrating Eq. (8.42) with respect to the volume $d\Gamma$ between the energy levels $\mathcal{E}_1$ and $\mathcal{E}_2$, we obtain

$$\frac{dN}{dt} = \int\limits_{\mathcal{E}_1 \leqslant \mathcal{E} \leqslant \mathcal{E}_2} A(\mathcal{E}) \frac{dI}{d\mathcal{E}}\, d\Gamma = \int\limits_{\mathcal{E}_1 \leqslant \mathcal{E} \leqslant \mathcal{E}_2} A(\mathcal{E}) \frac{dI}{d\mathcal{E}}\, d\mathcal{E} \int \frac{d\Gamma}{d\mathcal{E}}. \tag{8.46}$$

Since the right side of this expression must have the form $I(\mathcal{E}_1) - I(\mathcal{E}_2)$, then

$$A(\mathcal{E}) \int \frac{d\Gamma}{d\mathcal{E}} = 1.$$

Thus, $A(\mathcal{E})$ is actually the reciprocal volume of configuration space for a specified energy $\mathcal{E}$:

$$A(\mathcal{E}) = \left[\int \frac{d\Gamma}{d\mathcal{E}} \right]^{-1} = \left\{ \int \delta\, [\mathcal{E} - \mathcal{E}(v, r)]\, d^3v\, d^3r \right\}^{-1}. \tag{8.47}$$

Taking into consideration that we are investigating motion in a central force field U(r), we find

$$A(\mathcal{E}) = \left\{ \int \delta \left[\mathcal{E} - \frac{mv^2}{2} - U(r) \right] d^3v\, d^3r \right\}^{-1} = \frac{1}{16\pi^2} \left\{ \int v^2 r^2 \delta \left[\mathcal{E} - \frac{mv^2}{2} - U(r) \right] dv\, dr \right\}^{-1}$$
$$= \frac{m^{3/2}}{16\pi^2 \sqrt{2}} \left[\int \sqrt{\mathcal{E} - U(r)}\, r^2\, dr \right]^{-1}. \tag{8.48}$$

The integration in the last expression is carried out over the domain of r in which $\mathcal{E} - U(r) \geq 0$. For $\mathcal{E} < 0$ the coefficient $A(\mathcal{E})$ differs from zero only for particles which are being attracted, that is, only when $U(r) < 0$. If the function U(r) has a region where it falls off more rapidly than $1/r^2$, then bounded motion is also possible for positive energies $\mathcal{E} > 0$ (see §32). In this case, however, the boundary of the regions of bounded and unbounded motions, determined by conditions (8.11), depends both on the energy as well as on the angular momentum of the particle. If U(r) always falls off more slowly than $1/r^2$, then bounded motion is possible only for $\mathcal{E} < 0$ [see Eq. (8.7)].

Let us consider, for example, a Coulomb center with charge eZ. Then

$$U(r) = -\frac{e^2 Z}{r}. \tag{8.49}$$

It follows from Eq. (8.48), upon use of (8.49), that

$$A(\mathcal{E}) = \frac{m^{3/2} |\mathcal{E}|^{5/2}}{\sqrt{2}\, \pi^3 Z^3 e^6}. \tag{8.50}$$

Here the identity

$$\int_0^1 \sqrt{1-x}\, x^{3/2}\, dx = \frac{\pi}{16}.$$

has been taken into consideration. Similarly, the coefficient $A(\mathcal{E}, \mathcal{M})$ in the general equation (8.45) has the form

$$A(\mathcal{E}, \mathcal{M}) = \left[\int \frac{d^2\Gamma}{d\mathcal{E}\, d\mathcal{M}}\right]^{-1} = \frac{m^{5/2}}{8\sqrt{2}\,\pi^2\mathcal{M}} \left[\int \frac{dr}{\sqrt{\mathcal{E} - U(r) - \frac{\mathcal{M}^2}{2mr^2}}}\right]^{-1}. \tag{8.51}$$

The integration here is carried out over that range of values of r in which the expression under the radical sign is greater than zero. In particular, for the Coulomb field (8.49) when $\mathcal{E} < 0$, we find

$$\int \frac{dr}{\sqrt{\mathcal{E} + \frac{e^2Z}{r} - \frac{\mathcal{M}^2}{2mr^2}}} = \frac{\pi e^2 Z}{2|\mathcal{E}|^{3/2}}. \tag{8.52}$$

Consequently, in this case

$$A(\mathcal{E}, \mathcal{M}) = \frac{m^{5/2}|\mathcal{E}|^{3/2}}{4\sqrt{2\pi^3}Ze^2\mathcal{M}}. \tag{8.53}$$

c. Determination of the Coefficient $D(\mathcal{E})$. In order to clarify the meaning of the coefficient D in Eq. (8.44) we substitute the distribution function f describing a monoenergetic particle with energy $\mathcal{E}$ into it, that is, we assume that

$$f = f_{\mathcal{E}} = A(\mathcal{E})\,\delta\left[\mathcal{E} - \frac{mv^2}{2} - U(r)\right], \tag{8.54}$$

where $A(\mathcal{E})$ is the normalization coefficient defined above. From formula (8.47) defining $A(\mathcal{E})$ it is seen that $f_{\mathcal{E}}$ is normalized to one particle, that is, $\int f_{\mathcal{E}}\, d\Gamma = 1$.

Now we multiply the left and right sides of Eq. (8.42) by $(\Delta\mathcal{E})^2$, where

$$(\Delta\mathcal{E})^2 = [\mathcal{E}(v, r) - \mathcal{E}]^2, \tag{8.55}$$

and we integrate it with respect to $d\Gamma$. As a result, on the left we obtain

$$\frac{\partial}{\partial t}\left[\int f_{\mathcal{E}} (\Delta\mathcal{E})^2\, d\Gamma\right] = \frac{\partial\,\overline{(\Delta\mathcal{E})^2}}{\partial t}, \tag{8.56}$$

i.e., the square of the change of the energy of a particle per unit time, averaged over the distribution (8.54). The right side of this equation gives

$$\int (\Delta\mathcal{E})^2 A(\mathcal{E}) \frac{\partial}{\partial\mathcal{E}}\left[\frac{D(\mathcal{E})}{2A(\mathcal{E})}\left(\frac{\partial f_{\mathcal{E}}}{\partial\mathcal{E}} + \frac{1}{\varkappa T} f_{\mathcal{E}}\right)\right] d\Gamma = D(\mathcal{E}), \tag{8.57}$$

if it is integrated by parts twice and the fact that the function $f_{\mathcal{E}}$ is defined with the aid of (8.54) is taken into consideration. Thus,

$$D(\mathcal{E}) = \frac{\partial\,\overline{(\Delta\mathcal{E})^2}}{\partial t}, \tag{8.58}$$

i.e., $D(\mathscr{E})$ is the average of the square of the change per unit time of a particle's energy caused by collisions. In other words, $D(\mathscr{E})$ is the coefficient for the diffusion of the particle in energy space.

In order to calculate the quantity $\overline{\partial(\Delta\mathscr{E})^2}/\partial t$, it is now necessary to specifically examine the collision processes of particles which are trapped in finite orbits. We shall assume, as already mentioned above, that the captured particles are electrons which collide with neutral particles. In this case, the collision of an electron is a fast process in comparison with the frequency of revolution of an electron around the center of force, i.e., around the body.* Therefore, one can regard the electron as a free particle and immediately write $\overline{\partial(\Delta E)^2}/\partial t$ as the average of the product of the square of the energy change $(\Delta\mathscr{E})^2$ during one collision times the number per unit time, that is

$$\frac{\overline{\partial(\Delta\mathscr{E})^2}}{\partial t}=n_0\overline{\left[v_e\int\overline{(\Delta\mathscr{E})_n^2}\,d\sigma\right]}_e, \tag{8.59}$$

where n_0 is the concentration of neutral particles, v_e is the velocity of the electron, $d\sigma$ is the effective differential cross section for the scattering of an electron by an atom. In formula (8.59) the first averaging is carried out with respect to the velocities of the neutral particles, and the second is carried out with respect to the velocities of the electrons. Here it is also considered that the electron's velocity v_e is much larger than the velocity v_n of a neutral particle.

We now consider that the magnitude of the relative velocity of an electron does not change upon collision (see p. 60 of [27])

$$(\mathbf{v}_e-\mathbf{v}_n)^2=(\mathbf{v}_e'-\mathbf{v}_n')^2,$$

where $\mathbf{v}_e$ and $\mathbf{v}_n$ are the velocities of these particles after the collision. Therefore

$$v_e'^2-v_e^2=2(\mathbf{v}_e'\mathbf{v}_n'-\mathbf{v}_e\mathbf{v}_n)\simeq 2\mathbf{v}_n(\mathbf{v}_e'-\mathbf{v}_e),$$

if it is considered that the velocity of the neutral particle generally changes little upon collision with an electron.

Thus,

$$(\Delta\mathscr{E})^2=\frac{1}{4}m^2(v_e'^2-v_e^2)^2\simeq m^2[(\mathbf{v}_n\mathbf{v}_e)^2+(\mathbf{v}_n\mathbf{v}_e')^2-2(\mathbf{v}_n\mathbf{v}_e)(\mathbf{v}_n\mathbf{v}_e')]. \tag{8.60}$$

It is now necessary to average this expression with respect to the velocities of the atom. Taking into consideration that $\overline{v_n^2}=3\varkappa T/M$ and

$$\overline{(\mathbf{v}_n\mathbf{v}_e)_n^2}=\overline{(v_nv_e\cos\theta_1')_n^2}=\frac{1}{3}\overline{v_n^2}v_e^2=\frac{\varkappa T}{M}v_e^2,\quad \overline{[(\mathbf{v}_n\mathbf{v}_e)(\mathbf{v}_n\mathbf{v}_e')]_n}=\frac{1}{3}\overline{v_n^2}(\mathbf{v}_e\mathbf{v}_e')=\frac{\varkappa T}{M}(\mathbf{v}_e\mathbf{v}_e'),$$

we have

$$\overline{(\Delta\mathscr{E})_n^2}=\frac{2\varkappa Tm^2}{M}(v_e^2-\mathbf{v}_e\mathbf{v}_e')=\frac{2\varkappa Tm^2}{M}v_e^2(1-\cos\theta_2), \tag{8.61}$$

where θ_2 is the angle between the velocities $\mathbf{v}_e$ and $\mathbf{v}_e'$. Now substituting $\overline{(\Delta\mathscr{E})_n^2}$ into expression (8.59), we find that

$$D(\mathscr{E})=\frac{\overline{\partial(\Delta\mathscr{E})^2}}{\partial t}=\frac{2\varkappa Tm^2n_0}{M}\overline{[\sigma^*(v_e)v_e^3]}_e, \tag{8.62}$$

where $\sigma^*(v_e)$ is the effective transport cross section for the collision of an electron with an atom, that is

$$\sigma^*(v_e)=\int(1-\cos\theta_2)\,d\sigma. \tag{8.63}$$

*If a is the dimension of the neutral particle and R is the radius of the electron's orbit, then this condition is always satisfied for $R\gg a$. Since R is not smaller than the size of the body R_0, then for any macroscopic body $R_0\gg a$ and then the condition $R\gg a$ is automatically fulfilled.

The average in expression (8.62) is taken with respect to the distribution of electrons (8.54):

$$\overline{[\sigma^*(v_e)\,v_e^3]}_e=\int v^3\sigma^*(v)\,A(\mathscr{E})\,\delta\left[\mathscr{E}-\frac{mv^2}{2}-U(r)\right]d\Gamma=16\pi^2A(\mathscr{E})\int v^5\sigma^*(v)\,r^2\delta\left[\mathscr{E}-\frac{mv^2}{2}-U(r)\right]dv\,dr. \tag{8.64}$$

Assuming for simplicity that the field U(r) is the Coulomb field (8.49) and that the transport cross section σ^* does not depend on the electron's velocity v, we have (for $\mathscr{E}<0$)

$$\overline{[\sigma^*(v)\,v^3]}_e=\sigma^*\,\overline{[v^3]}_e=\frac{64\pi^2}{3}\,\frac{e^6Z^3\sigma^*}{m^3|\mathscr{E}|}\,A(\mathscr{E}). \tag{8.65}$$

Substituting this expression into (8.62) and taking (8.50) into consideration, we find that

$$D(\mathscr{E})=\frac{64\sqrt{2}}{3\pi}\,\frac{m^{1/2}\varkappa T\,|\mathscr{E}|^{3/2}\sigma^*n_0}{M}=\frac{64\sqrt{2}}{3\pi}\,\frac{m^{1/2}\varkappa T\,|\mathscr{E}|^{3/2}}{Ml}. \tag{8.66}$$

In the last expression it is considered that the mean free path of the electron is given by

$$l=\frac{1}{\sigma^*n_0}.$$

Thus, the coefficient $D(\mathscr{E})$ substantially depends on the nature of the collisions of the captured particles. In the general case, the diffusion equation (8.45) in the space of energy $\mathscr{E}$ and angular momentum $\mathscr{M}$ is now characterized by not one but by three diffusion coefficients $D_{\mathscr{E}\mathscr{E}}$, $D_{\mathscr{M}\mathscr{M}}$, $D_{\mathscr{E}\mathscr{M}}$, which one can find by the same method, evaluating $\overline{(\Delta\mathscr{E})^2}$, $\overline{(\Delta\mathscr{M})^2}$ and $\overline{(\Delta\mathscr{E}\,\Delta\mathscr{M})}$.

d. Solution of the Kinetic Equation. The kinetic equation (8.30), describing the distribution of the particles undergoing bounded motion in a weakly ionized plasma around a positive Coulomb center, is written in the form (8.44)

$$\overline{Y}(f)=A(\mathscr{E})\,\frac{\partial}{\partial\mathscr{E}}\left[\frac{D(\mathscr{E})}{2A(\mathscr{E})}\left(\frac{\partial f}{\partial\mathscr{E}}+\frac{1}{\varkappa T}f\right)\right]=0. \tag{8.44}$$

The coefficients $A(\mathscr{E})$ and $D(\mathscr{E})$ are determined by formulas (8.50) and (8.66). On the boundary of the region of bounded motion, where $\mathscr{E}=0$, the distribution function of the bounded particles must coincide with the distribution function (8.25) of the unbounded particles

$$f(0)=N_0\left(\frac{m}{2\pi\varkappa T}\right)^{3/2}. \tag{8.67}$$

It was already indicated above in §33 that the distribution function of the bounded particles is a solution of Eq. (8.30) in the presence of a stationary particle flux, caused by collisions, from the region of unbounded motion into the region of bounded motion. The general steady-state solution of Eq. (8.44) with nonvanishing total flux I has the form

$$f=I\exp\left(-\frac{\mathscr{E}}{\varkappa T}\right)\int_C^{\mathscr{E}}\frac{2A(\mathscr{E})}{D(\mathscr{E})}\exp\left(\frac{\mathscr{E}}{\varkappa T}\right)d\mathscr{E}, \tag{8.68}$$

where C is an arbitrary constant. It is necessary to use the boundary conditions in order to determine this constant. Since the entire investigation is carried out here under the assumption that the body's surface completely absorbs the particles incident on it, then, as already indicated above in §33, the distribution function of bounded particles, whose trajectories pass through the body's surface, is identically equal to zero. Consequently, if $U(R_0) = -e^2Z/R_0$ is the potential of the body's surface, then for energies $\mathcal{E} < U(R_0)$ the distribution function $f(\mathcal{E})$ is equal to zero. In other words, the constant C in formula (8.68) is equal to $U(R_0)$. We note that such a boundary condition is analogous to the boundary condition of the usual diffusion equation near an absorbing surface (see, for example, [38]). In this connection, if the potential $U(R_0)$ on the surface of the body is sufficiently large, $|U(R_0)| \gg \varkappa T$, then in the region of energies $\mathcal{E} > U(R_0)$ upon fulfillment of the condition $\mathcal{E} - U(R_0) \gg \varkappa T$ the quantity $U(R_0)$ only has an exponentially weak influence on the distribution function, and consequently on the value of interest to us of the concentration and flux of particles. Therefore, one can set $U(R_0) = -\infty$. Considering the boundary condition (8.67), we than obtain from (8.68) that the total particle flux at the center is given by

$$I = \frac{N_0}{2}\left(\frac{m}{2\pi\varkappa T}\right)^{3/2}\left\{\int_{-\infty}^{0} \exp\left(\frac{\mathcal{E}}{\varkappa T}\right)\frac{A(\mathcal{E})}{D(\mathcal{E})}\,d\mathcal{E}\right\}^{-1} = \frac{m^{3/2}N_0}{4\sqrt{2}\pi^{3/2}(\varkappa T)^{3/2}\int_0^{\infty}\frac{A(\mathcal{E})}{D(\mathcal{E})}\exp\left(-\frac{|\mathcal{E}|}{\varkappa T}\right)d|\mathcal{E}|}. \tag{8.69}$$

e. Concentration and Flux of the Particles Undergoing Bounded Motion. Taking into consideration that in a weakly ionized plasma the coefficient $D(\mathcal{E})$ for the electrons has the form (8.66), from (8.68) we find that the distribution function of the electrons undergoing bounded motion in the neighborhood of a Coulomb center with charge eZ is given by the expression

$$f = N_0\left(\frac{m}{2\pi\varkappa T}\right)^{3/2}\left(\frac{|\mathcal{E}|}{\varkappa T} + 1\right). \tag{8.70}$$

Here, of course, we neglect terms of order $\exp[U(R_0)/\varkappa T]$.

Integrating the distribution function (8.70), we find the concentration of bounded particles at the point r

$$N_b(r) = \int f\,d^3v = \frac{4}{3\sqrt{\pi}}N_0\left(\frac{e^2Z}{r\varkappa T}\right)^{3/2}\left(\frac{2}{5}\frac{e^2Z}{r\varkappa T} + 1\right). \tag{8.71}$$

The concentration of unbounded particles under the same conditions is given by expression (8.31). Substituting $U(r) = -e^2Z/r$ into it, we have

$$N^{+}_{unb}(r) = N_0\left\{\frac{2}{\sqrt{\pi}}\sqrt{\frac{e^2Z}{r\varkappa T}} + \left[1 - \Phi\left(\sqrt{\frac{e^2Z}{r\varkappa T}}\right)\right]\exp\left(\frac{e^2Z}{r\varkappa T}\right)\right\}. \tag{8.72}$$

Comparing formulas (8.71) and (8.72), one can convince oneself that in a weak field, when $e^2Z/r\varkappa T \ll 1$, the concentration of unbounded particles exceeds the concentration of bounded particles. With increase of the field, the concentration of bounded particles, however, increases considerably faster, so that in strong fields when $e^2Z/r\varkappa T \gg 1$ it is already considerably larger than the concentration of unbounded particles. For $e^2Z/r\varkappa T \gg 1$

$$\frac{N_b(r)}{N^{+}_{unb}(r)} = \frac{4}{15}\left(\frac{e^2Z}{r\varkappa T}\right)^2 = \frac{4}{15}\left(-\frac{U(r)}{\varkappa T}\right)^2.$$

Here

$$U(r) = -e^2Z/r$$

is the potential of the field at the point r.

It is necessary to emphasize that the concentration of bounded particles in the presence of an absorbing center is radically different from the equilibrium concentration (8.1). From here it is clear that no matter how small the body may be, the absorption of particles on its surface has a strong influence on the character of the steady-state distribution of the bounded particles even in the far region of the body. This assertion, proved here for a specific case, is general. The form of the steady-state particle distribution function in the vicinity of the body is determined by the character of the collisions of particles. For a power law dependence of the collision cross section on the particle's energy, the distribution function is also only a power law function of the energy, as is clear from Eq. (8.68). Accordingly, the concentration of bounded particles increases with increase of the potential of the field only according to a power law, but not according to the exponential law (8.1).

The total flux on a Coulomb center of the electrons, undergoing bounded motion in a weakly ionized plasma, is given by

$$I_b = \frac{N_0 m^{3/2}}{4\sqrt{2}\pi^{3/2}(\varkappa T)^{3/2}\int\limits_0^\infty \frac{A(|\mathscr{E}|)}{D(|\mathscr{E}|)}\exp\left(-\frac{|\mathscr{E}|}{\varkappa T}\right)d|\mathscr{E}|} = \frac{16\sqrt{2\pi}}{3}\,\frac{e^6 Z^3 m^{1/2} N_0}{M(\varkappa T)^{5/2}\, l}. \tag{8.73}$$

Here expression (8.66) is used for the coefficient $D(|\mathscr{E}|)$. According to Eqs. (10.30), (10.38), and (10.41), the total flux on the body of the electrons being attracted and undergoing bounded motion equals

$$I_b = \pi R_0^2 N_0 \sqrt{\frac{8\varkappa T}{\pi m}}\left(1 + \frac{e^2 Z}{R_0 \varkappa T}\right), \tag{8.74}$$

where R_0 is the radius of the absorbing surface (here it is assumed that $U(R_0) = -e^2Z/R_0$). Comparing (8.73) and (8.74) for $e^2Z/\varkappa TR_0 = -U(R_0)/\varkappa T \gg 1$, we have

$$\frac{I_b}{I_{unb}} = \frac{8}{3}\frac{R_0}{l}\frac{m}{M}\left(\frac{e^2Z}{\varkappa T R_0}\right)^2 = \frac{8}{3}\frac{R_0}{l}\frac{m}{M}\left(-\frac{U(R_0)}{\varkappa T}\right)^2. \tag{8.75}$$

It follows from (8.75) that the flux of bounded electrons is small in comparison with the flux of unbounded electrons, since $R_0/l \ll 1$, $m/M \ll 1$. However, with increase of the potential of the field on the absorbing surface or with decrease of the dimensions of the surface, the flux of bounded particles increases more rapidly, so that when

$$R_0 < \left(\frac{e^2Z}{\varkappa T}\right)^2 \frac{m}{Ml}$$

it becomes larger than the flux of unbounded particles.

CHAPTER IX

DISTURBANCE OF THE PLASMA AND THE ELECTRIC FIELD IN THE VICINITY OF A CHARGED BODY AT REST

§ 36. Basic Equations

Let us consider a sphere of radius R_0 at rest in an infinite (without boundaries) rarefied plasma. As always, the size of the body will be assumed to be small in comparison with the mean free path of the particles. Let the surface of the sphere have a potential φ_0. In this connection, the plasma in the vicinity of the body is, of course, disturbed since space charge is formed which screens the field of the body. In order to investigate the disturbed region it is necessary to determine the total field created by the body and by the space charge, and also the distribution of electrons and ions, that is, to find the corresponding distribution functions and the potential of the field $\varphi(r)$.

In virtue of the spherical symmetry of the problem, the distribution functions f_i (of the ions) and f_e (of the electrons) only depend on r, v_r, and v_θ (see §33) and are determined by the kinetic equations (8.15):

$$v_r \frac{\partial f_i}{\partial r} + \left(\frac{v_\theta^2}{r} + \frac{e}{M_i}\frac{\partial \varphi}{\partial r}\right)\frac{\partial f_i}{\partial v_r} - \frac{v_\theta v_r}{r}\frac{\partial f_i}{\partial v_\theta} + Y(f_i) = 0, \tag{9.1}$$

$$v_r \frac{\partial f_e}{\partial r} + \left(\frac{v_\theta^2}{r} - \frac{e}{m}\frac{\partial \varphi}{\partial r}\right)\frac{\partial f_e}{\partial v_r} - \frac{v_\theta v_r}{r}\frac{\partial f_e}{\partial v_\theta} + Y(f_e) = 0, \tag{9.2}$$

where v_r and v_θ are the radial and tangential components of the particle's velocity, M_i is the mass of the ion, m is the mass of an electron, $-e$ is the electron's charge, e is the ion's charge, $Y(f_i)$ and $Y(f_e)$ are the collision integrals for ions and electrons. In virtue of the spherical symmetry, the equation (1.30) for the potential of the field has the form

$$\frac{1}{r^2}\frac{d}{dr}\left(r^2 \frac{d\varphi}{dr}\right) = -4\pi e\left[\int f_i\, d^3v - \int f_e\, d^3v\right]. \tag{9.3}$$

It is necessary to solve Eqs. (9.1)-(9.3) with account of the boundary conditions on the body's surface and at infinity (see §3). For the potential of the field on the surface of the body we have

$$\varphi(R_0) = \varphi_0; \tag{9.4}$$

and at infinity

$$\varphi_{r\to\infty} = 0. \tag{9.5}$$

At infinity the distribution functions are the Maxwellian distributions given by Eqs. (1.25) and (1.26). On the surface of the body, if all the electrons are absorbed and the ions are neutralized upon collision with the body, the function $f(v_r, v_\theta, r)_{r=R_0} = 0$ where for particles undergoing unbounded motion this condition should be satisfied only for $v_r > 0$ (see §33). If the particles are reflected from the body, then the boundary condition on the surface of the body essentially depends on the character of the reflection, that is, on whether specular reflection occurs, diffuse scattering, and so forth (see §3). Everywhere below (with the exception of §41) we shall consider the case of a totally absorbing body.

One can look for the solution of equations (9.1)-(9.3) as follows: find the distribution functions f_i and f_e in an arbitrary central force field, calculate the ion and electron concentrations, $N_i(r, \varphi)$ and $N_e(r, \varphi)$, and then carry out the integration of Eq. (9.3) for the potential of the field.

Expressions for the concentration of attracted and repelled particles in an arbitrary central force field in the presence of an absorbing body were obtained in Chapter VIII, where it is also shown that it is convenient to write the distribution function, and therefore the concentration of attracted particles, separately for particles undergoing bounded and unbounded motion. Thus, in the general case, Eq. (9.3) has the form

$$\frac{1}{r^2}\frac{d}{dr}\left(r^2\frac{dU}{dr}\right)=-4\pi e^2[N_b^+(r,U)+N_{unb}^+(r,U)-N^-(r,U)], \tag{9.3a}$$

where $U(r)=-e|\varphi|$, N^+ is the concentration of particles being attracted, N^- is the concentration of particles being repelled. It is considered that only particles being attracted can perform bounded motion. What kind of particles are attracted – electrons or ions – depends, of course, on the sign of the charge on the body itself.

In Chapter VIII it was shown that the concentration of particles being repelled, at a given point r, depends only on the value of the potential of the field at that point and on the surface of the body. However, the concentration of attracted particles depends on the overall behavior of the potential of the field in the vicinity of the body. Here it is very important how the field falls off with increasing distance from the body – faster or slower than $1/r^2$. For example, if the field away from the surface of the body always falls off faster than $1/r^2$, then there are no finite trajectories which do not pass through the body's surface, and consequently in this case the concentration $N_b^+(r, U)$ in Eq. (9.3a) is equal to zero.

In a plasma over distances smaller than the Debye radius $D=(\varkappa T/4\pi e^2N_0)^{1/2}$, the potential of the field falls off according to Coulomb's law, that is, more slowly than $1/r^2$. Over distances which are large in comparison with D, Debye screening plays an important role, which leads to a much more abrupt decrease of the field. From here it follows that the behavior of the field and the structure of the plasma in the vicinity of large or small bodies, that is, when $R_0 \gg D$ or $R_0 \ll D$, differ substantially. Both of these cases are considered below. It is important to emphasize that at large distances from an absorbing body, distances which exceed its dimensions and the Debye radius, i.e., where $r \gg R_0$ and $r \gg D$, the potential φ of the electric field always falls off exactly like $1/r^2$, as shown in § 16. It is also easy to understand such a dependence of φ on r from simple considerations. In fact, in the regions $r \gg D$ Debye screening plays a decisive role, so that the plasma is quasi-neutral and the concentrations of electrons and ions are close to each other, that is, $N_i \approx N_e$. Here the potential of the field is proportional to the disturbance of the concentration (see § 14), so that $\varphi \sim \delta N = |N - N_0|$, where N_0 is the undisturbed concentration of electrons or ions. Since the disturbance is a consequence of the absorption of particles on the body's surface, so at large distances from the body when $r \gg R_0$ the disturbance of the concentration of particles is proportional to the angle which the body subtends as seen from a given point r, i.e., proportional to R_0^2/r. Thus, at large distances from the body, φ is always proportional to $\delta N \sim 1/r^2$, that is

$$\varphi=\varphi_\infty\frac{R_0^2}{r^2}, \tag{9.6}$$

where φ_∞ is the asymptotic value of the potential, characterizing the behavior of the field at large distances from the body.

§ 37. Electric Field and Structure of the Plasma in the Vicinity of a Large Body

a. Transformation of the Initial Equations. For not very large values of the potential on the surface of a large body, the field in its neighborhood falls off faster than $1/r^2$, so that the concentration N_b^+ of bounded particles is equal to zero. In this case Eq. (9.3a) is written in the form

$$\frac{1}{r^2}\frac{d}{dr}\left(r^2\frac{dU}{dr}\right)=-4\pi e^2[N_{unb}^+(r,U)-N^-(r,U)]. \tag{9.7}$$

In Eq. (9.7) N_{unb}^+ denotes the concentration of attracted particles performing unbounded motion. It is determined by expression (8.39)

$$N^{+}_{\text{unb}}(r,\ U)=\frac{N_0}{2}\Big\{\Big[1-\Phi\Big(\sqrt{-\frac{U(r)}{\varkappa T}}\Big)\Big]\exp\Big[-\frac{Ur}{\varkappa T}\Big]$$

$$+\frac{2}{\sqrt{\pi}}\Big(\sqrt{-\frac{U(r)}{\varkappa T}}-\sqrt{-\frac{U(r)}{\varkappa T}-\frac{e\varphi_\infty}{\varkappa T}\frac{R_0^2}{r^2}}\Big)+\sqrt{1-\frac{R_0^2}{r^2}}\times$$

$$\times\Bigg[1-\Phi\Bigg(\sqrt{\frac{R_0}{2\varkappa T}\frac{dU}{dr}(R_0)-\frac{U(r)-U(R_0)}{\varkappa T\Big(1-\frac{R_0^2}{r^2}\Big)}}\Bigg)\Bigg]\quad\exp\Big[\frac{R_0^2U(R_0)-r^2U(r)}{(r^2-R_0^2)\varkappa T}\Big]+X(r,\ R_0)-X(\infty,\ r)\Big\}. \tag{9.8}$$

Here the function X(a, b) is defined by the formula

$$X(a,\ b)=-\frac{1}{2\sqrt{\pi\varkappa T}}\int\limits_b^a\frac{\Big(1-\frac{r_0^2}{r^2}\Big)\Big(3\frac{dU}{dr_0}+r_0\frac{d^2U}{dr_0^2}\Big)\exp A}{\Big[U(r_0)-U(r)-\frac{r_0}{2}\frac{dU}{dr_0}\Big(\frac{r_0^2}{r^2}-1\Big)\Big]^{1/2}}dr_0,$$

where

$$A=\Big[-\frac{U(r_0)}{\varkappa T}-\frac{r_0}{2\varkappa T}\frac{dU}{dr_0}\Big].$$

In connection with the derivation of this formula, it was assumed that the potential of the field falls off more rapidly than $1/r^2$ in the vicinity of the body, and exactly like $1/r^2$ at large distances from the body (see §34). In addition, it is assumed that in the case of a large body, $R_0 \gg D$, and therefore $R_0 > r_K$, $r > r_K$, where r_K is the distance which characterizes the variation of the potential U(r) of the field (see Eq. (8.8)); under these conditions, as shown in §32, $\rho_0(R_0)=R_0$ and $\rho_0(r)=r$.

According to Eq. (8.41), the concentration of repelled particles is given by

$$N^{-}(r,\ U)=\frac{N_0}{2}\Big\{1+\Phi\Big(\sqrt{\frac{U(r)-U(R_0)}{\varkappa T}}\Big)+\sqrt{1-\frac{R_0^2}{r^2}}$$

$$\times\Big[1-\Phi\Big(\sqrt{\frac{r^2U(r)-R_0^2U(R_0)}{(r^2-R_0^2)\varkappa T}}\Big)\Big]\ \exp\Big[\frac{R_0^2U(r)-R_0^2U(R_0)}{(r^2-R_0^2)\varkappa T}\Big]\Big\}\exp\Big[\frac{U(r)}{\varkappa T}\Big]. \tag{9.9}$$

Here it is considered that the potential energy of the repelled particles is $e|\varphi| = -U(r)$.

We see that with account of expressions (9.8) and (9.9), Poisson's equation (9.7) in the case investigated is extremely complicated. In terms of the dimensionless variables $\xi = r/R_0$, $\varphi^* = -U/\varkappa T = e|\varphi|/\varkappa T$, it can be rewritten in the form

$$\frac{1}{\xi^2}\frac{d}{d\xi}\Big(\xi^2\frac{d\varphi^*}{d\xi}\Big)=\Big(\frac{R_0}{D}\Big)^2\Big\{\frac{N^{+}_{\text{unb}}(\xi,\ \varphi^*)}{N_0}-\frac{N^{-}(\xi,\ \varphi^*)}{N_0}\Big\}. \tag{9.10}$$

Since only a large body, $R_0 \gg D$, is being considered here, a large parameter stands in front of the right side of Eq. (9.10). Hence it is clear that everywhere, where the derivative $d\varphi^*/d\xi$ is not very large, so that

$$\frac{d\varphi^*}{d\xi}\ll\frac{R_0}{D}\varphi^*, \tag{9.11}$$

the solution of Eq. (9.10) correct to small terms of order $[(D/R_0\varphi^*)(d\varphi^*/d\xi)]^2$ coincides with the solution of the equation of quasi-neutrality (compare with §14):

$$N^{+}_{\text{unb}}(\xi,\ \varphi^*)=N^{-}(\xi,\ \varphi^*). \tag{9.12}$$

In the vicinity of a large body for not too large values of φ_0, condition (9.11) is not satisfied only in the double layer, of thickness on the order of the Debye radius, near the surface of the body. Since $D \ll R_0$, then it is clear that the structure of the plasma in the neighborhood of the body is generally determined by Eq. (9.12).

It is extremely important, however, that the expressions for N^+_{unb} and N^- depend on the value of the potential of the field and even on the intensity of the field on the surface of the body. Moreover, there is a double layer at the surface of the body, in which the potential of the field abruptly changes from the value φ_0 to some other value $\varphi_1(\varphi_0)$. The potential φ_1, of course, is the solution of Eq. (9.12) when $r \to R_0$, that is, when $\xi \to 1$. Taking into consideration that in expression (9.8) for N^+_{unb}, the last integral term $X(r, R_0)$ and the last nonintegral term, proportional to $(1 - R_0^2/r^2)^{1/2}$, vanish when $r \to R_0$, we find that the potential φ_1 is determined by the equation

$$e^{-\varphi_1^*}\left[1+\Phi\left(\sqrt{\varphi_0^*-\varphi_1^*}\right)\right] = e^{\varphi_1^*}\left[1-\Phi\left(\sqrt{\varphi_1^*}\right)\right] + \frac{2}{\sqrt{\pi}}\sqrt{\varphi_1^*} - C_1(1, \varphi_1^*). \tag{9.13}$$

Here $\varphi_0^* = e|\varphi_0|/\varkappa T$, $\varphi_1^* = e|\varphi_1|/\varkappa T$, and the constant $C_1(1, \varphi_1^*)$ is given by

$$C_1(1, \varphi_1^*) = -\frac{1}{2\sqrt{\pi}} \int_1^\infty \frac{(1-\xi_0^2)\left(3\frac{d\varphi^*}{d\xi_0} + \xi_0 \frac{d^2\varphi^*}{d\xi_0^2}\right)(1-\exp A)}{\left[\varphi_1^* - \varphi^*(\xi_0) + \frac{\xi_0}{2}(\xi_0^2-1)\frac{d\varphi^*}{d\xi_0}\right]^{1/2}} d\xi_0, \tag{9.14}$$

where

$$A = \left[\varphi^*(\xi_0) + \frac{\xi_0}{2}\frac{d\varphi^*}{d\xi_0}\right].$$

For calculation of (9.14) it is taken into consideration that

$$-\frac{1}{2\sqrt{\pi}} \int_\xi^\infty \frac{\left(1-\frac{\xi_0^2}{\xi^2}\right)\left(3\frac{d\varphi^*}{d\xi_0} + \xi_0\frac{d^2\varphi^*}{d\xi_0^2}\right) d\xi_0}{\left[\varphi^*(\xi) - \varphi^*(\xi_0) + \frac{\xi_0}{2}\left(\frac{\xi_0^2}{\xi^2}-1\right)\frac{d\varphi^*}{d\xi_0}\right]^{1/2}} = \frac{2}{\sqrt{\pi}}\sqrt{\varphi^*(\xi) - \varphi_\infty^*}, \tag{9.14a}$$

where $\varphi_\infty^* = e\varphi_\infty/\varkappa T$, and φ_∞ is determined from Eq. (9.6). Equations (9.13) and (9.14) show that the boundary value of the potential φ_1^* itself essentially depends on the overall behavior of the field in the plasma in the neighborhood of the body.

Thus, when $D/R_0 \to 0$, the potential on the boundary of the body jumps and consequently has two boundary values φ_0^* and φ_1^*. It is now necessary to explain which of these values of φ^* appears in the expressions for $N^+_{unb}(\xi, \varphi^*)$ and $N^-(\xi, \varphi^*)$. From formula (8.37) in §34 it follows that in the presence of a thin double layer of thickness D on the boundary of the body, that is, a layer in which the potential of the field abruptly increases, the concentration of attracted particles outside the layer changes extremely weakly in its dependence on the behavior of the field inside the layer – only terms of order $(D/R_0)^2 \ln^2(R_0\varphi_0^*/D)$ are added. As $D/R_0 \to 0$ the concentration of attracted particles outside the layer generally does not depend on the behavior of the potential inside the layer. It is easy to understand this conclusion directly, since if a thin layer, in which the potential of the field changes very abruptly, and consequently there is a large force of attraction, exists at the boundary of the body, then every particle hitting the layer also hits the surface of the body and is absorbed on it. The presence of such a layer is equivalent, therefore, to an increase of the dimensions of the body by an amount D, so that when $D \ll R_0$ the presence of the layer does not manifest itself in the distribution of the attracted particles outside the layer. For the particles which are repelled, the picture, of course, is completely different: the sharp increase of the force, repelling the particles in the double layer, leads to the result that the majority of these particles are reflected, without having reached the surface of the body. The boundary value of the field for them is, of course, the value of the potential on the surface of the body.

Thus, in the expression for N^- in Eq. (9.12), $\varphi^*(R_0) = \varphi_0^*$, where φ_0^* is the specified value of the potential on the surface of the body, and in the expression for N^+_{unb} in the same equation, $\varphi^*(R_0) = \varphi_1^*$, where φ_1^* is the boundary value of the potential, determined by relation (9.13). Of course, it is possible to determine φ_1^* and $d\varphi^*/d\xi\,(1)$ only from the solution of Eq. (9.12). With account of the indicated conditions on the boundary

($r = R_0$, that is $\xi = 1$), Eq. (9.12) becomes closed: by solving it one can determine the electric field in the vicinity of a large body, that is, for $R_0 \gg D$.

Inside the double layer the concentrations of electrons and ions differ substantially, and here it is necessary to now solve the complete equation (9.10). In this connection, however, since the question is only about very small distances from the surface of the body ($r - R_0 \sim D \ll R_0$), so the curvature of the surface is unimportant, thanks to which expressions (9.8) and (9.9) for the concentrations of attracted and repelled particles are substantially simplified*

$$N^+_{\text{unb}}(\varphi^*) = \frac{N_0}{2}\left\{e^{\varphi^*}\left[1-\Phi\left(\sqrt{\varphi^*}\right)\right] + \frac{2}{\sqrt{\pi}}\left(\sqrt{\varphi^*} - \sqrt{\varphi^* - \varphi^*_\infty}\right) + C(\varphi^*)\right\}, \tag{9.15}$$

$$N^-(\varphi^*) = \frac{N_0}{2} e^{-\varphi^*}\left\{1 + \Phi\left(\sqrt{\varphi^*_0 - \varphi^*}\right)\right\}. \tag{9.16}$$

Here the function $C(\varphi^*)$ is given by

$$C(\varphi^*) = -\frac{1}{2\sqrt{\pi}} \int_1^\infty \frac{(1-\xi_0^2)\left(3\frac{d\varphi^*}{d\xi_0} + \xi_0 \frac{d^2\varphi^*}{d\xi_0^2}\right)\exp A}{\left[\varphi^* - \varphi^*(\xi_0) + \frac{\xi_0}{2}(\xi_0^2 - 1)\frac{d\varphi^*}{d\xi_0}\right]^{1/2}} d\xi_0, \tag{9.15a}$$

where

$$A = \left[\varphi^*(\xi_0) + \frac{\xi_0}{2}\frac{d\varphi^*}{d\xi_0}\right].$$

Thus, Eq. (9.7) in the double layer can be rewritten in terms of the dimensionless variables $\varphi^* = e|\varphi|/\varkappa T$ and $\zeta = (r - R_0)/D$ in the form

$$\frac{d^2\varphi^*}{d\zeta^2} = \frac{1}{2}\left\{e^{\varphi^*}\left[1-\Phi\left(\sqrt{\varphi^*}\right)\right] + \frac{2}{\sqrt{\pi}}\left(\sqrt{\varphi^*} - \sqrt{\varphi^* - \varphi^*_\infty}\right) + C(\varphi^*) - e^{-\varphi^*}\left[1 + \Phi\left(\sqrt{\varphi^*_0 - \varphi^*}\right)\right]\right\}. \tag{9.17}$$

The boundary conditions are: for $\zeta = 0$, $\varphi^* = \varphi_0^*$, and as $\zeta \to \infty$, $\varphi^* \to \varphi_1^*$, where φ_1^* is the value of the potential of the field on the boundary of the double layer, determined by relation (9.13). Of course, when $\varphi^* \to \varphi_1^*$, the right side of Eq. (9.13) vanishes.

It is necessary to emphasize that we consider here only the case of a large body, that is, a body whose dimension is much larger than the thickness of the double layer. In spite of this, the specific shape of the body still influences the concentration and, consequently, the potential of the electric field in the double layer:

*We note that it is easy to generalize formula (9.16) to the case when the particles being repelled are partially reflected upon collision with the body's surface. Thus, if the reflection is specular, then

$$N^- = \frac{N_0}{2} e^{-\varphi^*}\left\{1 + \Phi\left(\sqrt{\varphi_0^* - \varphi^*}\right) + R\left[1 - \Phi\left(\sqrt{\varphi_0^* - \varphi^*}\right)\right]\right\}, \tag{9.16a}$$

where R is the coefficient for the reflection of repelled particles from the body's surface. In the case of total absorption, R = 0, and formula (9.16a) coincides with (9.16). For total reflection, R = 1, and the concentration of repelled particles is, of course, determined by the Boltzmann formula.

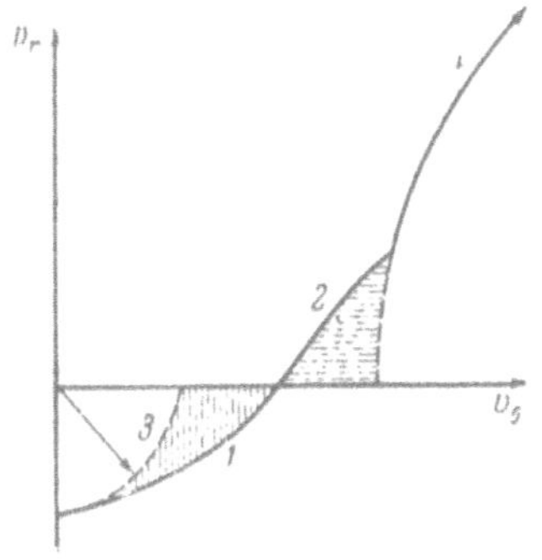

Fig. 59. Domain of integration in velocity space.

the form of the function $C(\varphi^*)$ varies depending on the shape of the body's surface. The reason for this is that the electric field, which exists not only in the double layer but also in the quasi-neutral plasma in the vicinity of the body, depends, of course, on the shape of the body. The function $C(\varphi^*)$ takes the influence of this field on the concentration of attracted particles in the double layer into consideration.

b. Solution of the Equation of Quasi-Neutrality. The potential of the field in the vicinity of a body is determined by Eq. (9.12). Its solution, of course, can only be obtained numerically. First let us consider for simplicity the case of a strongly charged body, when $\varphi^* = e|\varphi_0|/\varkappa T \gg 1$. In this case, expression (9.9) for the concentration of the repelled particles simplifies to

$$N^- = N_0 e^{-\varphi^*}, \tag{9.18}$$

and Eq. (9.12) assumes the form

$$\varphi^* = \ln \frac{2}{A}, \tag{9.19}$$

where

$$A = e^{\varphi^*}\left[1 - \Phi\left(\sqrt{\varphi^*}\right)\right] + \frac{2}{\sqrt{\pi}}\left(\sqrt{\varphi^*} - \sqrt{\varphi^* - \frac{\varphi^*_\infty}{\xi^2}}\right)$$

$$+ \sqrt{1 - \frac{1}{\xi^2}}\left[1 - \Phi\left(\sqrt{\frac{\varphi^* - \varphi_1^*}{1 - 1/\xi^2} - \frac{1}{2}\frac{d\varphi^*}{d\xi}(1)}\right)\right] \exp\left[\frac{\xi^2\varphi^* - \varphi_1^*}{\xi^2 - 1}\right] + X(r, R_0) - X(\infty, r).$$

Here X denotes the integral terms of (9.8), φ_1 is the value of φ_1^* for $\xi = 1$, and φ^*_∞ is the asymptotic value determined by the behavior of the function $\varphi^*(\xi)$ as $\xi \to \infty$ [see Eq. (9.6)].

In order to solve Eq. (9.19) it is natural to use the method of iterations. The success of this method essentially depends on the choise of the initial function $\varphi_0^*(\xi)$. In order to determine $\varphi_0^*(\xi)$, instead of (9.19) we write a simple approximate equation and find its solution. For this, it is necessary to replace the complicated integral terms in Eq. (9.19), that is, in formula (9.8) for $N^+_{\text{unb}}(\xi, \varphi^*)$, by simple approximate expressions.

Expression (9.8) for the concentration of the attracted particles was obtained in §34 as a result of integrating the Maxwellian distribution function in velocity space over the region whose boundary is indicated by the bold-face curve in Fig. 59. The first and second integral terms appear as the result of integration over the regions whose boundaries are marked by the numbers 1 and 2 on the figure. In order to obtain an approximate expression for the concentration of attracted particles, it is natural to continue the circle 3 (see Eq. (8.7)) and the hyperbola 4 (see Eq. (8.28)) up to the intersection with the v_θ axis. Then it is found that

$$N^+_{(0)\,\text{unb}} = \frac{N_0}{2}\left\{e^{\varphi^*}\left[1 - \Phi\left(\sqrt{\varphi^*}\right)\right] + \frac{2}{\sqrt{\pi}}\sqrt{\varphi^*} + \sqrt{1 - \frac{1}{\xi^2}}\exp\left(\frac{\xi^2\varphi^* - \varphi_1^*}{\xi^2 - 1}\right)\right\}. \tag{9.20}$$

Formula (9.20) is the desired simple approximate expression for N^+_{unb}.

Equation (9.19) can be rewritten in the form

$$\varphi^* = \ln \frac{2}{B - C_1[\xi, \varphi^*(\xi)] + C_2[\xi, \varphi^*(\xi)]}, \tag{9.21}$$

where

$$B = e^{\varphi^*}\left[1 - \Phi\left(\sqrt{\varphi^*}\right)\right] + \frac{2}{\sqrt{\pi}}\sqrt{\varphi^*} + \sqrt{1 - \frac{1}{\xi^2}}\exp\left(\frac{\xi^2\varphi^* - \varphi_1^*}{\xi^2 - 1}\right),$$

TABLE XXII. Dependence of the Potential $\varphi^* = e|\varphi|/\varkappa T$ on Distance $\xi = r/R_0$ for Successive Iterations

ξ	$\varphi^*_{(0)}$	$\varphi^*_{(1)}$	$\varphi^*_{(2)}$	$\varphi^*_{(3)}$	$\varphi^*_{(4)}$	φ^*
1.0	0.439	0.506	0.553	0.594	0.633	0.690
1.001	0.435	0.498	0.538	0.570	0.602	0.658
1.002	0.432	0.492	0.530	0.559	0.587	0.638
1.004	0.426	0.484	0.520	0.542	0.565	0.610
1.006	0.421	0.476	0.508	0.527	0.544	0.591
1.01	0.411	0.463	0.493	0.509	0.520	0.569
1.02	0.392	0.437	0.462	0.477	0.484	0.536
1.04	0.369	0.397	0.417	0.429	0.436	0.480
1.06	0.335	0.366	0.383	0.393	0.399	0.430
1.1	0.295	0.318	0.331	0.338	0.343	0.362
1.2	0.227	0.241	0.248	0.252	0.256	0.261
1.4	0.151	0.157	0.161	0.162	0.164	0.164
1.6	0.110	0.113	0.115	0.116	0.118	0.117
2.0	0.066	0.0068	0.069	0.069	0.070	0.068
2.5	0.041	0.041	0.042	0.042	0.043	0.041
3.0	0.028	0.028	0.028	0.028	0.029	0.028
φ^*_∞	0.219	0.226	0.231	0.233	0.235	0.237

and $C_1[\xi, \varphi^*(\xi)]$ and $C_2[\xi, \varphi^*(\xi)]$ are correction terms, arising as the result of integration over the cross-hatched regions, namely:

$$C_1[\xi, \varphi^*(\xi)] = -\frac{1}{2\sqrt{\pi}} \int_\xi^\infty \frac{\left(1-\frac{\xi_0^2}{\xi^2}\right)\left(3\frac{d\varphi^*}{d\xi_0}+\xi_0\frac{d^2\varphi^*}{d\xi_0^2}\right)\left(1-\exp\left[\varphi^*(\xi_0)+\frac{\xi_0}{2}\frac{d\varphi^*}{d\xi_0}\right]\right)}{\left[\varphi^*(\xi)-\varphi^*(\xi_0)+\frac{\xi_0}{2}\left(\frac{\xi_0^2}{\xi^2}-1\right)\frac{d\varphi^*}{d\xi_0}\right]^{1/2}} d\xi_0, \tag{9.22}$$

$$C_2[\xi, \varphi^*(\xi)] = -\frac{1}{2\sqrt{\pi}} \int_\xi^1 \frac{\left(1-\frac{\xi_0^2}{\xi_0^2}\right)\left(3\frac{d\varphi^*}{d\xi_0}+\xi_0\frac{d^2\varphi^*}{d\xi_0^2}\right)\exp\left[\varphi^*(\xi_0)+\frac{\xi_0}{2}\frac{d\varphi^*}{d\xi_0}\right]}{\left[\varphi^*(\xi)-\varphi^*(\xi_0)+\frac{\xi_0}{2}\left(\frac{\xi_0^2}{\xi^2}-1\right)\frac{d\varphi^*}{d\xi_0}\right]^{1/2}} d\xi_0 -$$

$$-\sqrt{1-\frac{1}{\xi^2}}\,\Phi\left(\sqrt{\frac{\varphi^*-\varphi_1^*}{1-1/\xi^2}-\frac{1}{2}\frac{d\varphi^*}{d\xi}(1)}\right)\exp\left(\frac{\xi^2\varphi^*-\varphi_1^*}{\xi^2-1}\right). \tag{9.23}$$

Equation (9.21) with account of expressions (9.22) and (9.23) for $C_1[\xi, \varphi^*]$ and $C_2[\xi, \varphi^*]$ (also see Eq. (9.14a)) identically coincides with Eq. (9.19).

The solution of Eq. (9.21) is found by the method of iterations, as already indicated above. In the zero order approximation, we assume C_1 and C_2 to be equal to zero. It is then easy to find the function $\varphi^*_{(0)}(\xi)$ as the solution of the transcendental equation (9.21) for $C_1 = 0$, $C_2 = 0$. It is given in Table XXII and in Figs. 60a and 60b. Then the function $\varphi^*_{(0)}(\xi)$ is substituted into expressions (9.22) and (9.23) for C_1 and C_2 and the correction functions

$$C_{10}(\xi) = C_1[\xi, \varphi^*_{(0)}(\xi)], \quad C_{20}(\xi) = C_2[\xi, \varphi^*_{(0)}(\xi)].$$

are calculated. The functions $C_{10}(\xi)$ and $C_{20}(\xi)$ are substituted into Eq. (9.21). By now solving the transcendental equation obtained for φ^*, we find the function $\varphi^*_{(1)}(\xi)$ of the first-order approximation. It is also given in Table XXII and in Fig. 60. It is clear from the table and from the figure that the deviation between $\varphi^*_{(0)}(\xi)$ and $\varphi^*_{(0)}(\xi)$ is not large, and therefore the zero-order approximation selected by us is correct.

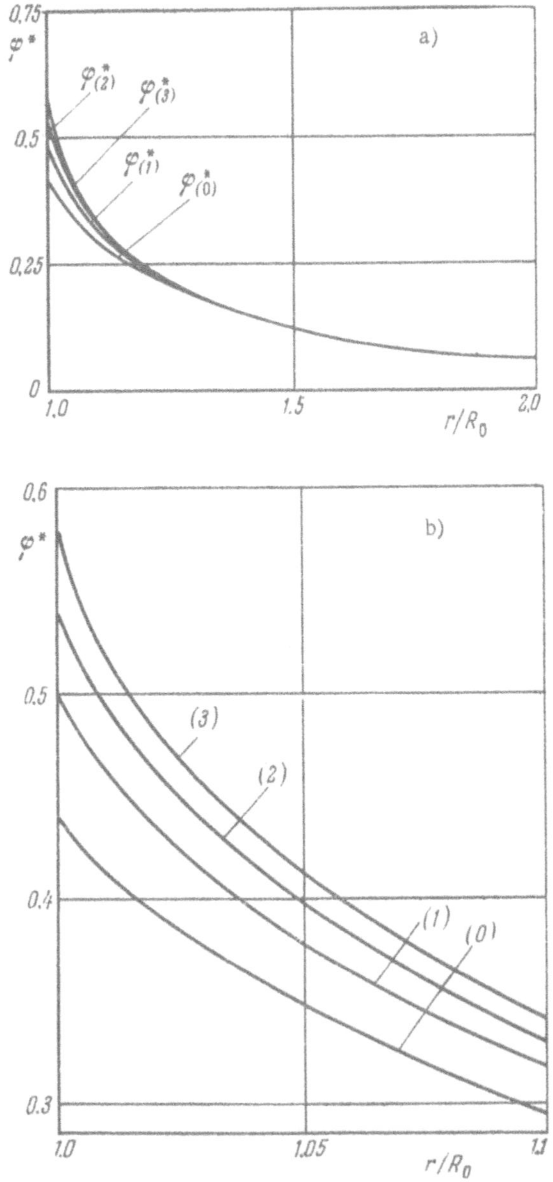

Fig. 60. a) Variation of the potential of the field in the vicinity of a body when $e\varphi_0/\varkappa T \gg 1$. The various curves are successive iterations. b) The same curves in the zone nearest to the body.

Now substituting $\varphi^*_{(1)}(\xi)$ into expressions (9.22) and (9.23), we find the correction functions $C_{11}(\xi)$ and $C_{21}(\xi)$:

$$C_{11}(\xi)=C_1[\xi,\ \varphi^*_{(1)}(\xi)],\quad C_{21}(\xi)=C_2[\xi,\ \varphi^*_{(1)}(\xi)].$$

Substituting the functions $C_{11}(\xi)$ and $C_{21}(\xi)$ into Eq. (9.21) and solving it, we find the second-order approximation function $\varphi^*_{(2)}(\xi)$,* and so forth.

This method leads to a converging sequence of functions $\varphi^*_{(0)}(\xi)$, $\varphi^*_{(1)}(\xi)$, $\varphi^*_{(2)}(\xi)$... It is necessary, however, to carry out a large number of iterations before one is able, by means of extrapolation, to determine the final solution; the determination of the final values of the function $\varphi^*(\xi)$ near $\xi = 1$ precisely constitutes the fundamental difficulty of these calculations because of the large gradient of the function $\varphi^*(\xi)$ near the point $\xi = 1$. The results of the first four iterations and the final values of the function $\varphi^*(\xi)$ are presented in Table XXII, and also in Fig. 60a.

It is easy to determine the asymptotic behavior of the function $\varphi^*(\xi)$ for large values of ξ. In fact, under these conditions the potential $\varphi^*(\xi)$ and its derivative $d\varphi^*/d\xi$ are small. Therefore, expanding expression (9.21) in a series in powers of φ^* and $1/\xi^2$, we find that as $\xi\to\infty$ the potential φ^* decreases like $1/\xi^2$, as it should.

Here

$$\varphi^*_{\xi\to\infty}=\frac{\varphi^*_\infty}{\xi^2}=\frac{1}{\xi^2}\frac{1}{4}\left\{\frac{te^{-t^2}}{\sqrt{\pi}}+\left(\varphi^*_1+\frac{1}{2}\right)[1-\Phi(t)]+C_0\right\}, \tag{9.24}$$

where $t=\sqrt{\varphi^*_1-\frac{1}{2}\frac{d\varphi^*}{d\xi}(1)}$, and

---

*For a numerical calculation of the functions $C_1[\xi,\ \varphi^*(\xi)]$ and $C_2[\xi,\ \varphi^*(\xi)]$, it is apparently more convenient to rewrite the integral terms in the form

$$C_1[\xi,\ \varphi^*(\xi)]=\frac{2}{\sqrt{\pi}}\int\limits_0^{\sqrt{\varphi^*(\xi)-(\varphi^*_\infty/\xi^2)}}\left[\left(1-\exp-\tau_\xi^2+\varphi^*(\xi)+\frac{\xi_0^3}{2\xi^2}\frac{d\varphi^*(\xi_0)}{d\xi_0}\right)\right]d\tau_\xi,$$

$$C_2[\xi,\ \varphi^*(\xi)]=\frac{2}{\sqrt{\pi}}\int\limits_0^{\tau_\xi(1)}\exp\left[-\tau_\xi^2+\varphi^*(\xi)+\frac{\xi_0^3}{2\xi^2}\frac{d\varphi^*(\xi_0)}{d\xi_0}\right]d\tau_\xi$$

$$-\sqrt{1-\frac{1}{\xi^2}}\,\Phi\left(\sqrt{\frac{\varphi^*-\varphi^*_1}{1-1/\xi^2}-\frac{1}{2}\frac{d\varphi^*}{d\xi}(1)}\right)\exp\left(\frac{\xi^2\varphi^*-\varphi^*_1}{\xi^2-1}\right),$$

where

$$\tau_\xi(\xi_0)=\sqrt{\varphi^*(\xi)-\varphi^*(\xi_0)+\frac{\xi_0^3}{2\xi^2}\frac{d\varphi^*}{d\xi_0}\left(1-\frac{\xi^2}{\xi_0^2}\right)}$$

and $\xi_0(\tau_\xi)$ is the function whose inverse is the function $\tau_\xi(\xi_0)$. Such a transformation of the integral gets rid of the necessity to twice differentiate the function $\varphi^*(\xi)$, which is difficult to do in a numerical calculation. We also note that as $\xi\to 1$ the derivative $(d\varphi^*/d\xi)$ tends to infinity. However, this increase is extremely weak: $d\varphi/d\xi\sim\ln(\xi-1)$. For a numerical calculation it is therefore convenient to use the expression (9.23) given above for $C_2[\xi,\ \varphi^*(\xi)]$ which considers that the derivative $d\varphi^*/d\xi$ is finite as $\xi\to 1$. In this connection, the calculation of the integral in (9.23) is taken only up to a value ξ_0 close to 1 (for example, up to $\xi_0 = 1.0001$), and the value of $d\varphi^*/d\xi$ for $\xi = 1.0001$ is substituted as the value of the derivative $d\varphi^*/d\xi$ for $\xi = 1$. The result, of course, does not depend on the choise of the boundary value ξ_0, provided this value is sufficiently close to unity. It is also possible to use the transformation presented in this footnote for evaluation of the constant in formulas (9.24) and (9.30).

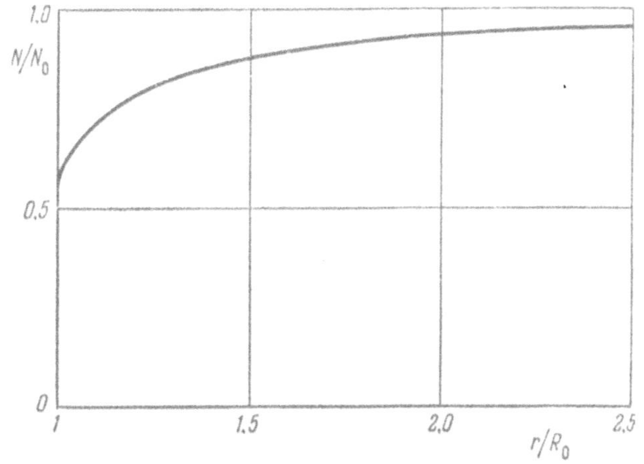

Fig. 61. Curve showing the variation of the concentration of particles in the vicinity of the body when $e\varphi_0/\varkappa T \gg 1$.

$$C_0 = -\frac{1}{4\sqrt{\pi}} \int_1^\infty \frac{\xi_0 \frac{d\varphi^*}{d\xi_0}\left(3\frac{d\varphi^*}{d\xi_0}+\xi_0\frac{d^2\varphi^*}{d\xi_0^2}\right)\exp\left(\varphi^*(\xi_0)+\frac{\xi_0}{2}\frac{d\varphi^*}{d\xi_0}\right)}{\left[-\varphi^*(\xi_0)-\frac{\xi_0}{2}\frac{d\varphi^*}{d\xi_0}\right]^{1/2}}\, d\xi_0. \tag{9.25}$$

In the case considered here, $\varphi^*_\infty = 0.237$ and, consequently,

$$\varphi_{\xi\to\infty} = 0.237/\xi^2. \tag{9.26}$$

Using the results of the calculations of the potential of the field in the vicinity of the body, we now determine the concentrations of ions and electrons with the aid of formula (9.18)

$$N = N^- = N^-_{\text{unb}} = N_0 \exp[-\varphi^*(\xi)].$$

The corresponding dependence of N on $\xi = r/R_0$ is shown in Fig. 61. From the figure it is clear that near the body's surface the disturbance of the charged particle concentrations is considered. At large distances from the body, $-\delta N = N_0 - N$ decreases in proportion to $1/r^2$, namely:

$$-\delta N_{\xi\to\infty} = N_0\frac{\varphi^*_\infty}{\xi^2} = \frac{0{,}237}{\xi^2}N_0.$$

The intensity of the electric field in the vicinity of the body is shown in Fig. 62.

c. The Disturbed Zone in the Vicinity of a Body for an Arbitrary Potential on Its Surface. Above the problem was solved under the assumption that the potential on the surface is large: $\varphi_0^* = e|\varphi_0|/\varkappa T \gg 1$. For arbitrary values of φ_0^*, Eq. (9.12) is written in the form

$$\varphi^* = \ln\left\{\frac{A}{B - C_1[\xi, \varphi^*(\xi)] + C_2[\xi, \varphi^*(\xi)]}\right\}, \tag{9.27}$$

where

$$A = 1 + \Phi\left(\sqrt{\varphi_0^* - \varphi^*}\right) + \sqrt{1-\frac{1}{\xi^2}}\left[1 - \Phi\left(\sqrt{\frac{\varphi_0^*-\varphi^*}{1-1/\xi^2}}\right)\right]\exp\left(\frac{\varphi_0^*-\varphi^*}{\xi^2-1}\right),$$

$$B = e^{\varphi^*}\left[1-\Phi\left(\sqrt{\varphi^*}\right)\right] + \frac{2}{\sqrt{\pi}}\sqrt{\varphi^*} + \sqrt{1-\frac{1}{\xi^2}}\exp\left(\frac{\xi^2\varphi^* - \varphi_1^*}{\xi^2-1}\right),$$

$\varphi_1^* = \varphi^*(1)$, and C_1 and C_2 are defined as before by expressions (9.22) and (9.23).

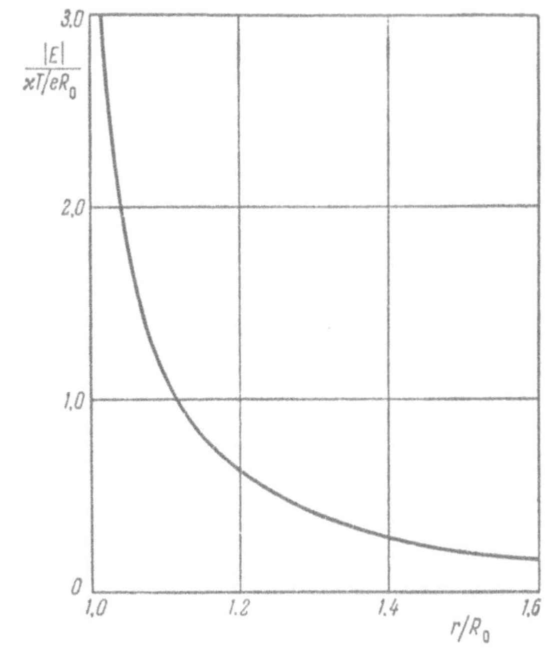

Fig. 62. Curve of the variation of the field intensity in the vicinity of the body for $e\varphi_0/\varkappa T \gg 1$.

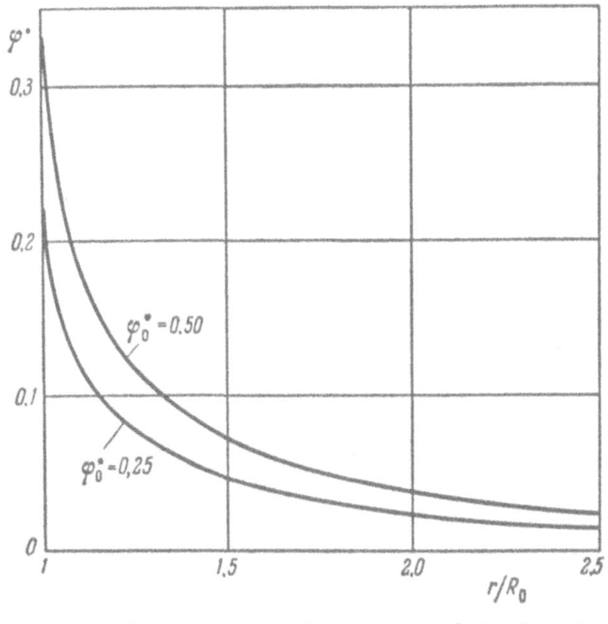

Fig. 63. Curves showing the variation of the electric field's potential in the vicinity of a body for $|\varphi_0| = 0.25(\varkappa T/e)$ and $|\varphi_0| = 0.5(\varkappa T/e)$.

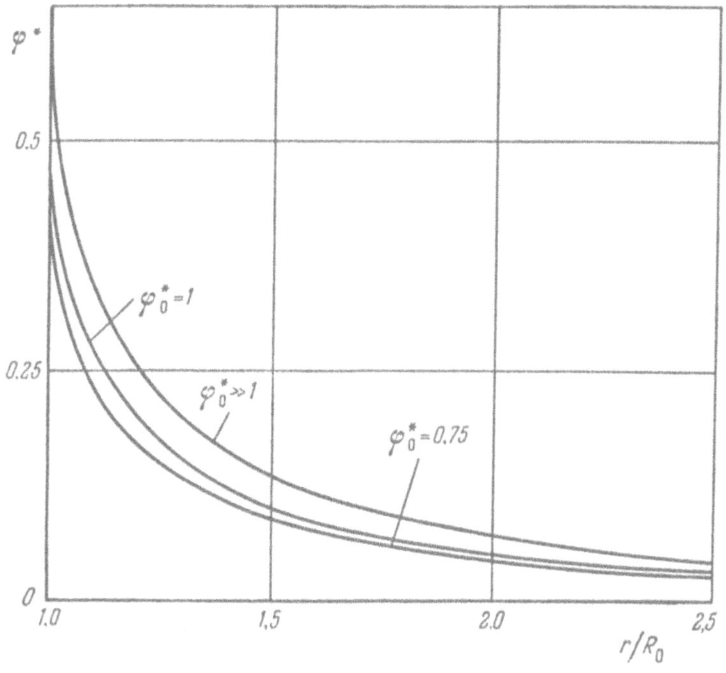

Fig. 64. Curves showing the variation of the electric field's potential in the vicinity of a body for $|\varphi_0| = 0.75(\varkappa T/e)$, $\varkappa T/e$, and $|\varphi_0| \gg (\varkappa T/e)$.

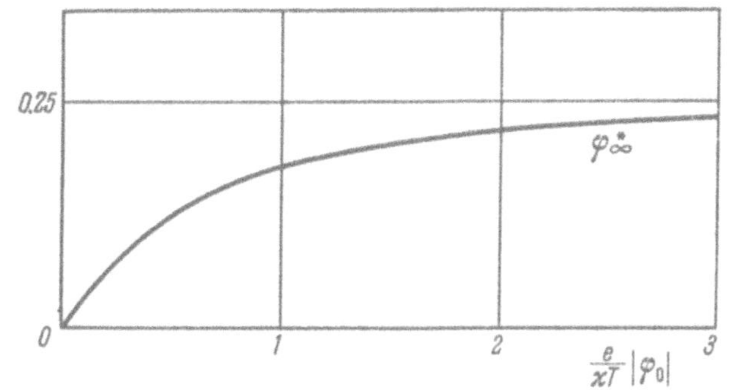

Fig. 65. Dependence of the asymptotic value (φ_∞^*), characterizing the behavior of the field at large distances from the body, on the potential on the body's surface.

TABLE XXIII. Basic Constants, Characterizing the Electric Field and the Particle Concentration, for Different Values of the Potential on the Body's Surface

φ_0^*	φ_1^*	φ_∞^*	δN_∞
0	0	0	0.25
$\varphi_0^* \ll 1$	φ_0^*	$\varphi_0^*/2$	0.25
0.25	0.22	0.082	0.25
0.50	0.37	0.13	0.25
0.75	0.47	0.16	0.25
1.0	0.53	0.18	0.24
$\varphi_0^* \gg 1$	0.69	0.24	0.24

The solution of Eq. (9.27) for arbitrary values of φ_0^* is found in quite analogous fashion to the solution of Eq. (9.21) considered above. The results of the corresponding calculation of the function $\varphi^*(\xi)$ for values of $\varphi_0^* = 0.25$, 0.50, 0.75, 1.0, and for $\varphi_0^* \gg 1$ are shown in Figs. 63 and 64. Using the values of the function $\varphi^*(\xi)$ found, with the aid of formula (9.9) it is not difficult to determine the concentrations of electrons and ions in the disturbed zone. In particular, at large distances from the body,

$$N^+_{\text{unb}} = N_0 \left\{ 1 + \varphi^* - \frac{1}{2\xi^2} \left[\left(\varphi_1^* + \frac{1}{2} \right) (1 - \Phi(t)) + \frac{te^{-t^2}}{\sqrt{\pi}} + C_0 \right] \right\} \tag{9.28}$$

and

$$N^- = N_0 \left\{ 1 - \varphi^* + \frac{1}{2\xi^2} \left[\left(\varphi_0^* - \frac{1}{2} \right) \left(1 - \Phi\left(\sqrt{\varphi_0^*}\right)\right) - \frac{\sqrt{\varphi_0^*}}{\sqrt{\pi}} e^{-\varphi_0^*} \right] \right\}. \tag{9.29}$$

as follows from expressions (9.8) and (9.9).

Here $t = \sqrt{-\varphi_1^* - \frac{1}{2} \frac{d\varphi^*}{d\xi}(1)}$, as before, and the constant C_0 is defined by expression (9.25). Consequently, for $\xi \gg 1$, $\varphi^* = \varphi_\infty^*/\xi^2$, where

$$\varphi_\infty^* = \frac{1}{4} \left\{ \left(\varphi_1^* + \frac{1}{2} \right) [1 - \Phi(t)] + \frac{te^{-t^2}}{\sqrt{\pi}} + C_0 + \left(\varphi_0^* - \frac{1}{2} \right) \left[1 - \Phi\left(\sqrt{\varphi_0^*}\right)\right] - \frac{\sqrt{\varphi_0^*}\, e^{-\varphi_0^*}}{\sqrt{\pi}} \right\}. \tag{9.30}$$

The disturbance of the particle concentration is here given by

$$N_0 - N_{\xi \to \infty} = -\delta N_{\xi \to \infty} = \frac{N_0}{4\xi^2} \left\{ \left(\varphi_1^* + \frac{1}{2} \right) [1 - \Phi(t)] + \frac{te^{-t^2}}{\sqrt{\pi}} + C_0 - \left(\varphi_0^* - \frac{1}{2} \right) \left[1 - \Phi\left(\sqrt{\varphi_0^*}\right)\right] + \frac{\sqrt{\varphi_0^*}}{\sqrt{\pi}} e^{-\varphi_0^*} \right\}.$$

The dependence of the values of φ_∞^* and $\delta N_\infty = -\xi^2 \delta N_{\xi \to \infty}/N_0$ on the value φ_0 of the potential on the body's surface is shown in Table XIII (also see Fig. 65). Also the values of the potential φ_1^* on the external boundary of the double layer are listed there. We note that for small values of the potential of the field on the body's surface ($\varphi_0^* \ll 1$) its value φ_1^* on the outer boundary of the double layer coincides with φ_0^*, that is,

$$\frac{\varphi_0^* - \varphi_1^*}{\varphi_0^*} \to 0 \text{ for } \varphi_0^* \to 0.$$

§ 38. Double Layer Around the Surface of a Large Body

The variation of the potential in the double layer around the surface of a large body is determined by Eq. (9.17)

$$\frac{d^2\varphi^*}{d\zeta^2} = \frac{1}{2} \left\{ e^{\varphi^*} \left[1 - \Phi\left(\sqrt{\varphi^*}\right)\right] + \frac{2}{\sqrt{\pi}} \left(\sqrt{\varphi^*} - \sqrt{\varphi^* - \varphi_\infty^*} \right) + C(\varphi^*) - e^{-\varphi^*} \left[1 + \Phi\left(\sqrt{\varphi_0^* - \varphi^*}\right)\right] \right\}. \tag{9.17}$$

Here $\zeta = (r - R_0)/D$, $\varphi_0^* = e|\varphi_0|/\varkappa T$ is the potential on the surface of the body (for $\zeta = 0$), and $\varphi_\infty^* = e\varphi_\infty/\varkappa T$ is the calculated above asymptotic value of the potential (9.6), characterizing the behavior of the field at large distances from the body. The function $C(\varphi^*)$ in (9.17) is defined by expression (9.15a).

The form of the function $C(\varphi^*)$ depends on the nature of the variation of the potential in the vicinity of the body, that is, it depends on the function $\varphi^*(\xi)$ which was determined in the previous section for different values of φ_0^*. Using the function $\varphi^*(\xi)$ it is possible to calculate the function $C(\varphi^*)$ with the aid of formula (9.15a).

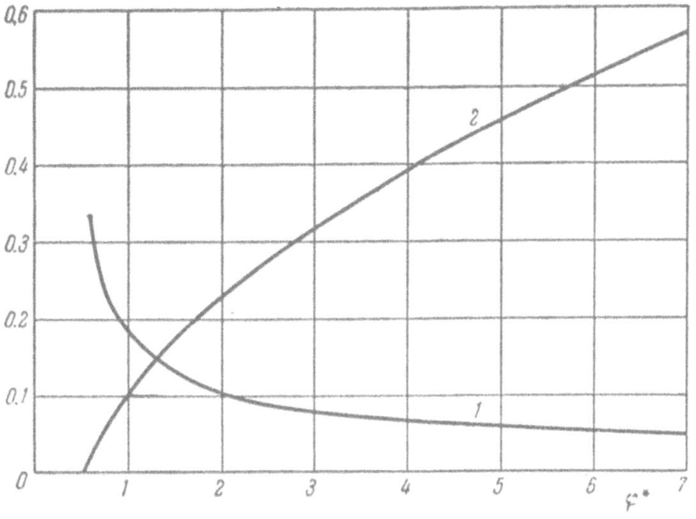

Fig. 66. The function $C(\varphi^*)$ (curve 1) and $\int_{\varphi_1^*}^{\varphi^*} C(\varphi^*)\, d\varphi^*$ (curve 2).

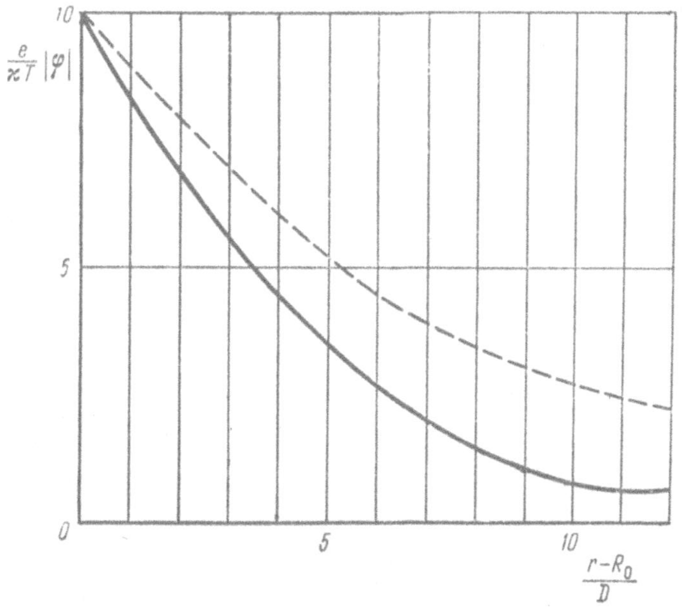

Fig. 67. Dependence of the potential of the field inside the double layer around the surface of a spherical body for $|\varphi_0| = 10\ \varkappa T/e$ on the distance from the body's surface. The dashed curve – the same, when the influence of the field in the quasi-neutral plasma is not taken into consideration.

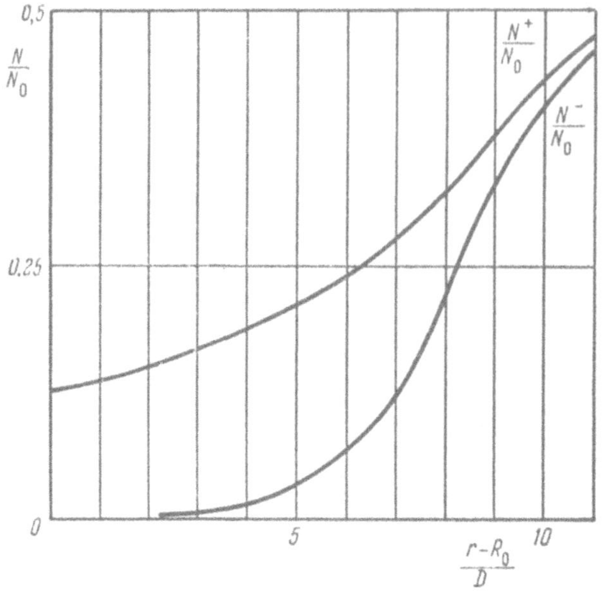

Fig. 68. Dependence of the concentrations of attracted and repelled particles in the double layer on distance from the body's surface, for $|\varphi_0| = 10\varkappa T/e$.

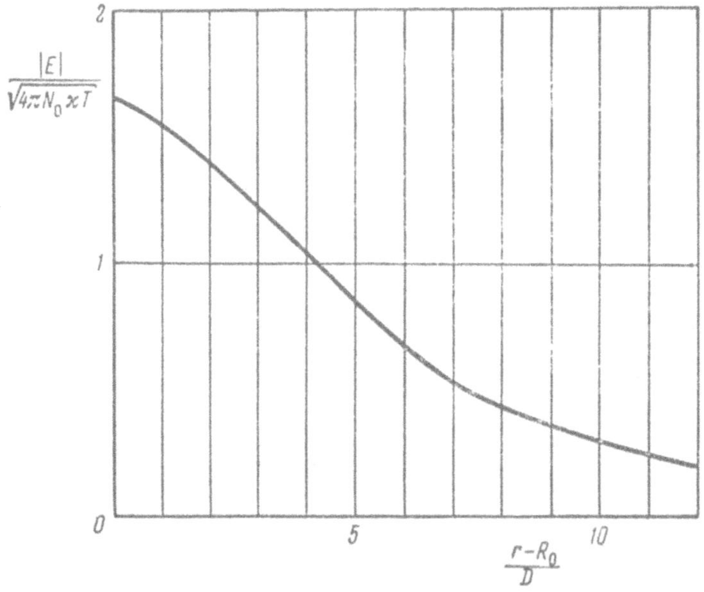

Fig. 69. Dependence of the electric field intensity inside the double layer on distance from the body's surface, for $|\varphi_0| = 10\varkappa T/e$.

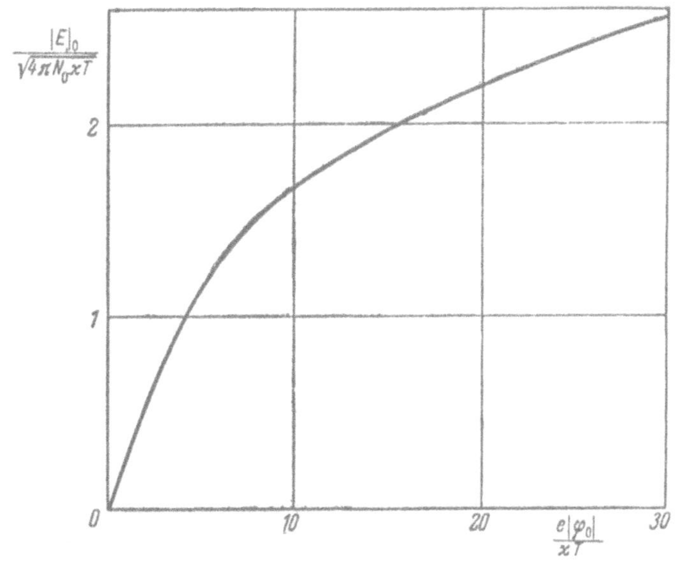

Fig. 70. Intensity of the field on a body's surface as a function of the surface potential.

The results of the corresponding calculation are given in Fig. 66 for $\varphi^* \gg 1$ (curve 1). The dependence of $\int C(\varphi^*)\,d\varphi^*$ (curve 2) on φ^* is also plotted in this same figure; this dependence is needed for the calculation of the potential in the double layer [see Eq. (9.32)].

On the external boundary of the double layer, that is, as $\zeta \to \infty$, the potential φ^* of the field should, of course, coincide with the boundary value of the function $\varphi^*(\xi)$. In other words,

$$\varphi^*_{\zeta\to\infty}(\zeta) = \varphi^*_{\xi\to 1}(\xi) = \varphi^*_1,$$

where φ_1^* is the calculated above value of the potential of the field on the external boundary of the double layer. For $\varphi^*(\zeta) \to \varphi_1^*$ the right side of Eq. (9.17) vanishes, of course, i.e., the plasma on the external boundary of the double layer is quasi-neutral for $\zeta \to \infty$, as it should be.

Taking the boundary conditions

$$\varphi^* = \varphi_0^* \quad \text{for} \quad \zeta = 0$$

and

$$\varphi^* = \varphi_1^* \quad \text{for} \quad \zeta \to \infty,$$

into consideration, after integration of the solution of Eq. (9.17) (see § 14) we find in implicit form

$$\zeta = \int\limits_{\varphi^*}^{\varphi_0^*} \frac{d\varphi^*}{\sqrt{2\int\limits_{\varphi_1^*}^{\varphi^*} f(\varphi^*)\,d\varphi^*}}\,.$$

Here the function $f(\varphi^*)$ is the right side of Eq. (9.17). Substituting $f(\varphi^*)$, we find that the integral under the radical sign in formula (9.31) is

$$Z(\varphi^*) = 2\int_{\varphi_1^*}^{\varphi^*} f(\varphi^*)\,d\varphi^* = e^{-\varphi^*}[1+\Phi(\sqrt{\varphi_0^*-\varphi^*})] + e^{\varphi^*}[1-\Phi(\sqrt{\varphi^*})] + \frac{4}{3\sqrt{\pi}}[\varphi^{*3/2}-(\varphi^*-\varphi_\infty^*)^{3/2}]$$

$$+\frac{2\sqrt{\varphi^*}}{\sqrt{\pi}} - \frac{2e^{-\varphi_0^*}\sqrt{\varphi_0^*-\varphi^*}}{\sqrt{\pi}} - \int_{\varphi_1^*}^{\varphi^*} C(\varphi^*)\,d\varphi^* - \Big\{e^{-\varphi_1^*}[1+\Phi(\sqrt{\varphi_0^*-\varphi^*})] + e^{\varphi_1^*}[1-\Phi(\sqrt{\varphi_1^*})] + \frac{4}{3\sqrt{\pi}}[\varphi_1^{*3/2}$$

$$-(\varphi_1^*-\varphi_\infty^*)^{3/2}] + \frac{2\sqrt{\varphi_1^*}}{\sqrt{\pi}} - \frac{2e^{-\varphi_0^*}\sqrt{\varphi_0^*-\varphi_1^*}}{\sqrt{\pi}}\Big\}. \tag{9.32}$$

Thus, the solution of Eq. (9.17) has the form

$$\zeta = \int_{\varphi^*}^{\varphi_0^*} \frac{d\varphi^*}{\sqrt{Z(\varphi^*)}}. \tag{9.33}$$

One can accomplish the integration of (9.33) only by numerical methods. The results of the corresponding calculation of the dependence of the potential in a double layer on distance from the surface of the body are given in Fig. 67 for the value $\varphi_0^* = 10$.

The dependence of the concentrations of attracted and repelled particles on distance inside the double layer, calculated with the aid of formulas (9.15) and (9.16), is shown in Fig. 68 for the same value of the potential on the body's surface. It is clear from the figures that not only the concentration of repelled particles, but also the concentration of attracted particles decreases as one approaches the surface of the body. Of course, the concentration of repelled particles decreases with increasing φ^* much more rapidly than the concentration of attracted particles does. In fact, from formulas (9.15) and (9.16) it follows that for large values of $\varphi^* \gg 1$

$$N^- = e^{-\varphi^*}\left[\frac{1}{2}+\frac{1}{2}\Phi(\sqrt{\varphi_0^*-\varphi^*})\right] \leqslant e^{-\varphi^*}, \tag{9.15b}$$

$$N_{\text{unb}}^+ = \frac{C}{\sqrt{\varphi^*}},$$

where

$$C = \frac{1}{2\sqrt{\pi}}\left\{1+\varphi_\infty^* + \frac{1}{2}\int_1^\infty (\xi_0^2-1)\left(3\frac{d\varphi^*}{d\xi_0} + \xi_0\frac{d^2\varphi^*}{d\xi_0^2}\right)\exp\left[\varphi^*(\xi_0) + \frac{\xi_0}{2}\frac{d\varphi^*}{d\xi_0}\right]d\xi_0\right\}. \tag{9.15c}$$

The physical meaning of expression (9.15b) is quite clear. In fact, a certain flux of attracted particles is incident on the external boundary of the double layer. Under the influence of the attractive field, the velocity of these particles in the double layer increases, but their flux here, of course, remains constant. Consequently, the concentration of attracted particles in the double layer decreases as their average velocity increases. For large values $\varphi^* \gg 1$ the average velocity if proportional to $\sqrt{\varphi^*}$ and, therefore, $N_{\text{unb}}^+ \sim 1/\sqrt{\varphi^*}$. The constant C in expression (9.15c) is equal to 0.415.

The electric field intensity in the double layer, as is clear from Eq. (9.33), is determined by the expression

$$|E| = -\frac{\varkappa T}{eD}\frac{d\varphi^*}{d\zeta} = \sqrt{4\pi N_0 \varkappa T}\sqrt{Z(\varphi^*)}, \tag{9.34}$$

where $Z(\varphi^*)$ is a dimensionless function [see Eq. (9.32)]. The sign of the electric field intensity in formula (9.34) depends, of course, on the sign of the charge of the body itself: if the body is negatively charged, then $E < 0$, and if its is positively charged, then $E > 0$. The variation of the electric field intensity in the double layer as a function of the distance from the body's surface, calculated with the aid of formula (9.34) for the value $\varphi_0^* = 10$, is shown in Fig. 69. The dependence of the electric field intensity, on the surface of the body itself, on the surface potential φ_0^* is shown in Fig. 70. For large values of the potential, when $\varphi^* \gg 1$, as is clear from (9.32) and (9.15c),

$$Z(\varphi^*) = 4C\sqrt{\varphi^*} - C_0 \approx 1.66\sqrt{\varphi^*} - 2.83, \tag{9.32a}$$

where C is defined by formula (9.15c); in the last expression it is taken into consideration that when $\varphi_0^* \geq \varphi^* \gg 1$ the constant C is equal to 0.415. Consequently, for $\varphi^* \gg 1$

$$|E| = 4.57\sqrt{N_0 \varkappa T}\sqrt[4]{e|\varphi|/\varkappa T}. \tag{9.34a}$$

From here it is seen that the electric field intensity on the surface of the body increases very slowly with increase of the surface potential φ_0^*, namely in proportion to $(\varphi^*)^{1/4} = (e|\varphi_0|/\varkappa T)^{1/4}$. Correspondingly the thickness of the double layer, that is, the thickness of the layer on the surface of the body in which the potential of the electric field is considerably diminished, increases with increase of the surface potential in proportion to $(e|\varphi_0|/\varkappa T)^{3/4}$. For $e|\varphi_0|/\varkappa T \sim 1$ the thickness of the double layer is the order of the Debye radius.

Now for comparison let us consider the form of the double layer in the case when it is possible to neglect the function $2(\varphi^*/\pi)^{1/2} - C_1(1, \varphi^*)$, describing the influence of the field in the vicinity of the body on the motion of the attracted particles. In this case the potential is described by the equation

$$\frac{d^2\varphi^*}{d\zeta^2} = \frac{1}{2}\left\{e^{\varphi^*}\left[1 - \Phi\left(\sqrt{\varphi^*}\right)\right] - e^{-\varphi^*}\left[1 + \Phi\left(\sqrt{\varphi_0^* - \varphi^*}\right)\right]\right\}. \tag{9.17a}$$

Equation (9.17a) approximately describes the behavior of the potential inside the double layer around an arbitrary surface and is integrated in the same way as Eq. (9.17). The result of a calculation of the potential in this case for $\varphi^* = 10$ is represented by the dashed curve in Fig. 67. From the figure it is clear that the behavior of the potential inside the double layer in the case considered is appreciably different from the behavior of the potential around a sphere: Also the intensities of the electric field on the body's surface are different. Thus, for large values of the potential, $\varphi_0^* \gg 1$, the field intensity on the surface of a sphere turns out to be $(4\pi C)^{1/4} \approx 1.21$ times larger. For small values of the potential on the body's surface, when $\varphi_0^* \ll 1$, the solutions of Eqs. (9.17) and (9.17a) differ in a qualitative manner. Thus, according to the exact solution, around a spherical surface, as already mentioned above, for small values of φ_0^* the potential of the field on the external boundary of the double layer is equal to φ_0^*, that is, in the linear (with respect to φ_0^*) approximation, the double layer around the surface of a sphere is absent. According to the approximate equation (9.17a) the double layer survives for small values of φ_0^*; the value of the potential on the external boundary of the double layer is equal to $\varphi_0^*/2$. This immediately follows from Eq. (9.17a). In fact, the value of the potential on the external boundary of the double layer is determined from condition (9.12) for quasi-neutrality of the plasma, that is, from the condition that the right side of Eq. (9.17a) vanishes. For small values of φ_0^* and φ^*, the latter is equal to $(\varphi^*/\pi)^{1/2} - [(\varphi_0^* - \varphi^*/\pi)]^{1/2}$, and therefore $\varphi^* = \varphi_0^*/2$ on the external boundary of the double layer.

Thus, we see that the electric field in the quasi-neutral plasma in the vicinity of the body substantially influences the structure of the double layer, although the thickness of the double layer was here assumed by us to be arbitrarily small in comparison with the radius of the sphere. The double layer around the surface of a large body consequently depends on the specific shape of the body. This is also understandable from simple considerations, since the structure of the double layer essentially depends on the flux of particles arriving on its outer boundary. The flux now depends on the shape of the body, since the overall field distribution in the vicinity of the body influences it.

§39. Structure of the Plasma and the Electric Field for a Very High Potential on the Surface of a Large Body

In the preceding sections it was assumed that the thickness of the double layer, that is, the size of the region where the condition of quasi-neutrality is substantially violated, is much smaller than the size of the body. We saw at the same time that the thickness of the double layer increases with increasing surface potential φ_0 in proportion to $(e|\varphi_0|/\varkappa T)^{3/4}$. For very large values of the potential φ_0, that is when

$$|\varphi_0| \gtrsim \frac{\varkappa T}{e}\left(\frac{R_0}{D}\right)^{4/3}, \tag{9.35}$$

the thickness of the double layer becomes comparable with the dimensions of the body. Below the case of very high potentials on the surface of the body will be considered, when condition (9.35) is satisfied. Here, as before, the case of a large body, $R_0 \gg D$, is considered.

For the solution of Poisson's equation (9.3a)

$$\frac{1}{r^2}\frac{d}{dr}\left(r^2\frac{dU}{dr}\right) = -4\pi e^2\left[N_b^+(r, U) + N_{unb}^+(r, U) - N^-(r, U)\right] \tag{9.3a}$$

($U = -e|\varphi|$) one can now isolate two regions: the quasi-neutral plasma region.

In the quasi-neutral plasma region, where the value of the potential φ^* is not large ($\varphi^* \lessgtr 1$), Eq. (9.3a) is equivalent to Eq. (9.12) whose solution was obtained above in §37. The only difference is that in §37 the thickness of the double layer on the body's surface was assumed to be small in comparison with the body's radius R_0, and therefore the boundary of the quasi-neutral region was considered to be the surface of the body, $r = R_0$. Now the size R_c of the space charge may be larger, and even considerably larger than R_0. However, if the potential in the region of space charge falls off sufficiently sharply, so that the condition

$$|\varphi_0|\frac{R_0^2}{R_c^2} \gg \frac{\varkappa T}{e}, \tag{9.36}$$

is satisfied, then every attracted particle incident on the space charge region also hits the surface of the body and is absorbed here. The sphere of radius R_c then plays the role of the absorbing surface in the equation of quasi-neutrality. Therefore, it is sufficient for us to replace the radius R_0 by R_c in the equation of quasi-neutrality (9.12) and in its solution, which was obtained in §37b for $|\varphi_0| \gg \varkappa T/e$. In other words, the potential of the field in the region of quasi-neutrality, $r \geq R_c$, is described as before by the solution of Eq. (9.12), where the dimensionless variable ξ is equal to

$$\xi = \frac{r}{R_c}.$$

The radius R_c of the space charge region depends, of course, on the radius R_0 of the body and on the potential φ_0 on its surface; it will be determined in what follows. Comparing condition (9.36) with condition (9.35), we convince ourselves that condition (9.36) is automatically satisfied if

$$\frac{R_c}{R_0} \lessgtr \left(\frac{R_0}{D}\right)^{2/3},$$

that is, up to very large values of the radius of the space charge region. In fact, condition (9.36) is always satisfied, as will be shown below.

TABLE XXIV. Dependence of the Potential $\varphi^{(*)} = (e|\varphi|/\varkappa T)(D/R_c)^{4/3}$ on the Distance $1/\eta = R_c/r$ in the Region of Space Charge ($R_c > r$)

$1/\eta$	$\varphi^{(*)}$	$1/\eta$	$\varphi^{(*)}$	$1/\eta$	$\varphi^{(*)}$	$1/\eta$	$\varphi^{(*)}$
1.00	0	1.80	0.632	3.00	1.85	5.4	4.42
1.05	0.018	1.90	0.727	3.2	2.06	5.6	4.64
1.10	0.046	2.00	0.823	3.4	2.27	5.8	4.86
1.15	0.078	2.10	0.922	3.6	2.48	6.0	5.08
1.20	0.112	2.20	1.02	3.8	2.69	6.5	5.62
1.25	0.148	2.30	1.12	4.0	2.90	7.0	6.16
1.30	0.186	2.40	1.22	4.2	3.12	7.5	6.71
1.35	0.226	2.50	1.32	4.4	3.33	8.0	7.26
1.40	0.269	2.60	1.43	4.6	3.55	8.5	7.81
1.45	0.311	2.70	1.53	4.8	3.76	9.0	8.36
1.50	0.355	2.80	1.64	5.0	3.98	9.5	8.91
1.60	0.445	2.90	1.74	5.2	4.20	10.0	9.46
1.70	0.538						

In the space charge region ($r \le R_c$) it is now necessary to solve the complete equation (9.3a). We assume, however, that the potential φ is of the order of φ_0 for $r < R_c$ and consequently $|\varphi| \gg \varkappa T/e$. In this case it is possible to substantially simplify the right side of Eq. (9.3a). First of all, the concentration N^- of repelled particles [see Eq. (9.9)] is exponentially small for $|\varphi| \gg \varkappa T/e$ and one can neglect it. In addition, expression (8.39) for the concentration of attracted particles undergoing unbounded motion, $N^+_{unb}(r)$, can be substantially simplified.* In fact, the integral terms in the region where $|U(r_0)| \gg \varkappa T$ give an exponentially small contribution because of the influence of the exponential terms $\exp\left[-\frac{U(r_0)}{\varkappa T} - \frac{r_0}{2\varkappa T}\frac{dU}{dr_0}\right]$. Consequently, only the region where $|U(r_0)/\varkappa T| \lessdot 1$ turns out to be important, that is, the quasi-neutral plasma region, $r_0 \ge R_c$. Taking into account that here $|U(r)| \gg |U(r_0)|$, we obtain the following expression for the first integral term in formula (8.39):

$$X[\varrho(r), \infty] = \frac{1}{2\sqrt{-\pi\varkappa T U(r)}} \int_{R_c}^{\infty} \left(1 - \frac{r_0^2}{r^2}\right)\left[3\frac{dU}{dr_0} + r_0\frac{d^2U}{dr_0^2}\right] \exp\left[-\frac{U(r_0)}{\varkappa T}\right.$$

$$\left. - \frac{r_0}{2\varkappa T}\frac{dU}{dr_0}\right] dr_0 = \frac{1}{\sqrt{-\pi U(r)/\varkappa T}}\left\{\frac{2}{r^2}\int_{R_c}^{\infty} r_0\left(1 - \exp\left[-\frac{U(r_0)}{\varkappa T} - \frac{r_0}{2\varkappa T}\frac{dU}{dr_0}\right]\right) dr_0 + \left(\frac{R_c^2}{r^2} - 1\right)\right\}.$$

Upon fulfillment of condition (9.36), the second integral term is equal to zero since the region of integration with respect to dr_0 in this case does not include the quasi-neutral plasma, that is, the region $\rho(R_0) > R_c$. Also the term proportional to $(1 - R_0^2/r^2)^{1/2}$ turns out to be exponentially small. Simplifying the remaining terms in formula (8.39) (considering that $-U(r)/\varkappa T \gg 1$), we finally find that

$$N^+_{unb}(r) = \frac{N_0}{2\sqrt{\pi}} \frac{R_c^2}{r^2\sqrt{-U(r)/\varkappa T}} \left\{1 + \varphi^*_\infty + \frac{2}{R_c^2}\int_{R_c}^{\infty} r_0\left(1 - \exp\left[-\frac{U(r_0)}{\varkappa T} - \frac{r_0}{2\varkappa T}\frac{dU}{dr_0}\right]\right) dr_0\right\}$$

$$= \frac{0.415 R_c^2}{r^2\sqrt{-U(r)/\varkappa T}} N_0. \qquad (9.37)$$

*In the case considered here the field may fall off both more rapidly as well as more slowly than $1/r^2$, and therefore it is necessary to use the general expression (8.39) for the density of attracted particles undergoing unbounded ("infinite") motion.

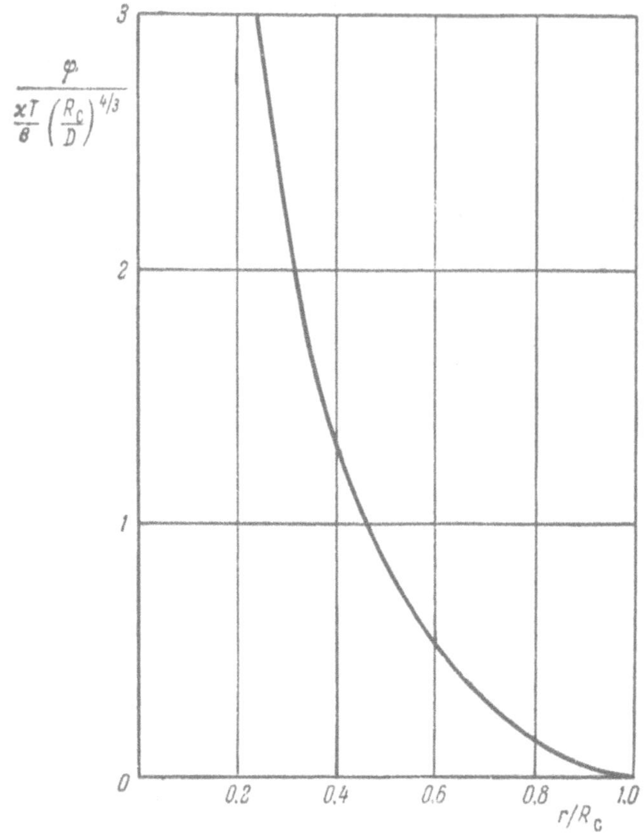

Fig. 71. Dependence of the potential in the region of space charge on the distance.

In the last expression it is considered that, in the quasi-neutral plasma region, the potential $\varphi^*(\xi)$ coincides with the result obtained in §37b and, therefore

$$\frac{1}{2\sqrt{\pi}}\left\{1+\varphi_\infty^*+2\int_1^\infty \xi_0\left[1-\exp\left(\varphi^*+\frac{\xi_0}{2}\frac{d\varphi^*}{d\xi_0}\right)\right]d\xi_0\right\}=0.415$$

(see §38). Expression (9.37) has the simple physical meaning of conservation of flux of attracted particles, arriving at the external boundary of the space charge region.

Thus, Eq. (9.3a) in the space charge region $r \le R_c$ has the form

$$\frac{1}{r^2}\frac{d}{dr}\left(r^2\frac{dU}{dr}\right)=-4\pi e^2\left\{\frac{0.415R_c^2N_0}{r^2\sqrt{-U(r)/\varkappa T}}+N_b^+\right\}. \tag{9.38}$$

In terms of the dimensionless variables

$$\eta=\frac{r}{R_c},\quad \varphi^{(*)}=\left[-\frac{U(r)}{\varkappa T}\left(\frac{D}{R_c}\right)^{4/3}\right]=\frac{e|\varphi|}{\varkappa T}\left(\frac{D}{R_c}\right)^{4/3} \tag{9.39}$$

Eq. (9.3a) can be rewritten in the form

$$\frac{1}{\eta^2}\frac{d}{d\eta}\left(\eta^2\frac{d\varphi^{(*)}}{d\eta}\right)=\frac{0.415}{\eta^2\sqrt{\varphi^{(*)}}}+\frac{N_b^+}{N_0}\left(\frac{R_c}{D}\right)^{2/3}. \tag{9.3b}$$

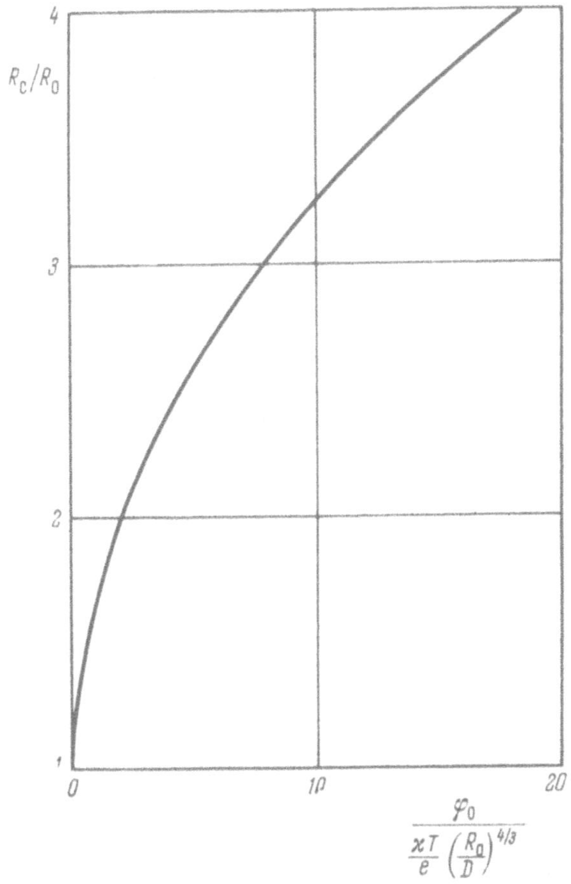

Fig. 72. Dependence of the radius of the space charge region on the value of the potential on the body's surface.

On the surface of the body, for $r = R_0$, the potential of the field is equal to φ_0, that is

$$\varphi^{(*)}\big|_{\eta=\eta_0=R_0/R_c} = \varphi_0^{(*)} = \frac{e\,|\varphi_0|}{\varkappa T}\left(\frac{D}{R_c}\right)^{4/3}. \tag{9.40}$$

On the boundary of the space charge region, for $r = R_c$, the potential of the field is of order $\varkappa T/e$, and consequently

$$\varphi^{(*)}\big|_{\eta\to 1} \to 0. \tag{9.41}$$

On the boundary of the space charge region and the quasi-neutral plasma, one can distinguish a transition layer in which it is still impossible to neglect the particles which are being repelled. The thickness of this layer, naturally, is of the order of the Debye radius D. In this layer the field is described by Eq. (9.17), and, consequently, its solution coincides with the solution for the double layer, obtained above in §38. Here, in the same way as in §38, the intensity of the field for $e\,|\varphi|/\varkappa T \gg 1$ increases in proportion to $(e\,|\varphi|/\varkappa T)^{1/4}$ [more precisely, see Eq. (9.34a)].

Rewriting this expression in terms of the variables (9.39), we obtain

$$\left.\frac{d\varphi^{(*)}}{d\eta}\right|_{\eta=1} = 1.29\varphi^{(*)1/4} = 0.$$

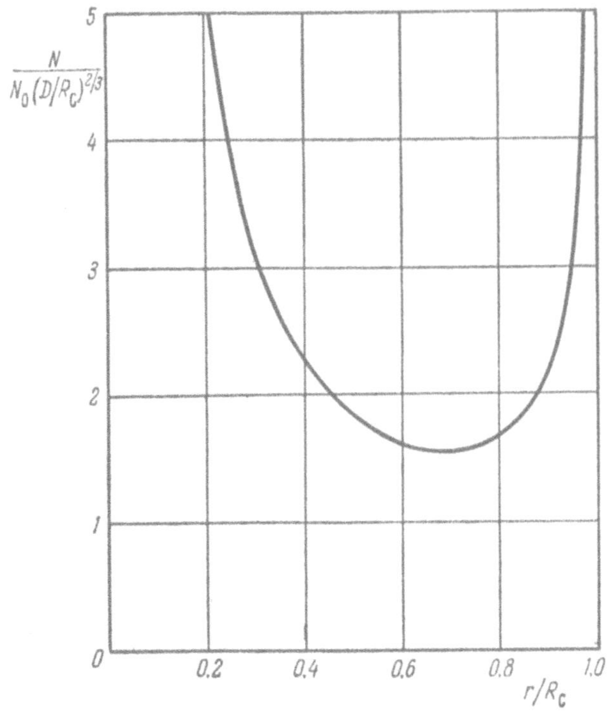

Fig. 73. Dependence of the concentration of attracted particles in the space charge region on distance.

Therefore, the condition

$$\left.\frac{d\varphi^{(*)}}{d\eta}\right|_{\eta\to 1} \to 0. \tag{9.42}$$

must still be satisfied for $\eta \to 1$. Thus, the solution of Eq. (9.3b) must satisfy the three boundary conditions (9.40), (9.41), and (9.42) at once. This is possible since the radius R_c [or the parameter $\eta_0 = R_0/R_c$ in boundary condition (9.40)] is an arbitrary parameter. The boundary conditions (9.40), (9.41), and (9.42) determine, therefore, not only the solution of Eq. (9.3b), but the radius of the space charge region as well. With account of these conditions, the problem becomes closed, and only the integration of Eq. (9.3b) remains.

First let us assume that it is possible to neglect the particles undergoing finite motion. Equation (9.3b) then assumes the form

$$\frac{d}{d\eta}\left(\eta^2 \frac{d\varphi^{(*)}}{d\eta}\right) = \frac{0.415}{\sqrt{\varphi^{(*)}}}.$$

An equation of this type with boundary conditions (9.41) and (9.42) was solved by Langmuir and Blodgett [39]. Using their results, we immediately obtain the solution $\varphi^{(*)}(\eta)$ presented in Table XXIV and in Fig. 71. The boundary conditions (9.41) and (9.42) are always satisfied. And the boundary condition (9.40),

$$\varphi^{(*)}|_{\eta=\eta_0} = \varphi_0^{(*)}$$

determines the value of η_0, that is, the radius of the space charge region,

$$R_c = \frac{R_0}{\eta_0}. \tag{9.43}$$

The solution $\varphi^{(*)}(\eta)$ is given in Table XXIV only for values of $\eta = r/R_c \geq 0.1$. For $\eta < 0.1$,

$$\varphi^{(*)} = 1.67 \cdot \frac{1}{\eta} - 2.47.$$

correct to within 0.5%. From here it is evident that for very large values of the ratio R_c/r, that is, for very large values of the potential of the field on the body's surface, when $|\varphi_0| \gg (\varkappa T/e)(R_c/D)^{4/3}$, the space charge near the body does not play an essential role, and the field falls off according to Coulomb's law

$$\varphi = \varphi_0 \cdot \frac{R_0}{r}.$$

In this case

$$R_c = \left(0.598 R_0 D^{4/3} \frac{e|\varphi_0|}{\varkappa T}\right)^{3/7} = 0.803 R_0 \left[\frac{e|\varphi_0|}{\varkappa T}\left(\frac{D}{R_0}\right)^{4/3}\right]^{3/7}. \tag{9.44}$$

Thus, for very large values of the potential on the surface of the body, the radius of the space charge region increases with increasing φ_0 in proportion to $|\varphi_0|^{3/7}$. For relatively small values of the potential on the surface, when

$$\varphi_0^{(*)} = \frac{e|\varphi_0|}{\varkappa T}\left(\frac{D}{R_0}\right)^{4/3} \ll 1, \tag{9.45}$$

the values of η_0, determined by the function $\varphi^{(*)}(\eta)$ presented in Table XXIV, are close to unity. In other words, upon fulfillment of condition (9.45) the space charge region $(R_c - R_0)$ is small in comparison with the body's radius R_0. In this case, the space charge region is a thin charged layer near the body's surface; such a situation was already investigated in the preceding section. The dependence of the radius R_c on the surface potential φ_0 is shown in Fig. 72.

The concentration of attracted particles in the space charge region is naturally much smaller than the concentration of particles in the undisturbed plasma, $N^+_{unb} \sim N_0(D/R_c)^{2/3}$, since the velocity of the particles being attracted is much larger in the space charge region than the thermal velocity $v \sim (e\varphi/M)^{1/2} \sim (\varkappa T/M)^{1/2}(R_c/D)^{2/3}$, and the total particle flux is conserved. However, for small values of r the potential increases slowly, since in this region the concentration of attracted particles increases with decreasing r (in virtue of the same reason – conservation of the total flux). The minimum value $N_{min} = 1.57\,(D/R_c)^{2/3} N_0$ is reached when $r/R_c \approx 0.67$. The behavior of the concentration of attracted particles in the space charge region, determined according to formula (9.37), is shown in Fig. 73.

Above we neglected the particles performing bounded motion. This is rigorously valid in the case when the potential always falls off faster than $1/r^2$. This is realized, as indicated in §33, if the radius R_0 of the body is larger than the distance r_K, where r_K is determined from formula (8.8)

$$3\frac{dU}{dr}(r_K) + r_K \frac{d^2U}{dr^2}(r_K) = 0. \tag{8.8}$$

In terms of the dimensionless variables $\varphi^{(*)}$ and η, formula (8.8) can be rewritten in the form

$$3\frac{d\varphi^{(*)}}{d\eta}(\eta_K) + \eta_K \frac{d^2\varphi^{(*)}}{d\eta}(\eta_K) = 0. \tag{9.46}$$

Using the values of the function $\varphi^{(*)}(\eta)$ listed in Table XXIV, we find that

$$\eta_K = 0.846, \quad \varphi^{(*)}(\eta_K) = 0.101, \quad \frac{d\varphi^{(*)}}{d\eta}(\eta_K) = 0.70 \tag{9.47}$$

or

$$R_{cK} = 1.183 R_0, \quad \varphi_{0K} = 0.127 \frac{\varkappa T}{e} \left(\frac{R_0}{D}\right)^{4/3}. \tag{9.48}$$

Thus, if the value of the potential on the surface of the body is less than the value φ_{0K} [see formula (9.48)], then bounded trajectories not passing through the body's surface are not present. In this case, the concentration of bounded particles is equal to zero (for a sufficiently large mean free path), and the solution obtained is the exact solution of Eq. (9.3b).

It is important that for $\varphi_0 > \varphi_{0K}$, the concentration of bounded particles is also exponentially small. In fact, let us assume that the particles interact according to Coulomb's law, so that the distribution function of the bounded particles is described by the Fokker-Planck equation in the space of energy and angular momentum (see §3). In the region of unbounded motion the distribution function is determined by the condition that the particle flux from the region of unbounded motion to the region of bounded motion and then onto the body's surface (see §33) is constant. With respect to order of magnitude, this leads to the following expression for the particle concentration in the region of bounded motion (see §35):

$$N_b^+ \sim (\varphi^*)^\lambda \exp\left(-\frac{\mathscr{E}_0}{\varkappa T}\right) N_0. \tag{9.49}$$

In (9.49), $\varphi^* = e|\varphi|/\varkappa T$, the exponent λ is the order of unity and depends on the characteristics of the collisions of the bounded particles, and $\mathscr{E}_0$ is the minimum value of a bounded particle's energy on the boundary of the regions of finite and infinite motions. For $\varphi_0 = \varphi_{0K}$, it is obvious that

$$\mathscr{E}_0 = \mathscr{E}(r_K) = U(r_K) + \frac{r_K}{2} \frac{dU}{dr}(r_K),$$

or, taking (9.47) into account,

$$\frac{\mathscr{E}_0}{\varkappa T} = 0.195 (R_c/D)^{4/3}.$$

Consequently, in this case, the concentration of bounded particles is exponentially small. Furthermore, one can prove that with increase of φ_0 the ratio $\mathscr{E}_0/\varkappa T$ only increases. In particular, for $\varphi_0 \gg \varphi_{0K}$

$$\frac{\mathscr{E}_0}{\varkappa T} = \frac{2.79}{\varphi_0^{(*)}} \left(\frac{R_c}{D}\right)^{4/3} = 2.15 \left(\frac{R_0}{D}\right)^{8/7} \left(\frac{e|\varphi_0|}{\varkappa T}\right)^{1/7}. \tag{9.50}$$

Thus, for $R_0/D \gg 1$, N_b^+ is always exponentially small. This conclusion is quite understandable. The point is that an unbounded particle which can be captured in a finite orbit must possess a very large angular momentum

$$\mathscr{M} \sim \sqrt{\frac{\mathscr{E}_0}{M}} M R_0 \gtrsim M\sqrt{\varkappa T/M} R_0 \left(\frac{R_0}{D}\right)^{2/3}.$$

The number of such particles is exponentially small.

Thus, in the case considered here of a large body ($R_0 \gg D$), one can neglect the concentration N_b^+. The solution of Eq. (9.3b) obtained above is therefore exact. We also note that if formula (9.50) for R_c is taken into consideration, then it is easy to convince oneself that condition (9.36), essentially used by us earlier, is satisfied for arbitrary values of the potential on the surface of the body.

a. Electric Field. Now let us consider the case when the size of the body is small in comparison with the Debye radius, $R_0 \ll D$. Under these conditions, the space charge generally has much less effect on the behavior of the electric field in the vicinity of the body than when $R_0 \gg D$. This enables us to comparatively simply determine the field distribution in the vicinity of a small body for not very large values of the potential on the surface.

In fact, at small distances from the body's surface, no space charge exists, and the field falls off according to Coulomb's law, $\varphi(r) = \varphi_0 R_0 / r$. Space charge plays an important role only beginning with distances on the order of the Debye radius, which are large in comparison with the size of the body R_0. We shall assume that at such large distances the field is now weak, that is, for $r \gtrsim D$

$$|\varphi(r)| \ll \frac{\varkappa T}{e}. \tag{9.51}$$

For sufficiently large values of D this condition is always satisfied. In this case, simple expressions are valid for the concentrations of attracted and repelled particles in that region where they have a substantial influence on the field.

Upon fulfillment of condition (9.51), that is, for $\varphi^* = e|\varphi|/\varkappa T \ll 1$, the concentration of repelled particles is determined by formula (9.29)

$$N^- = N_0\left\{1 - \varphi^* + \frac{1}{2\xi^2}\left[\left(\varphi_0^* - \frac{1}{2}\right)\left(1 - \Phi\left(\sqrt{\varphi_0^*}\right)\right) - \frac{\sqrt{\varphi_0^*}}{\sqrt{\pi}} e^{-\varphi_0^*}\right]\right\}, \tag{9.29}$$

where, as before, $\xi = r/R_0$. The concentration of attracted particles undergoing unbounded motion is determined by expression (8.37), as indicated in § 34. For $r \gg R_0$ and $\varphi^* \ll 1$ in the linear approximation this expression simplifies substantially, and we have

$$N_{\text{unb}}^+(\xi) = N_0\left[1 + \varphi^* - \frac{1}{\xi^2}\left(\varphi_0^* + \frac{1}{2}\right)\right]. \tag{9.52}$$

In the same approximation, the concentration of bounded particles is equal to zero.*

Thus, Eq. (9.10) for the electric field in the case considered of a small body ($R_0 \ll D$) can be written in the form

$$\frac{1}{\xi^2}\frac{d}{d\xi}\left(\xi^2 \frac{d\varphi^*}{d\xi}\right) = 2\left(\frac{R_0}{D}\right)^2\left(\varphi^* - \frac{\varphi_\infty^*}{\xi^2}\right), \tag{9.53}$$

where

$$\varphi_\infty^* = \frac{1}{8}\left[4\varphi_0^* - 2\varphi_0^*\Phi\left(\sqrt{\varphi_0^*}\right) + \Phi\left(\sqrt{\varphi_0^*}\right) - \frac{2\sqrt{\varphi_0^*}}{\sqrt{\pi}} e^{-\varphi_0^*}\right]. \tag{9.54}$$

The boundary conditions are: $\varphi^* = \varphi_0^*$ for $\xi = 1$ and $\varphi^* \to 0$ for $\xi \to \infty$.

After the substitution

$$\varphi^*(\xi) = \frac{1}{\xi}\varphi_1^*(\xi)\exp\left(-\frac{\sqrt{2}R_0}{D}\xi\right)$$

*For $e|\varphi| \ll \varkappa T$, the concentration of bounded particles is proportional to $[e|\varphi|/\varkappa T]^{3/2}$ (see, for example, § 35).

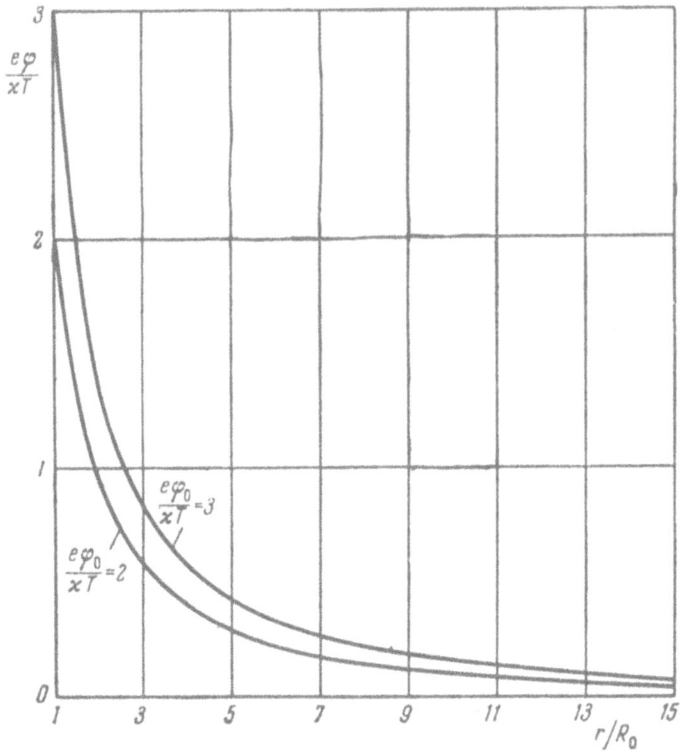

Fig. 74. Curves showing the variation of the potential in the vicinity of a small body for $D/R_0 = 14.2$; the values of φ_0 are indicated in the figure.

the solution of Eq. (9.53) is found by means of an elementary integration and has the form

$$\varphi^* = \frac{\varphi_0^*}{\xi}\exp\left[-\frac{\sqrt{2}\,R_0}{D}(\xi-1)\right] + \frac{\varphi_\infty^* R_0}{\sqrt{2}\,D\xi}\left\{-\exp\left[-\frac{\sqrt{2}\,R_0}{D}\xi\right]Ei\left[-\frac{\sqrt{2}\,R_0}{D}\xi\right]\right.$$

$$\left.+\left(\overline{Ei}\left[\frac{\sqrt{2}\,R_0}{D}\xi\right]-\overline{Ei}\left[\frac{\sqrt{2}\,R_0}{D}\right]\right)\exp\left(-\frac{\sqrt{2}\,R_0}{D}\xi\right)+Ei\left[-\frac{\sqrt{2}\,R_0}{D}\xi\right]\exp\left[\frac{\sqrt{2}\,R_0}{D}(2-\xi)\right]\right\}, \tag{9.55}$$

where $R_0/D = (4\pi e^2 N_0 R_0^2/\varkappa T)^{1/2}$, and Ei and $\overline{Ei}$ are integrals of the exponential function

$$Ei[-x] = -\int_x^\infty \frac{e^{-t}}{t}\,dt, \quad \overline{Ei}[x] = \int_{-\infty}^x \frac{e^t}{t}\,dt.$$

At comparatively small distances from the body, where $r < D$, the field described by expression (9.55) falls off according to Coulomb's law; for $D < r < (D/\sqrt{2})\ln(2D/R_0\varphi_0^*)$ the field falls off exponentially, and at very large distances when $r > D\ln(2D/R_0\varphi_0^*)$ the field decreases like $1/r^2$, that is

$$\varphi_{r\to\infty} = \frac{\varkappa T \varphi_\infty^*}{e}\,\frac{R_0^2}{r^2}, \tag{9.56}$$

where φ_∞^* is determined by formula (9.54). The dependence of the potential $\varphi^* = e|\varphi|/\varkappa T$ on $\xi = r/R_0$ for $D \approx 14R_0$ is shown in Fig. 74 for two values of φ_0^*.

b. Concentration of Particles. The concentration of attracted unbounded particles is given by expression (8.37). Let us substitute in it the potential $\varphi^*(\xi)$ from formula (9.55). The lower limit of varia-

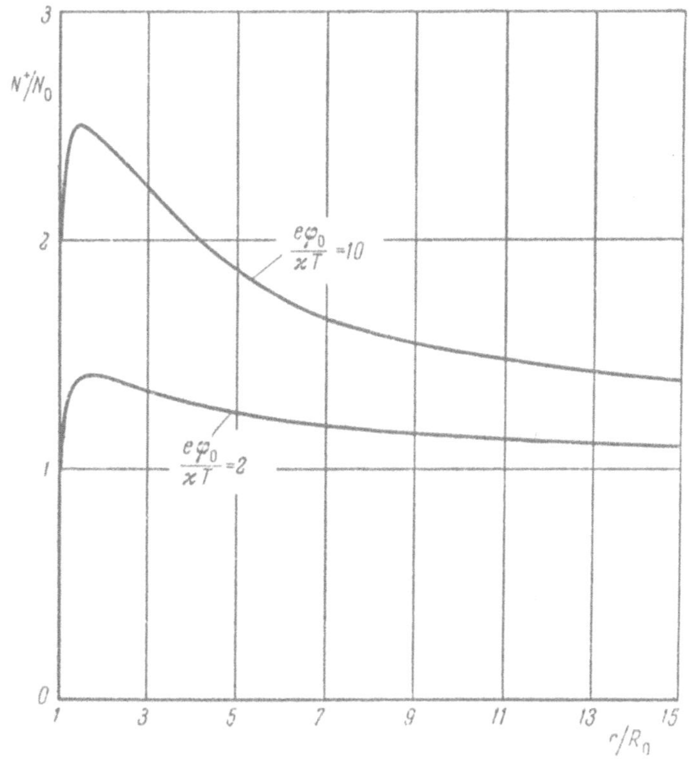

Fig. 75. Curves showing the variation of the concentration of attracted particles in the vicinity of a small body for $D/R_0 = 14.2$; values of φ_0 are indicated on the figure.

tion of the parameter r_0 in the integral terms in expression (8.37) is equal to $\rho_0(R_0)$, and in our case it exceeds the value of D, since $r_K \sim D$ (see §32). Consequently, over the entire region of variation of r_0 the potential $\varphi^*(r_0) \lessgtr \varphi^*(D) \ll 1$, so that at not very large distances from the body, when $r < D$,

$$N^+_{\text{unb}}(\xi) = N_0\left\{\frac{\sqrt{\varphi_0^*}}{\sqrt{\pi\xi}}\left(1+\sqrt{1-\frac{1}{\xi}}\right)+\frac{1}{2}\left[1-\Phi\left(\sqrt{\frac{\varphi_0^*}{\xi}}\right)\right]\exp\left(\frac{\varphi_0^*}{\xi}\right)\right.$$
$$\left.+\frac{1}{2}\sqrt{1-\frac{1}{\xi^2}}\left[1-\Phi\left(\sqrt{\frac{\varphi_0^*}{\xi+1}}\right)\right]\exp\left(\frac{\varphi_0^*}{\xi+1}\right)\right\}. \tag{9.57}$$

This expression is valid,correct to terms of order $(R_0/D\varphi_0^*)^2$. It coincides with the expression obtained directly from the formula which determines the concentration of unbounded particles in a slowly decreasing field [see formula (8.32)], which is as it should be. At large distances from the body, when $\varphi^* \ll 1$, the concentration of attracted particles is given simply by expression (9.52), where φ^* is determined in accordance with (9.55). These two formulas for $\varphi^*(\xi)$ overlap in a considerable region for $\varphi_0^* < \xi < D/R_0$, that is, they overlap where the field is already small ($\varphi^* < 1$), but still a Coulomb field. Consequently, the concentration of unbounded attracted particles is known over the entire disturbed region. It is shown in Fig. 75 for $D \approx 14R_0$ and $\varphi_0^* = 2$ and 10. From the figure it is clear that $N^+_{\text{unb}}(r)$ increases appreciably in the vicinity of the body.

The concentration of the particles being repelled is determined by expression (9.9). At small distances from the body, that is, $r \ll D$ ($\xi \ll D/R_0$), taking (9.55) into account, we obtain

$$N^-(\xi) = \frac{N_0}{2}\left\{1+\Phi\left(\sqrt{\varphi_0^*\left(1-\frac{1}{\xi}\right)}\right)+\right.$$

$$+\sqrt{1-\frac{1}{\xi^2}}\left[1-\Phi\left(\sqrt{\frac{\varphi_0^*\xi}{\xi+1}}\right)\right]\exp\left[\frac{\varphi_0^*}{\xi(\xi+1)}\right]\Bigg\}\exp\left(-\frac{\varphi_0^*}{\xi}\right). \tag{9.58}$$

At large distances from the body, the concentration N^- is determined by formula (9.29). The behavior of $N^-(\xi)$ in the vicinity of a small body is shown in Fig. 76. It is clear from the figure that, at distances of the order of the body's size, the concentration N^- strongly differs from its undisturbed value.

The concentration of particles undergoing bounded motion essentially depends on the character of their collisions. If the body is positively charged and the plasma is weakly ionized, that is, if the attracted particles are electrons, and they collide primarily with neutral particles, then the concentration of electrons captured in finite orbits is determined by expression (8.71),

$$N_b^+=\frac{4N_0}{3\sqrt{\pi}}\left(\frac{\varphi_0^*}{\xi}\right)^{3/2}\left(\frac{2}{5}\frac{\varphi_0^*}{\xi}+1\right). \tag{9.59}$$

as is shown in §35. This formula is valid only at sufficiently large distances from the body, when $r > (M/mR_0)^{1/2}$. At closer distances, the number of bounded particles is less than that determined by formula (9.59); in particular, as $r\to R_0$, $N_b^+\to 0$.

In the general case the concentration of bounded particles at the point r is given by

$$N_b^+(r)=\frac{4\pi}{M^2 r}\int\limits_{\mathscr{E}_{\min}(r)}^{\mathscr{E}_{\max}(r)} d\mathscr{E}\int\limits_{\sqrt{2MR_0^2(\mathscr{E}+\varkappa T\varphi_0^*)}}^{\sqrt{2Mr^2(\mathscr{E}+\varkappa T\varphi^*)}}\frac{\mathscr{M}f(\mathscr{E},\mathscr{M})\,d\mathscr{M}}{[2Mr^2(\mathscr{E}+\varkappa T\varphi^*)-\mathscr{M}^2]^{1/2}}. \tag{9.60}$$

In formula (9.60) $f(\mathscr{E},\mathscr{M})$ is the distribution function for the particles in finite orbits, and $\mathscr{E}_{\min}(r)$ and $\mathscr{E}_{\max}(r)$ are respectively the minimum and maximum energy of a bounded particle at the point r under the condition that its trajectory does not pass through the absorbing surface of the body. It is easy to see that

$$\mathscr{E}_{\min}(r)=-\varkappa T\frac{r^2\varphi^*-R_0^2\varphi_0^*}{r^2-R_0^2}=-\varkappa T\frac{\xi^2\varphi^*-\varphi_0^*}{\xi^2-1}. \tag{9.61}$$

The energy $\mathscr{E}_{\max}(r)$ is equal to zero providing that the potential φ falls off more slowly than $1/r^2$, which holds in our case.

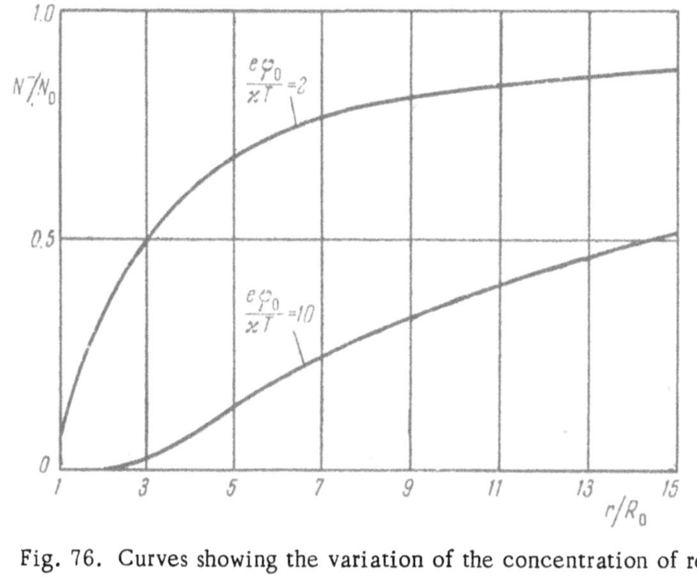

Fig. 76. Curves showing the variation of the concentration of repelled particles in the vicinity of a small body for $D/R_0 = 14.2$; values of φ_0 are indicated on the figure.

For large values of the absolute value of the energy $|\mathscr{E}|$, the distribution function of the electrons undergoing bounded motion in a weakly ionized plasma is proportional to $|\mathscr{E}|/\varkappa T$, as shown in §35. For the Coulomb interaction $f \sim (|\mathscr{E}|/\varkappa T)^3$. In order to estimate the concentration N_b^+ for large values of the potential φ, it is natural to assume that the distribution function for particles in finite orbits has the form

$$f \sim (|\mathscr{E}|/\varkappa T)^{\lambda}. \tag{9.62}$$

for large values of $|\mathscr{E}|$. In the general case, λ is a quantity of the order of unity, as indicated in §35e. Now substituting (9.62) into (9.60) and integrating, we find that

$$N_b^+(r) = \frac{\Gamma(\lambda+1)}{\Gamma\left(\lambda+\frac{5}{2}\right)} N_0 \sqrt{1-\frac{R_0^2}{r^2}} \left(\frac{|\mathscr{E}_{\min}|}{\varkappa T}\right)^{\lambda+\frac{3}{2}} = \frac{\Gamma(\lambda+1)}{\Gamma\left(\lambda+\frac{3}{2}\right)} N_0 \sqrt{1-\frac{1}{\xi^2}} \left(\frac{\xi^2\varphi^*-\varphi_0^*}{\xi^2-1}\right)^{\lambda+\frac{3}{2}}, \tag{9.63}$$

where $\Gamma(x) = \int\limits_0^\infty t^{x-1}e^{-t}\,dt$ is the gamma function. In particular, if the field is close to a Coulomb field, then

$$N_b^+(\xi) = \frac{\Gamma(\lambda+1)}{\Gamma\left(\lambda+\frac{3}{2}\right)} N_0 \varphi_0^{*\,\lambda+\frac{3}{2}} \frac{(\xi-1)^{1/2}}{\xi(\xi+1)^{\lambda+1}}. \tag{9.64}$$

Comparing (9.64), (9.63), and (9.59) with (9.57), we see that for sufficiently large values of the potential on the surface of the body the concentration of bounded particles in the vicinity of the body is generally larger than the concentration of unbounded particles.

c. Large Values of the Potential on the Surface of a Body. All of the investigation above was carried out under the assumption that the Debye radius is very large, so that the electric field at a distance D from the body's surface is weak because of the Coulomb fall off of the potential. This is valid, of course, only when the potential on the body's surface is bounded and satisfies the condition

$$|\varphi_0| \ll \frac{\varkappa T}{e}\frac{D}{R_0} \quad \text{or} \quad \varphi_0^* \ll \frac{D}{R_0}. \tag{9.65}$$

Now let us consider the opposite case of large values of φ_0, when the influence of the screening charge already becomes substantial at comparatively small distances from the body. It is important to emphasize here that the concentration of attracted unbounded particles increases with increase of the surface potential φ_0^* only like $\sqrt{\varphi_0^*}$. The concentration of bounded particles now increases more rapidly, and their role becomes decisive for larger values of the potential. Let us consider a region not very far removed from the body, so that the potential in this region is large: $\varphi^* \gg 1$. In this region, upon utilization of the approximate expression (9.63) for the particles undergoing bounded motion, Poisson's equation (9.3a) assumes the form

$$\frac{1}{\xi^2}\frac{d}{d\xi}\left(\xi^2\frac{d\varphi^*}{d\xi}\right) = \left(\frac{R_0}{D}\right)^2\left\{\frac{\Gamma(\lambda+1)}{\Gamma\left(\lambda+\frac{3}{2}\right)}\sqrt{1-\frac{1}{\xi^2}}\left(\frac{\xi^2\varphi^*-\varphi_0^*}{\xi^2-1}\right)^{\lambda+\frac{3}{2}} + \frac{\sqrt{\varphi^*}}{\sqrt{\pi}}\left(1+\sqrt{1-\frac{\varphi_0^*}{\xi^2\varphi^*}}\right)\right\}. \tag{9.66}$$

Here it is taken into account that when $\varphi^* \gg 1$ the role of the repelled particles is insignificant, and the concentration of attracted unbounded particles in a field which falls off faster than $1/r^2$ is given by the formula

$$N_{\text{unb}}^+ \approx \frac{N_0}{\sqrt{\pi}}\left(\sqrt{\varphi^*} + \sqrt{\varphi^* - \frac{\varphi_0^*}{\xi^2}}\right).$$

For very large values of φ_0^*, the nonlinear term describing the concentration of bounded particles begins to play a major role in Eq. (9.66). Therefore, it is convenient to rewrite Eq. (9.66) in the form

$$\varphi^* = \frac{\varphi_0^*}{\xi^2} + \left(1 - \frac{1}{\xi^2}\right)^{\frac{\lambda+1}{\lambda+\frac{3}{2}}} \left\{ \frac{\Gamma\left(\lambda+\frac{5}{2}\right)}{\Gamma(\lambda+1)} \left[\left(\frac{D}{R_0}\right)^2 \frac{1}{\xi^2} \frac{d}{d\xi}\left(\xi^2 \frac{d\varphi^*}{d\xi}\right) - \frac{\sqrt{\varphi^*}}{\sqrt{\pi}}\left(1 + \sqrt{1 - \frac{\varphi_0^*}{\xi^2 \varphi^*}}\right)\right]\right\}^{\frac{1}{\lambda+3/2}}. \tag{9.67}$$

It is obvious that the solution of this equation, obtained by the method of iterations, is represented in the form of the series

$$\varphi^* = \frac{\varphi_0^*}{\xi^2} + \left(1 - \frac{1}{\xi^2}\right)^{\frac{\lambda+1}{\lambda+\frac{3}{2}}} \left\{ \frac{\Gamma\left(\lambda+\frac{5}{2}\right)}{\Gamma(\lambda+1)} \left[2\left(\frac{D}{R_0}\right) \frac{\varphi_0^*}{\xi^4} - \frac{\sqrt{\varphi_0^*}}{\xi\sqrt{\pi}}\right]\right\}^{\frac{1}{\lambda+\frac{3}{2}}} + \ldots \tag{9.68}$$

The second term in this expression is much smaller than the first provided

$$\xi < \varphi_0^{*\,\frac{2\lambda+1}{4\lambda-2}} \left(\frac{R_0}{D}\right)^{\frac{4}{4\lambda-2}}. \tag{9.69}$$

Consequently, for sufficiently large values of φ_0^* in the considerable neighborhood of the body defined by condition (9.69), the electric field is described by the simple expression (9.68). Limiting ourselves to only the principal term of the series (9.68), we have

$$\varphi(r) = \varphi_0 \frac{R_0^2}{r^2}. \tag{9.70}$$

The simple formula

$$N_b^+(r) = \frac{R_0^2 |\varphi_0|}{2\pi e r^4}. \tag{9.71}$$

for the concentration of bounded particles is also obtained from Eq. (9.3a) under the same approximation. The concentration of unbounded particles under condition (9.69) is, of course, much smaller.

The physical meaning of the expression obtained for $\varphi(r)$ is quite clear. In fact, if the potential is sufficiently large, then the number of bounded particles is large and they screen the field of the body. It therefore falls off much more abruptly than a Coulomb field. However, if the field falls off faster than $1/r^2$, then bounded motion in the presence of an absorbing surface is impossible. Consequently, the number of bounded particles cannot increase without limit, but only as long as the field, because of screening, does not start to fall off like $1/r^2$.

It is important to stress that the asymptotic formulas (9.70) and (9.71) were obtained under very general assumptions about the nature of the particle distribution in finite orbits: it is only necessary that λ should be larger than $-1/2$, that is, the distribution function of the bounded particles should not decrease with increase of the absolute value of the energy. This condition is usually satisfied, as is clear from the specific example considered in §35, and also from an analysis of the general formula (8.68) (also see [71]).

Now let us consider the case of very high values of the potential φ_0. In this case the size of the strongly disturbed (by the field) region – the space charge region – is much larger than the Debye radius. It is therefore natural to expect that the nature of the field variation will be similar to the case of a large body, considered in §39. In §39 it was shown that the role of bounded particles in Eq. (9.66) is suppressed by the exponential factor exp $(-\mathcal{E}_0/\varkappa T)$, where, according to (9.50), the value of $\mathcal{E}_0/\varkappa T$ is given by

$$\frac{\mathcal{E}_0}{\varkappa T} = 2.15\left[\left(\frac{R_0}{D}\right)^8 \varphi_0^*\right]^{1/7}. \tag{9.50}$$

Consequently, for very high values of the potential,

$$\varphi_0^* \gg \left(\frac{D}{R_0}\right)^8 \qquad (9.72)$$

the concentration of bounded particles is exponentially small. Under the same conditions, the concentration of attracted particles undergoing unbounded motion is given by expression (9.37). In addition, all of them hit the surface of the body. Therefore, the behavior of the potential in the vicinity of a small body upon fulfillment of condition (9.72) is exactly described by the solution for a large body, which was given in §39. In this connection, the size of the space charge region (9.44) is given by

$$R_c = 0.803D\left[\frac{e\,|\varphi_0|}{\varkappa T}\,\frac{R_0}{D}\right]^{3/7}. \qquad (9.44)$$

Upon fulfillment of condition (9.72), of course, $R_c \gg D$.

d. Small Values of the Potential on the Surface of a Body. If the potential on the body's surface is small, that is

$$\varphi_0^* = \frac{e\,|\varphi_0|}{\varkappa T} \ll 1,$$

then it is also small everywhere in the vicinity of the body. In this case, in the linear (with respect to φ^*) approximation expressions (8.39) and (8.41) for the concentrations of attracted and repelled particles simplify considerably and assume the form

$$N^+_{\text{unb}} = \frac{N_0}{2}\left[1 + \sqrt{1-\frac{1}{\xi^2}} - \frac{\varphi_0^*}{\xi^2\sqrt{1-\frac{1}{\xi^2}}} + \varphi^*\left(1 + \frac{1}{\sqrt{1-\frac{1}{\xi^2}}}\right)\right], \qquad (9.73)$$

$$N^- = \frac{N_0}{2}\left[1 + \sqrt{1-\frac{1}{\xi^2}} + \frac{\varphi_0^*}{\xi^2\sqrt{1-\frac{1}{\xi^2}}} - \varphi^*\left(1 + \frac{1}{\sqrt{1-\frac{1}{\xi^2}}}\right)\right]. \qquad (9.74)$$

In the same approximation, the concentration of particles undergoing bounded motion is equal to zero, as was indicated in §37. Therefore, Eq. (9.3) is now rewritten in the form

$$\frac{1}{\xi^2}\frac{1}{d\xi}\left(\xi^2\frac{d\varphi^*}{d\xi}\right) = \left(\frac{R_0}{D}\right)^2\left(1 + \frac{1}{\sqrt{1-\frac{1}{\xi^2}}}\right)\left[\varphi^* - \frac{\varphi_0^*}{\xi^2\left(1+\sqrt{1-\frac{1}{\xi^2}}\right)}\right]. \qquad (9.75)$$

If the body is large, then the left side in Eq. (9.75) is unessential, and its solution has the form

$$\varphi^* = \frac{\varphi_0^*}{\xi^2\left(1+\sqrt{1-\frac{1}{\xi^2}}\right)}. \qquad (9.76)$$

For $\xi \to 1$, the value of φ^* is obviously equal to φ_0^*. This is understandable since for small values of φ_0^* the boundary value of the potential φ_1^*, determined by Eq. (9.13) correct to terms of order φ_0^{*2}, coincides with φ_0^* (this already indicated in §37). In this case the double layer around the surface of a large body is not present. At large distances when $\xi \gg 1$ the field falls off like $1/\xi^2$, and also the asymptotic value of the potential φ_∞^* coincides with the value obtained from formula (9.30).

In the case of a small body, when $R_0/D \ll 1$, the right side of Eq. (9.75) becomes important only at large distances from it, so that it is possible to expand $(1 - 1/\xi^2)^{1/2}$ in a series in powers of $1/\xi^2$. The solution of the equation then obtained naturally coincides with the solution (9.55).

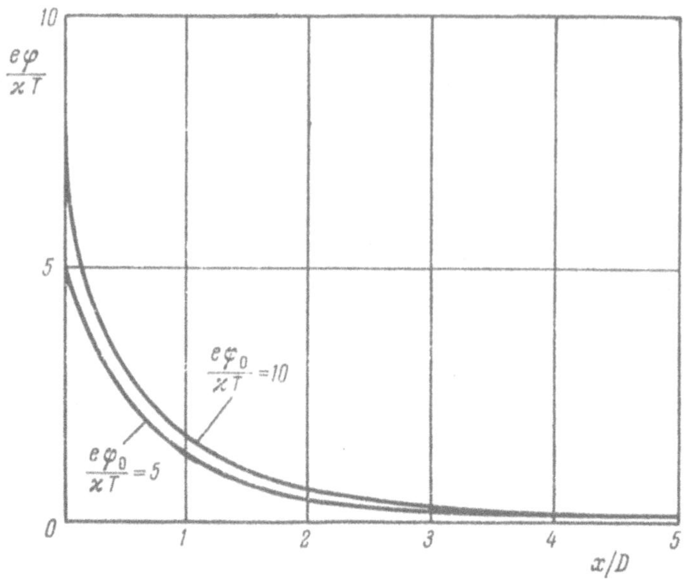

Fig. 77. Curves showing the variation of the potential near the surface of a perfectly reflecting body.

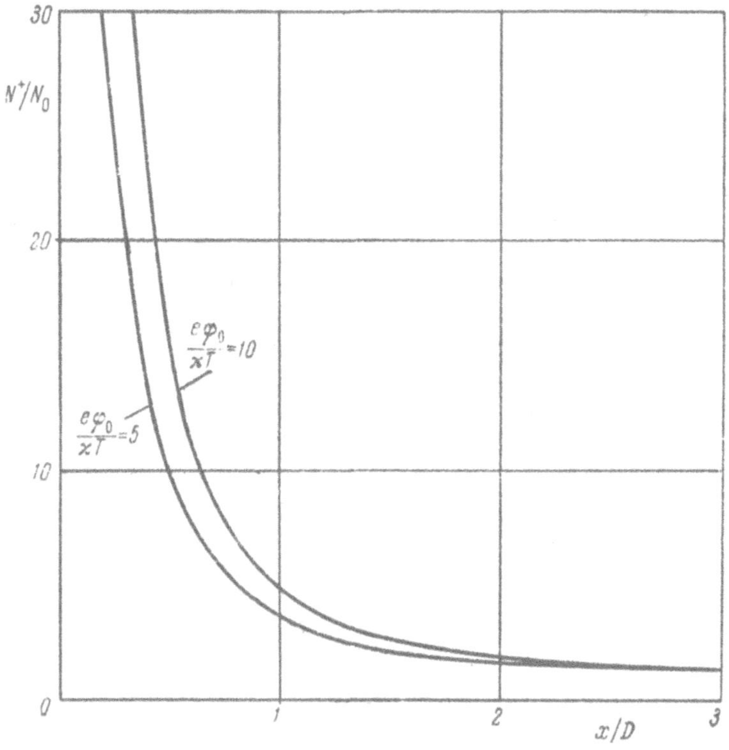

Fig. 78. Concentration of attracted particles near the surface of a perfectly reflecting body.

§41. Structure of the Disturbed Region in the Vicinity of a Perfectly Reflecting Body

Now let us consider the case when the body elastically reflects the particles incident on its surface. In this case an equilibrium Boltzmann distribution is established both for the repelled as well as for the attracted particles. Equation (9.3a) then has the simple form

$$\frac{1}{r^2}\frac{d}{dr}\left(r^2\frac{dU}{dr}\right) = -4\pi e^2\left[\exp\left(-\frac{U}{\varkappa T}\right) - \exp\left(\frac{U}{\varkappa T}\right)\right]. \tag{9.77}$$

The thickness of the disturbed layer near a large body is small in comparison with the size of the body. The curvature of the surface in this case is not essential, and Eq. (9.77) can be rewritten in terms of the dimensionless variables $\varphi^* = -U/\varkappa T$ and $\zeta = (r - R_0)/D = x/D$ in the form

$$\frac{d^2\varphi^*}{d\zeta^2} = 2\,\mathrm{sh}\,\varphi^*. \tag{9.78}$$

The boundary conditions are: $\varphi^* = \varphi_0^*$ for $\zeta = 0$ and $\varphi^* \to 0$ as $\zeta \to \infty$. Integrating Eq. (9.78), we find its solution in the form

$$\varphi^* = 2\ln\left[\frac{e^{\varphi_0^*/2} + 1 + (e^{\varphi_0^*/2} - 1)\,e^{-\sqrt{2}\zeta}}{e^{\varphi_0^*/2} + 1 - (e^{\varphi_0^*/2} - 1)\,e^{-\sqrt{2}\zeta}}\right]. \tag{9.79}$$

The behavior of the field around the body's surface, described by formula (9.79), is shown in Fig. 77 for the values $\varphi_0^* = 5$ and 10. From the figure it is seen that the field near the body falls off very abruptly. In this connection, the larger the value of φ_0^*, the more abrupt the decrease of the field, so that the thickness of the disturbed layer itself does not vary, always remaining approximately the order of the Debye radius. The intensity of the field on the body's surface here, of course, increases exponentially rapidly with increase of φ_0^*, since

$$\frac{d\varphi^*}{d\zeta}(0) = -\frac{eD}{\varkappa T}|E| = -2^{3/2}\,\mathrm{sh}\,\frac{\varphi_0}{2}. \tag{9.80}$$

The intensity of the field on the surface of an absorbing body increases less strongly with increase of φ_0^*, namely, it increases like $\varphi_0^{*1/4}$ [see (9.34a)].

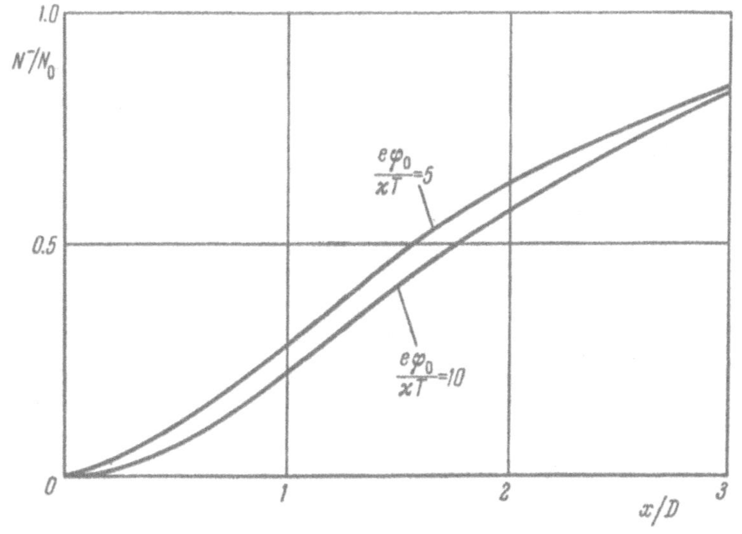

Fig. 79. Concentration of repelled particles near the surface of a perfectly reflecting body.

The concentrations of attracted and repelled particles are determined by the Boltzmann formula. Substituting expression (9.79) into it, for the potential we obtain

$$N^{+}=N_0 e^{\varphi^*}=N_0\left[\frac{e^{\varphi_0^*/2}+1+(e^{\varphi_0^*/2}-1)\,e^{-\sqrt{2}\zeta}}{e^{\varphi_0^*/2}+1-(e^{\varphi_0^*/2}-1)\,e^{-\sqrt{2}\zeta}}\right]^2, \tag{9.81}$$

$$N^{-}=N_0 e^{-\varphi^*}=N_0\left[\frac{e^{\varphi_0^*/2}+1-(e^{\varphi_0^*/2}-1)\,e^{-\sqrt{2}\zeta}}{e^{\varphi_0^*/2}+1+(e^{\varphi_0^*/2}-1)\,e^{-\sqrt{2}\zeta}}\right]^2. \tag{9.82}$$

The dependences of N^+ and N^- on ζ for two values of φ_0^* are shown in Figs. 78 and 79. For $\varphi_0^* \gg 1$ the concentration of attracted particles in the layer next to the surface is exponentially large. It decreases very abruptly with increasing distance from the surface. We note that particles undergoing bounded motion here play a major role. The concentration of repelled particles at the surface of the body is negligibly small; conversely, it abruptly increases with distance. For $r - R_0 \gtrsim D$ the concentration of attracted and repelled particles is always close to N_0.

Thus, both the field and the structure of the disturbed region in the case of a totally reflecting body and in the case of a completely absorbing body, considered in the preceding sections, differ radically. For partial reflection or absorption it is possible to obtain any intermediate case, depending on the value of the coefficient of reflection. However, in order that the structure of the layer should turn out to be close to the case of a totally reflecting body, it is necessary that the coefficient R for reflection of particles from the body's surface should be extremely close to unity; in any case it is necessary that

$$1-R \ll \frac{R_0}{l}, \frac{D}{l},$$

where l is the mean free path of the particles and R_0 is the size of the body. From here it is clear that the case close to the case of a completely absorbing body is apparently always realized in a rarefied plasma under actual conditions. If the absorption of particles is not total, then it is only necessary to correct the results obtained by taking into consideration the reflected unbounded particles. This does not qualitatively change the structure of the disturbed region.

CHAPTER X

PARTICLE FLUX IN THE VICINITY OF THE BODY

§42. Introductory Remarks

The disturbances produced by a moving body naturally lead to a variation of the particle flux in comparison with the flux in an undisturbed plasma. As we saw above (§14), it is necessary to know the fluxes of electrons and ions on the body's surface in order to determine the potential of the electric field on the surface of the body. On the other hand, the determination of the particle flux on variously oriented surface area elements at an arbitrary point in the vicinity of the body represents the starting point for an analysis of the result of various probe measurements.

As far as the flux of neutral particles is concerned, its calculation does not involve any major difficulties, since the neutral particle distribution function is known (see §7). In a number of specific cases, the calculation becomes difficult only because of the complexity of evaluating the corresponding integrals. We also note that since the trajectories of motion of the neutral particles remain straight lines, then for the calculation of the flux it often happens to be more convenient to start not from the solution of the kinetic equation, but from a geometrical picture of the motion of the particles. Such a method, of course, leads to the same results as the solution of the kinetic equation.

For a calculation of the flux of ions and electrons in the general case it is necessary to start from the solution of the kinetic equation. Since, however, a rigorous solution of the kinetic equations for electrons and ions simultaneously with the solution of the equation for the electric field is obtained above only in a limited number of cases, so an exact calculation of the charged particle fluxes can be accomplished only for those same cases.

One should emphasize that the difficulties basically arise in connection with the calculation of the flux of the particles being attracted to the body. Calculation of the flux of repelled particles usually does not cause any difficulties (see §44). Sometimes the field also has little influence on the attracted particles. In such cases one can determine the flux of attracted particles according to the formulas calculated for the flux of neutral particles. This, of course, is valid when the body is weakly charged, that is, $\varphi_0 < \varkappa T/e$. In addition, in front of a rapidly moving large body the flux of ions in the vicinity of the body and on its surface coincides with the flux of neutral particles, correct to terms of order v_i/V_0. The latter is a consequence of the fact that the kinetic energy, $M_i V_0^2/2$, of the ions incident on a rapidly moving body is extremely large. The potential energy of the ions in the electric field outside of the double layer, $e\varphi \sim \varkappa T$, is much smaller, as we saw above (see §14). Therefore, the influence of the field on the motion of the ions in front of the body is negligible, and for a calculation of their flux in the region of "compression," one can use the formulas for the flux of neutral particles. It is only necessary to take the difference in the coefficients of reflection (for reflection of ions and neutral particles from the surface of the body) into consideration.*

The flux of particles reflected from the body strongly depends, of course, on the nature of their scattering and on the geometrical configuration of the surface. In this connection, conditions may be realized in which the presence of a measuring apparatus near the body – a probe – can substantially influence the flux of particles reflected from the body (see §43). It is natural that the body itself will have a strong influence on the particle flux on the probe. It is particularly important to take this influence of the body on the particle

*Here it is assumed, in addition, that the body is not specially charged, so that the potential on its surface is also much smaller than $M_i V_0^2/2e$.

flux into consideration when the body is moving slowly (that is, $V_0 \ll v_i$) and its size is smaller than or comparable with the Debye radius. In this case the particle flux near the surface, because of the influence of the field created by the charged body, may change by a very large factor. Such conditions are realized in the layers of the ionosphere most distant from the earth, or in the region of interplanetary space nearest to the earth (see §§1 and 2). All of this indicates that an analysis of the results of various probing measurements, carried out with artificial satellites and rockets and pursueing the goal of obtaining information about the undisturbed medium, is extremely complicated. Interpretation of the results obtained in this connection requires a great deal of care.

§43. Flux of Neutral Particles

a. Particle Flux on the Body's Surface. The distribution function of the neutral particles incident on the body is Maxwellian [see Eq. (1.6)]. In order to determine the density of the particle flux on a unit element of area at an arbitrary point $\mathbf{r}_S$ on the surface of a body, it is only necessary to integrate it with respect to the velocities

$$j_n = -\overline{(\mathbf{n}\mathbf{v})} = -\int \mathbf{n}\mathbf{v} f_n \, d^3v. \tag{10.1}$$

Here $\mathbf{nv}$ is the projection of the velocity $\mathbf{v}$ of a particle onto the outward normal $\mathbf{n}$ to the surface of the body at the point $\mathbf{r}_S$, f_n is the distribution function of the particles incident on the body.

If the body is at rest, then

$$f_n = \left(\frac{M}{2\pi\varkappa T}\right)^{3/2} n_0 \exp\left(-\frac{Mv^2}{2\varkappa T}\right) = \frac{n_0}{\pi^{3/2} v_n^3} \exp\left[-\frac{M(v_x^2+v_y^2+v_z^2)}{v_n^2}\right], \tag{10.2}$$

where $v_n = (2\varkappa T/M)^{1/2}$ is the thermal velocity of the neutral particles. Substituting this expression for f_n into Eq. (10.1) and choosing the z axis along the direction of the inward normal to the surface of the body, we obtain

$$j_n = \frac{n_0}{\pi^{3/2} v_n^3} \int_0^\infty v_z dv_z \int_{-\infty}^{\infty} dv_x \int_{-\infty}^{\infty} dv_y \exp\left[-\frac{M(v_x^2+v_y^2+v_z^2)}{2\varkappa T}\right] = \frac{n_0 v_n}{2\sqrt{\pi}} = n_0 \sqrt{\frac{\varkappa T}{2\pi M}}. \tag{10.3}$$

The flux density j_n is the same at all points on the surface. Consequently, the total particle flux on the surface of the body is given by

$$J_0 = \oint_S j_n \, dS = n_0 S \sqrt{\frac{\varkappa T}{2\pi M}},$$

where S is the surface area of the body. In particular, if the body is a sphere of radius R_0, then

$$J_0 = \sqrt{\frac{8\pi\varkappa T}{M}} n_0 R_0^2. \tag{10.4}$$

If the body moves with velocity $\mathbf{V}_0$, then, as indicated above in Eq. (1.6),

$$f_n = \frac{n_0}{\pi^{3/2} v_n^3} \exp\left[-\frac{M(\mathbf{v}+\mathbf{V}_0)^2}{v_n^2}\right]. \tag{10.5}$$

Substituting this expression into Eq. (10.1) and choosing, as before, the z axis in the direction $-\mathbf{n}$ of the inward normal to the surface, and choosing the x axis in the direction perpendicular to the plane $\mathbf{V}_0\mathbf{n}$, we have

$$j_n = \frac{n_0}{\pi^{3/2} v_n^3} \int_0^{\infty} v_z \, dv_z \int_{-\infty}^{\infty} dv_x \int_{-\infty}^{\infty} dv_y \exp\left[-\frac{v_x^2 + (V_{0y}+v_y)^2 + (V_{0z}+v_z)^2}{v_n^2}\right]. \tag{10.6}$$

Here V_{0z} and V_{0y} are the projections of the velocity $\mathbf{V}_0$ on the z axis and on the y axis:

$$V_{0z} = nV_0, \quad V_{0y} = \sqrt{V_0^2 - (nV_0)^2}. \tag{10.7}$$

After elementary integration, we obtain

$$j_n = n_0 \left\{\frac{nV_0}{2}\left[1 + \Phi\left(\frac{nV_0}{v_n}\right)\right] + \frac{v_n}{2\sqrt{\pi}} \exp\left[-\frac{(nV_0)^2}{v_n^2}\right]\right\}, \tag{10.8}$$

where $\Phi(x)$ is the probability integral. The flux density is now already substantially different for different points on the surface of the body. For example, in the case of a rapidly moving body, the particles are primarily incident on its leading surface; in this connection

$$j_n \simeq n_0 (nV_0). \tag{10.9}$$

If the moving body is a sphere, then $\mathbf{n}\mathbf{V}_0 = V_0 \cos\theta'$, where θ' is the polar angle which determines the position of the point $\mathbf{r}_S$ (under consideration) on the surface of the sphere, and which is measured from the direction of $\mathbf{V}_0$. Integrating the flux density (10.8), we determine the total flux of particles on the surface of a sphere

$$J = \oint_S j_n \, dS = 2\pi R_0^2 \int_0^{\pi} \sin\theta' \, d\theta' \cdot j_n(\theta') = \pi R_0^2 n_0 \left\{V_0 \Phi\left(\frac{V_0}{v_n}\right)\left[1 + \frac{1}{2}\left(\frac{v_n}{V_0}\right)^2\right] + \frac{v_n}{\sqrt{\pi}} \exp\left(-\frac{V_0^2}{v_n^2}\right)\right\}. \tag{10.10}$$

For the case of a sphere at rest, this expression, of course, goes over into (10.4). In the case of a rapidly moving sphere,

$$J = \pi R_0^2 n_0 V_0 \left[1 + \frac{1}{2}\left(\frac{v_n}{V_0}\right)^2\right]. \tag{10.11}$$

correct to exponentially small terms. The physical meaning of the major term in this expression is quite clear. Since a body with cross section πR_0^2 and velocity V_0 is incident on a gas whose density is n_0, so the flux of particles on the surface of the body is $J \approx \pi R_0^2 V_0 n_0$. Affiliated with that argument, it is assumed that the particles of the gas are at rest, i.e., their thermal motion is neglected. It is clear from formula (10.11) that if $V_0 \gg v_n$, then the expressions for the flux obtained by neglecting the thermal motion are valid, correct to an additional term of order $(v_n/V_0)^2$.

b. Particle Flux in the Vicinity of the Body. The disturbances caused by a moving body also lead to a change of the particle flux in the vicinity of the body, in comparison with the flux in an undisturbed medium. It is clear from general considerations that, with increasing distance r from the body the disturbances of the flux fall off like $1/r^2$, that is, they decrease in the same way as the disturbances of the particle concentration (see §§ 16 and 36). The density of the particle flux on a unit element of area with normal $\mathbf{n}$, located at point A in the vicinity of the body (see Fig. 4), is in general composed of the flux j_{n1} of the "direct" particles, that is, the particles which do not collide with the body, and the flux j_{n2} of the reflected particles (compare with §7). The distribution function is well known in the general case (see §7); depending on the shape of the body's surface, only the function $w_n(\mathbf{v}, \mathbf{v}_1, \mathbf{r}_S)$ (see §3b) describing the probability of scattering on the surface of the body, and also the region of integration in velocity space for direct and reflected particles change. If the disturbing body is a sphere then, as shown in §7, the distribution function at the point $A(r, \theta)$ has the form

$$f_n = \begin{cases} f_0(v+V_0) = n_0\left(\frac{M}{2\pi\varkappa T}\right)^{3/2}\exp\left[-\frac{M(v+V_0)^2}{2\varkappa T}\right] & \text{for } \beta > \theta \\ -\frac{1}{(n_S v)}\int f_0(v_1+V_0)(n_S V_0)\, w_n\, d^3v_1 & \text{for } \beta < \theta \end{cases}, \tag{10.12}$$

where $\mathbf{n}_S$ is the outward normal to the surface of the sphere at the point of the particle's impact with the surface.

The flux of "direct" particles on a unit area element with outward normal **n**, situated in the vicinity of the sphere at the point A(r, θ), is determined by the expression

$$j_{n1} = -\int_0^{2\pi} d\varphi \int_\theta^{\pi} \sin\beta\, d\beta \int_0^{\infty} (nv)\, v^2 f_0(v+V_0)\, dv, \tag{10.13}$$

where φ is the azimuthal angle, and β is the polar angle in the space of velocities **v**, measured from the direction of **r**. Since one is usually interested in the flux on one side of the unit area element, so it is necessary to carry out the integration only over the domain of velocities in which **nv** is greater than zero.

For the flux of reflected particles we have

$$j_{n2} = \int_0^{2\pi} d\varphi \int_\theta^{\theta} \sin\beta\, d\beta \int_\theta^{\infty} \frac{nv}{(n_S v)} v^2\, dv \int d^3v_1 f_0(v_1+V_0)\,(n_S v_1)\, w_n. \tag{10.14}$$

In particular, for specular reflections from the surface of a sphere, when $w_n = \delta[\mathbf{v}_1 - \mathbf{v} + 2\mathbf{n}_S(\mathbf{v}\mathbf{n}_S)]$ (see §3b), after integration with respect to d^3v and with account of (2.60), we obtain

$$j_{n2} = -\int_0^{2\pi} d\varphi_1 \int_\theta^{\pi} \sin\beta_1\, d\beta_1 \int_0^{\infty} v_1^2\,[(nv_1) - 2(n_S n)(n_S v_1)]\; f_0(v_1+V_0)\,\frac{d^3r_1}{d^3r}\, dv_1, \tag{10.15}$$

where d^3r_1/d^3r is the change of the elementary volume element upon specular reflection from the surface [see Eq. (2.62)]. It is also possible to obtain a somewhat different expression for j_{n2} by first carrying out the integration with respect to d^3v_1 in Eq. (10.14) (see §7). Similarly it is not difficult to obtain a general expression for the flux of reflected particles in the case of diffuse reflection from the body's surface.

Now let us consider a rapidly moving sphere with velocity $V_0 \gg v_n$. In this case, as we saw above, in the first approximation one can neglect the thermal motion of the particles, that is, one can assume that

$$f(v+V_0) = \begin{cases} n_0\delta(v+V_0) & \text{for } z \geqslant 0, \text{ or for } x^2+y^2 > R_0^2, \text{ if } z < 0; \\ 0 & \text{when } z < 0, \text{ if } x^2+y^2 < R_0^2. \end{cases} \tag{10.16}$$

Here the shading (by the sphere) of the incident particle flux is taken into consideration. Substituting (10.16) into (10.13), we have

$$j_{n1} = \begin{cases} n_0(nV_0) & \text{for } nV_0 > 0 \text{ and the first condition of (10.16)}, \\ 0 & \text{for } nV_0 < 0 \text{ or for the second condition of (10.16)}. \end{cases} \tag{10.17}$$

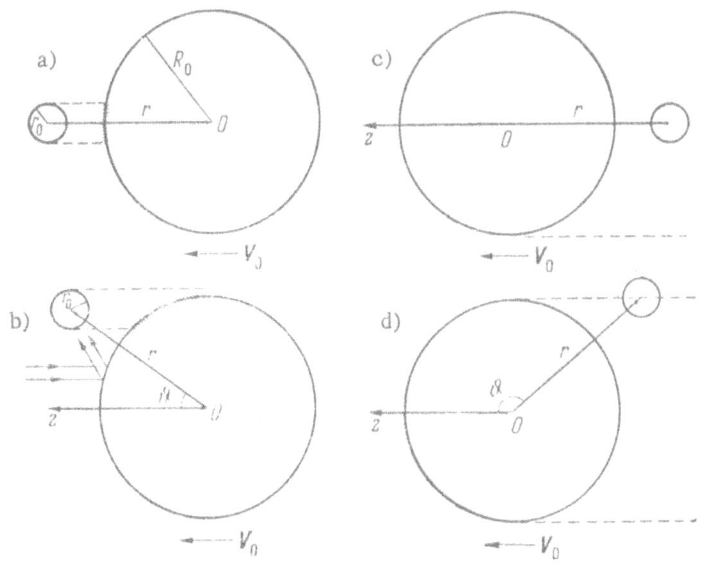

Fig. 80. Particle flux on a spherical probe situated near a rapidly moving body.

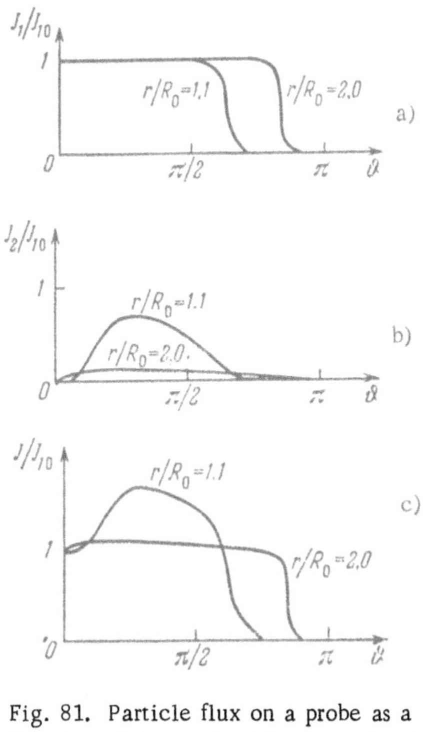

Fig. 81. Particle flux on a probe as a function of the angle ϑ between the direction to the probe and the direction of motion of the body; the dependence is plotted for $r_0/R_0 = 0.05$ and for the different values of r/R_0 indicated in the figure; $J_{10} = \pi r_0^2 n_0 V_0$. a) Direct particle flux; b) reflected particle flux (specular reflection); c) total flux.

The meaning of this formula is quite clear: the flux density j_{n1} everywhere coincides with the undisturbed value of the flux density (10.9) except in the region shaded by the sphere where, of course, the flux is equal to zero. Similarly from Eq. (10.14) we obtain

$$j_{n2} = \begin{cases} [2(\boldsymbol{n}\boldsymbol{n}_S)V_0\cos\theta' - \boldsymbol{n}V_0]\, n_0 \dfrac{R_0^2\sin^2\theta'\cos^2\theta'}{\varrho^2\left(1-\dfrac{R_0}{\varrho}\sin^3\theta'\right)} \\ \quad \text{for} \quad 2(\boldsymbol{n}\boldsymbol{n}_S)V_0\cos\theta' - \boldsymbol{n}V_0 > 0 \quad \text{and the first condition of (10.16)}, \\ 0 \quad \text{for} \quad 2(\boldsymbol{n}\boldsymbol{n}_S)V_0\cos\theta' - \boldsymbol{n}V_0 < 0 \\ \quad \text{or for the second condition of (10.16)}. \end{cases} \tag{10.18}$$

Here θ' is the angle between the normal to the sphere at the point of impact and the z axis, which is parallel to the velocity $\mathbf{V}_0$, that is, $\cos\theta' = \mathbf{n}_S\mathbf{V}_0/V_0$ (see Fig. 4). The angle θ' is related to the coordinates ρ and z of the considered point $A(\rho, z)$ by equation (2.35):

$$2z\cos\theta' + 2\varrho\sin\theta' - \frac{\varrho}{\sin\theta'} = R_0.$$

Upon integration of (10.15), expression (2.62) for d^3r_1/d^3r is taken into consideration. Formula (10.18) for the reflected particle flux j_{n2} has a simple meaning. We see that the current is equal to the product of the concentration of reflected particles,

$$n_2 = n_0 \frac{R_0^2 \sin^2\theta' \cos^2\theta'}{\varrho^2 \left(1 - \frac{R_0}{\varrho} \sin^3\theta'\right)} \tag{10.19}$$

[see Eq. (2.40)], times the projection of the velocity of the reflected particles on the inward normal, $-\mathbf{n}$, to the unit area element. It should be emphasized that since the formula for the concentration of reflected particles was tabulated earlier (see Table VI and Fig. 6), so one can obtain numerical values for the flux j_{n2} without difficulty. In fact, if the unit area element is oriented so that the direction of the inward normal to it coincides with the direction of the velocity of the reflected particles, then the flux is given by

$$j_{n2} = V_0 n_0 \frac{R_0^2 \sin^2\theta' \cos^2\theta'}{\varrho^2 \left(1 - \frac{R_0}{\varrho} \sin^3\theta'\right)} = V_0 n_2 (\varrho, z), \tag{10.20}$$

where $n_2(\rho, z)$ is the appropriate value of the concentration of reflected particles, taken from Table VI.

For any other orientation of the unit area element, it is necessary to multiply expression (10.20) by the cosine of the angle between the inward normal $-\mathbf{n}$ and the direction of the velocity of the reflected particles.

On the z axis in front of the body, $\cos\theta' = R_0/2z$, and formula (10.20) simplifies to

$$j_{n2}(0, z) = -n_0 (V_0 n) \frac{R_0^2}{(R_0 - 2z)^2}. \tag{10.21}$$

We also note that the flux density is equal to zero in the shaded region (the region of "rarefaction"), provided that the thermal motion of the particles is neglected. With account of thermal motion of the particle, the fluxes j_{n1T} and j_{n2T} are exponentially small in the shaded region; for example,

$$j_{n1T} \sim j_{n1} \exp\left[-\left(\frac{V_0 R_0}{v_n z}\right)^2\right],$$

where j_{n1} is the flux in the unshaded region.

<u>c. Particle Flux on a Probe Located Near a Rapidly Moving Body.</u> Now let us consider a small spherical probe of radius r_0 located not far from the surface of a rapidly moving sphere of radius R_0, as is shown in Fig. 80. We shall determine the particle flux on the surface of such a probe. This flux, of course, varies substantially, depending on the position of the probe.

If the probe is located in front of the body (Fig. 80a, b), then the "direct" particle flux, in the approximation in which their thermal motion is neglected, is of course determined by formula (10.11), that is,

$$J_{10} = \pi r_0^2 n_0 V_0. \tag{10.22}$$

Behind the body (Fig. 80c) $J_1 = 0$ in the shaded zone. On the boundary of the shaded zone (Fig. 80d) the flux is given by

$$J_1 = n_0 V_0 S_0, \tag{10.22a}$$

where S_0 is the unshaded (by the body) part of the probe's cross section. The dependence of the "direct" particle flux on the angle ϑ, which determines the position of the probe relative to the direction of motion of the body (see Fig. 80), is shown in Fig. 81a. The different curves in the figure, obtained with the aid of formulas (10.22) and (10.22a), correspond to different distances r between the center of the probe and the center of the sphere. As is clear from the figure, the change of the "direct" particle flux from J_{10} to 0 upon passage of the probe into the shaded region turns out to be smeared out over an angle $\delta\vartheta$, whose magnitude increases with increasing dimensions of the probe, namely $\delta\vartheta \approx 2r_0/r$.

In order to determine the reflected particle flux J_2 it is necessary to integrate the flux density j_{n2} (10.15) over the entire surface of the probe. If the reflection of a particle from the body's surface is specular and the dimensions r_0 of the probe are small in comparison with the dimensions R_0 of the body, so that the reflected particle density (10.19) varies little over the distance r_0, then in the approximation in which thermal motion is neglected

$$J_2 = \pi r_0^2 V_0 n_2(\rho, z) = \pi r_0^2 V_0 n_2(\vartheta, r), \tag{10.23}$$

where ρ and z are the coordinates of the center of the probe: $\rho = r \sin\vartheta$, $z = r\cos\vartheta$. Expressions for the concentration $n_2(\rho, z)$ and its numerical values are given in §6 (Table VI, Fig. 6).

Behind the body, in the shaded region, the concentration n_2 and consequently also the flux J_2 tend to zero. It is important, however, that the concentration n_2 and the flux J_2 also vanish when the probe is located exactly in front of the the body – for values of ϑ_2 close to zero. In fact, the expressions for n_2 given in §6 do not take the presence of the probe, which absorbs part of the particles incident on the body, into consideration. Now the probe itself creates a shaded region on the surface of the body, in which the incoming particles do not hit (the crosshatched region shown in Figs. 80a and b). In this connection it is clear that those particles, which should be reflected by the shaded (by the probe) region on the body's surface are generally not present in the reflected particle flux. Therefore, in the presence of a probe, the expression for $n_2(\rho, z)$ is changed somewhat. In particular, it is clear that if the probe is located exactly in front of the body ($\vartheta = 0$, Fig. 80a), then the region on the surface from which the particles incident on the probe must be reflected is completely shaded and $J_2 = 0$. If the angle ϑ is close to zero, or more precisely, if

$$\vartheta \leqslant \frac{r_0}{r - R_0}, \tag{10.24}$$

then the reflecting region is partially shaded. On the other hand, if $\vartheta > r_0/(r - R_0)$, then the presence of a shaded zone on the body's surface already has no influence on the flux J_2, since now only those particles which were reflected in the unshaded zone (see Fig. 80b) are incident on the probe. In this case, therefore, the reflected particle flux is determined by formula (10.23).

The dependence of the reflected particle flux J_2 on the angle ϑ, constructed with the aid of formula (10.23) with account of the shaded (by the probe) part of the body's surface, is shown in Fig. 81b. The total particle flux on the probe, $J = J_1 + J_2$, is shown in Fig. 81c. It is clear from the figure that the total particle flux differs considerably from $J_{10} = \pi r_0^2 n_0 V_0$ and varies substantially depending on the position of the probe. The difference between the total current J and J_{10} turns out to be even more considerable for diffuse scattering of particles on the surface of the body. Therefore it is clear that for a sufficiently precise measurement of the concentration n_0 with the aid of probes of the type considered, which are situated on a body which is not oriented in space, an extremely complicated analysis of the experimental data, and also knowledge of the laws governing the reflection of particles from the surface of the body, is required.

Above the thermal motion of the particles was not taken into consideration. In general this is valid for a rapidly moving body, when the probe is not very far removed from its surface, that is,

$$\frac{r_0}{r - R_0} \frac{V_0}{v_n} > 1. \tag{10.25}$$

In the opposite case the region shaded by the probe is smeared out because of the thermal motion, so that the reflected particle flux varies more smoothly as a function of the angle ϑ. Some calculations of the flux on a probe of direct and reflected particles, with account of their thermal motion, are carried out in article [26].

It should be indicated that at the present time calculations of the flux of charged particles in the vicinity of a rapidly moving body have been performed only for a limited number of cases and calculations which take the influence of a magnetic field into consideration are generally absent; the role of the magnetic field may turn out to be extremely important in those cases, for example, when the Larmor radius of the electrons is commensurable with or smaller than the dimensions of the body (probe).

Going ahead below to a calculation of the charged particle flux in the vicinity of a body at rest, we only note that some new results for a moving body were obtained for a number of special cases in article [78]. Thus, for example, in the case of a body charged to the potential φ_0, whose radius is much smaller than the Debye radius and which is moving with a velocity $V_0 \gg (2\varkappa T/M)^{1/2}$, the flux of attracted particles is given by

$$J^{+} = \pi R_0^2 N_0 V_0 \left(1 + \frac{2e\phi_0}{MV_0^2}\right),$$

and the flux of repelled particles (electrons) on the surface of the body, if $MV_0^2/2 > \varphi_0$, is given by

$$J^{-} = \frac{1}{2} \pi R_0^2 N_0 V_0 \left(1 - \frac{2e\phi_0}{MV_0^2}\right) \left\{1 + \phi\left[\frac{MV_0^2}{2\varkappa T}\left(1 - \frac{2e\phi_0}{MV_0^2}\right)\right]\right\}$$

for $MV_0^2/2 \leq e\varphi_0$, $J_e = 0$ in the approximation considered.

§44. Flux of Electrons and Ions in the Vicinity of an Absorbing Body

Now let us determine the fluxes of electrons and of ions in the vicinity of a sphere which is charged to a potential φ_0, at rest in a plasma. Space charge forms in the vicinity of the sphere, due to polarization of the plasma. In virtue of symmetry, the total field produced by the body and by the space charge is spherically symmetric, that is, $\varphi = \varphi(r)$. Therefore it is advisable to first obtain general formulas for the fluxes of attracted and repelled particles in a spherically symmetric field $U(r)$, and afterwards to calculate the flux in those specific cases when the exact form of this field is known (see Chapter IX).

Particles in an attractive field can perform both unbounded as well as bounded motion (see §32). The flux j^{+}_{unb} of unbounded attracted particles on a unit element of area with outward normal **n**, located at the point **r**, is determined by the expression

$$j^{+}_{\text{unb}} = -\int\limits_{\text{unb}} (\mathbf{nv}) f_{\text{unb}} d^3v = -\frac{N_0}{\sqrt{2\pi}}\left(\frac{M}{\varkappa T}\right)^{3/2} \int (\mathbf{nv}) \exp\left[-\frac{M(v_\theta^2 + v_r^2)}{2\varkappa T} - \frac{U(r)}{\varkappa T}\right] v_\theta \, dv_\theta \, dv_r. \tag{10.26}$$

Here it is assumed that the distribution function in a central force field depends only on r and on the velocities v_r and v_θ (see §33) and that it has the form (8.25) for unbounded particles. The integration in (10.26) is carried out, of course, over the region of unbounded motion, and also only with respect to those velocities **v** which satisfy the condition $\mathbf{nv} < 0$.

In a plasma the electric field may decrease with increasing distance both more rapidly as well as more slowly than $1/r^2$, but at infinity it always falls off exactly like $1/r^2$. The region of unbounded motion was investigated previously in §§32 and 34 for all of these cases, and it is determined by formulas (8.7), (8.11), and (8.28). In the simplest case, if the unit area element at the point **r** is oriented so that $\mathbf{n} \parallel \mathbf{r}$, then $\mathbf{nv} = v_r$ and the condition $\mathbf{nv} < 0$ can be written in the form

$$v_r < 0. \tag{10.27}$$

In this case the presence of an absorbing surface of radius R_0 in no way manifests itself on the particle flux at the point **r**. Expression (10.26) therefore assumes the form

$$j^{+}_{\text{unb}} = -\frac{N_0}{\sqrt{2\pi}}\left(\frac{M}{\varkappa T}\right)^{3/2} \int\limits_{\substack{\text{unb} \\ v_r < 0}} \exp\left[-\frac{M(v_r^2 + v_\theta^2)}{2\varkappa T} - \frac{U(r)}{\varkappa T}\right] v_r v_\theta \, dv_r \, dv_\theta$$
$$= -\frac{\sqrt{2}N_0}{\sqrt{\pi}}\left(\frac{M}{\varkappa T}\right)^{1/2} \int\limits_{-\infty}^{0} \exp\left[-\frac{M[v_r^2 + v_\theta^2(v_r)]}{2\varkappa T} - \frac{U(r)}{\varkappa T}\right] v_r dv_r, \tag{10.28}$$

where the function $v_\theta(v_r)$ is the limit up to which the integration with respect to the velocities v_θ goes for a given value of the velocity v_r. In this connection, if the magnitude of the velocity v_r is larger than $[-(2/M)U(r)]^{1/2}$, then $v_\theta(v_r)=0$; however, if $v_r < [-(2/M)U(r)]^{1/2}$, then the function $v_\theta(v_r)$ is the boundary between the regions of bounded and unbounded motions. Taking this into account, we go over, as usual (see § 34) in the integral (10.28), to a new variable r_0 which is related to v_r by Eq. (8.11). Then, in the general case when the field at infinity falls off (in absolute value) exactly like $1/r^2$, but for smaller values of r it first falls off faster than, and then more slowly than $1/r^2$, we obtain

$$j^+_{\text{unb}} = j_0\left\{1+\frac{R_0^2}{r^2}\varphi^*_\infty+\frac{1}{2}\int\limits_{\rho_0(r)}^{\infty}\left(\frac{r_0^2}{r^2}-1\right)\left(3\frac{d\varphi^*}{dr_0}+r_0\frac{d^2\varphi^*}{dr_0^2}\right)\exp\left[\varphi^*(r_0)+\frac{r_0}{2}\frac{d\varphi^*}{dr_0}\right]dr_0\right\}, \tag{10.29}$$

where j_0 is the particle flux (10.3) in the absence of a field

$$j_0 = N_0\sqrt{\frac{\varkappa T}{2\pi M}}, \tag{10.30}$$

and, as usual, $\varphi^* = -U(r)/\varkappa T$, and φ^*_∞ is a quantity characterizing the asymptotic behavior of the potential $U(r)$ as $r\to\infty$ (see § 34), that is,

$$-\frac{U(r)}{\varkappa T}_{r\to\infty} = \varphi^*_\infty\frac{R_0^2}{r^2}. \tag{10.31}$$

We note that the expression $\frac{1}{2}[3(d\varphi^*/dr_0)+r_0(d^2\varphi^*/dr_0^2)]$, standing inside the integral in (10.29) is the derivative of the argument of the exponential function appearing in this same integral. Therefore, one can carry out an integration by parts in (10.29). Then we obtain

$$j^+_{\text{unb}} = j_0\left\{\varphi^*_\infty\frac{R_0^2}{r^2}+\frac{\varrho_0^2(r)}{r^2}-\left(\frac{\varrho_0^2(r)}{r^2}-1\right)\exp\left[\varphi^*(\varrho_0)+\right.\right.$$
$$\left.\left.+\frac{\varrho_0}{2}\frac{d\varphi^*}{dr}(\varrho_0)\right]+\frac{2}{r^2}\int\limits_{\rho_0(r)}^{\infty}r_0\left[1-\exp\left(\varphi^*(r_0)+\frac{r_0}{2}\frac{d\varphi^*}{dr_0}\right)\right]dr_0\right\}. \tag{10.32}$$

The distance $\rho_0(r)$ is determined from Eq. (8.12)

$$U(\varrho_0)-U(r)-\frac{\varrho_0}{2}\frac{dU}{dr}(\varrho_0)\left(\frac{\varrho_0^2}{r^2}-1\right)=0, \tag{10.33}$$

where ρ_0 is the maximum root of this equation. We previously analyzed the formula for ρ_0 in § 32. In this connection, it was shown in particular that the potential $U(r)$ at the point ρ_0 always falls off faster than $1/r^2$ with increasing r. From formula (10.32) it is clear that the flux of attracted particles at the point r substantially depends not only on the value of the potential at this same point, but also on the overall behavior of $\varphi^*(r_0)$ in the region $r_0 \geq \rho_0$.

Now let us consider certain limiting expressions for the flux, which are obtained from formula (10.32). First let us assume that the condition

$$\left|\varphi^*(r_0)+\frac{r_0}{2}\frac{d\varphi^*}{dr_0}\right| \ll 1 \tag{10.34}$$

is satisfied everywhere in the region $r_0 > \rho_0(r)$. Then it is possible to expand the exponential term in a power series. Integrating by parts in (10.32) and taking relation (10.34) into consideration, we obtain

$$j^+_{\text{unb}}(r) = j_0\{1+\varphi^*(r)-A_2(r)\}, \tag{10.35}$$

where

$$A_2(r)=\frac{1}{r^2}\int\limits_{\rho_0}^{\infty} r_0\left[\varphi^*(r_0)+\frac{r_0}{2}\frac{d\varphi^*}{dr_0}\right]^2 dr_0+\frac{1}{2}\left(\frac{\varrho_0^2}{r^2}-1\right)\left[\varphi^*(\varrho_0)+\frac{\varrho_0}{2}\frac{d\varphi^*}{dr}(\varrho_0)\right]^2. \tag{10.36}$$

The last term in (10.35) is small, since

$$A_2(r)\sim|\varphi^*(\varrho_0)+\frac{\varrho_0}{2}\frac{d\varphi^*}{dr}(\varrho_0)|\ll 1, \tag{10.37}$$

and consequently, the simple expression of Langmuir and Mott-Smit [42],

$$j^+_{\text{unb}}(r)=j_0(1+\varphi^*)=j_0\left(1-\frac{U(r)}{\varkappa T}\right) \tag{10.38}$$

is valid for the particle flux in the case under consideration. One can immediately obtain this expression from formula (10.29) if the exponential term in it is set equal to unity. Since for all values of $r_0 \geq \rho_0$ the potential falls off not more slowly than $1/r^2$, so a negative quantity always appears in the exponent. Consequently, the flux at the point r for a given value of the potential U(r) in any case cannot exceed the flux given by (10.38). We note that formula (10.38) for the particle flux is rigorously valid if the potential always falls off with increasing r not faster than $1/r^2$. In fact, the boundary of the regions of bounded and unbounded motions in this case is determined by formula (8.7). Consequently, in formula (10.28) $v_\theta(v_r)=[-v_r^2-(2/M)U(r)]^{1/2}$ for $v_r<[-(2/M)U(r)]^{1/2}$ and $v_\theta(v_r)=0$ for $v_r>[-(2/M)U(r)]^{1/2}$. Taking this into consideration, we at once obtain expression (10.38) for the flux of attracted particles from formula (10.28). Thus, the simple formula (10.38) for the density of the flux of attracted particles on the surface of a sphere of radius R_0 is valid if the potential falls off not faster than $1/r^2$, up to large distances from the sphere, $r/R_0 \gg 1$.

Now let us consider the other limiting case when the potential decreases very abruptly with increasing distance from the surface of a sphere of radius R_0. For example, let

$$U(r)=U(R_0)\exp\left(-\frac{r-R_0}{D}\right), \tag{10.39}$$

where $U(R_0)$ is the potential on the surface of the sphere, and $D \ll R_0$. In this case $\varphi^*_\infty=0$ and $\rho_0(r)=r$ (see §32). Substituting expression (10.39) for U(r) into (10.32) and integrating, we find that

$$j^+_{\text{unb}}=j_0\left\{1+\frac{1}{2}\left[\frac{D}{R_0}\ln\left(-\frac{R_0}{D}\frac{U(R_0)}{\varkappa T}\right)\right]^2\right\}. \tag{10.40}$$

In other words, if the field is cut off very abruptly at the surface of the sphere, then it only very weakly influences the flux of attracted particles, since $D/R_0 \ll 1$. The flux density j^+_{unb} coincides in this case with the flux density j_0 of neutral particles. Such a field distribution occurs on the leading surface of a rapidly moving large body (see §14). Consequently, the flux of ions on the leading surface of a rapidly moving large body and in its vicinity is close to the neutral particle flux. This was already mentioned earlier (see §§14 and 42). We note that the total particle flux on a spherical surface of radius R_0 is, of course, given by

$$J_{\text{unb}}=4\pi R_0^2 j_{\text{unb}}(R_0). \tag{10.41}$$

The flux of bounded particles at the point r is also determined by the integral (10.26) over the region of bounded motion, using the distribution function of the bounded particles. However, since we are in fact interested only in the flux on a certain absorbing surface, so the presence of such a surface changes the distribution of bounded particles, since particles passing along trajectories which intersect this surface disappear. Thus, in the approximation which does not take collisions into consideration, the bounded particle flux on any absorbing surface is equal to zero. With account of collisions, this flux differs from zero, but it is always small in comparison with the flux of unbounded particles under conditions when the size R of the disturbed zone is much smaller than the mean free path l of the particles. In a weakly ionized plasma, the expression for the total flux of bounded electrons on an absorbing Coulomb center was obtained in §35 [see Eq. (8.73)]. Comparing it with the flux of unbounded particles (8.75), one can convince oneself that under conditions $R \ll l$ the flux of bounded particles is generally small.

In conclusion, we note that for repelled particles the Maxwell-Boltzmann distribution is always valid. Therefore, calculation of the flux of repelled particles on a unit area element at the point $\mathbf{r}$ with normal $\mathbf{n} \parallel \mathbf{r}$ leads to the obvious result

$$j^-(r) = -\int v_r f \, d^3v = j_0 \exp[-\varphi^*(r)], \tag{10.42}$$

where j_0 as usual is the particle flux (10.3) in the absence of any field.

§45. Theory of a Spherical Probe

Now let us consider the flux of charged particles on a metallic body of small size – a probe. The flux on the probe, of course, essentially depends on the potential of its surface and is given by

$$\mathcal{J} = e(J_i - J_e), \tag{10.43}$$

where J_i and J_e are the total fluxes of ions and of electrons on the surface of the probe. If the probe is positively charged, then ions are repelled, and electrons are attracted to it; if the probe is negatively charged, then the converse.

The flux of repelled particles is determined only by the potential of the probe (10.42); the attracted particle flux depends, as we saw above, on the overall behavior of the field $\varphi(r)$ in the vicinity of the probe (see Eq. (10.32)). In this connection, the cases of large ($R_0 \gg D$) and small ($R_0 \ll D$) probes are substantially different (see Chapter IX). Also the case of very high potentials on the surface of the probe is distinguished (see §§39 and 40).

a. Flux of Attracted Particles on the Surface of a Large Probe. The dependence of the potential φ on r in the case when the radius R_0 of the probe is much larger than the Debye radius $D = (\varkappa T/4\pi e^2 N_0)^{1/2}$ in the undisturbed plasma is obtained in §37 for the case of not very high values of the potential φ, that is, when

$$|\varphi_0| \ll \frac{\varkappa T}{e}\left(\frac{R_0}{D}\right)^{4/3}. \tag{10.44}$$

Since in this case $R_0 > r_K$ and, consequently, $\rho_0(r) = r$, then formula (10.32) for the flux of attracted particles simplifies and takes the form

$$J^+ = J_0\left\{1 + \varphi_\infty^* + 2\int_1^\infty \xi\left[1 - \exp\left(\varphi^* + \frac{\xi}{2}\frac{d\varphi^*}{d\xi}\right)\right] d\xi\right\}. \tag{10.45}$$

TABLE XXV. Dependence of J^+/J_0 on the Potential, $\varphi_0^* = e|\varphi_0|/\varkappa T$ on the Surface of the Probe

φ_0^*	0	$\varphi_0^* \ll 1$	0.25	0.50	0.75	1.0	$\varphi_0^* \gg 1$
J^+/J_0	1	$1+\varphi_0^*$	1.18	1.26	1.33	1.38	1.47

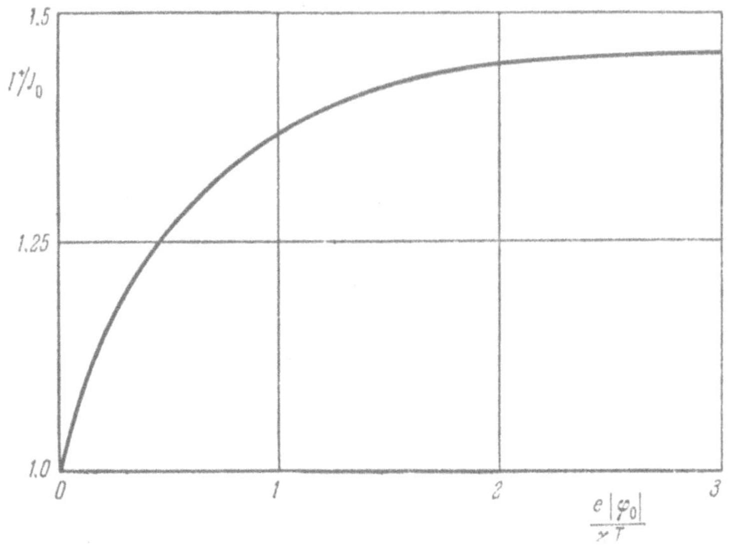

Fig. 82. Variation of the total flux of attracted particles as a function of the potential on the body's surface: $J_0 = (8\pi\varkappa T/M)^{1/2} R_0^2 N_0$.

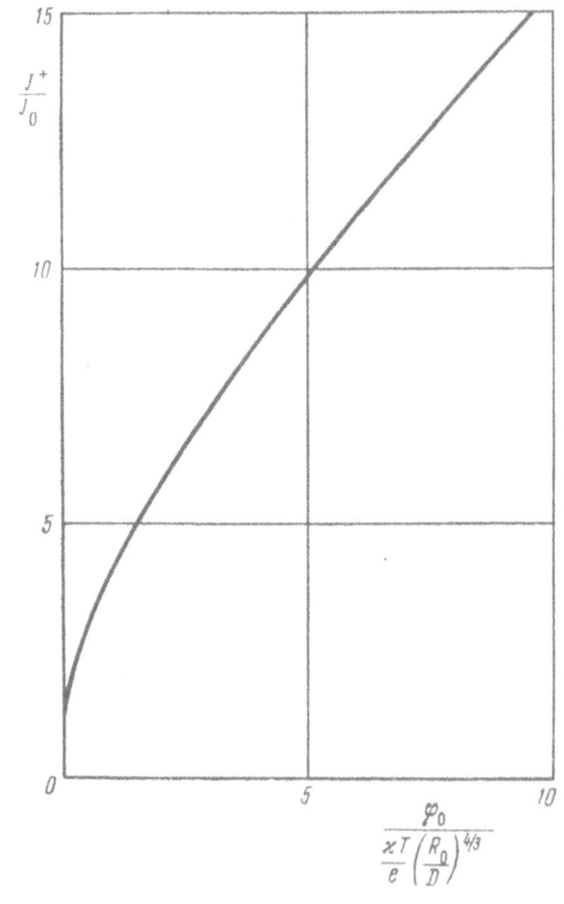

Fig. 83. Dependence of the total flux of the attracted particles on φ_0 for very high values of the potential φ_0 on the surface of the body.

TABLE XXVI. Dependence of J^+/J_0 on the Potential of the Surface of the Probe for Large Values of φ_0 [$\varphi_0^* = (e\,|\,\varphi_0\,|/\varkappa T)(D/R_0)^{4/3}$]

$\varphi_0^{(*)}$	J^+/J_0	$\varphi_0^{(*)}$	J^+/J_0	$\varphi_0^{(*)}$	J^+/J_0	$\varphi_0^{(*)}$	J^+/J_0
0	1.47	0.80	3.71	6.0	11.0	30	33.6
0.02	1.67	1.0	4.11	7.0	12.2	35	37.6
0.05	1.77	1.4	4.75	8.0	13.3	40	41.4
0.10	1.96	1.7	5.32	9.0	14.4	45	45.2
0.15	2.14	2.0	5.77	10.0	15.5	50	49.0
0.20	2.30	2.5	6.52	12.0	17.5	55	52.8
0.25	2.45	3.0	7.25	14.0	19.4	60	56.5
0.30	2.59	3.5	7.94	16.0	21.3	70	63.7
0.40	2.84	4.0	8.60	18.0	23.2	80	70.7
0.50	3.07	4.5	9.23	20	25.0	90	77.4
0.60	3.29	5.0	9.83	25	29.4	100	83.8

Here J_0 is the total flux of attracted particles in the absence of a field (10.30), (10.41):

$$J_0 = (8\pi\varkappa T/M)^{1/2} N_0 R_0^2, \tag{10.46}$$

and $\varphi^* = e\,|\,\varphi\,|/\varkappa T$, $\xi = r/R_0$, φ_∞^* is a constant characterizing the behavior of the field $\varphi^*(\xi)$ as $\xi \to \infty$, R_0 is the radius of the probe, N_0 is the concentration of particles in the undisturbed plasma, T and M are their temperature and mass.

Using the potential $\varphi^*(\xi)$, calculated in §37 for different values of the surface potential φ_0, in formula (10.45) and integrating (numerically), we obtain values of the flux J^+ for different values of φ_0. These values are listed in Table XXV.

A curve of the dependence of J^+/J_0 on φ_0^* is shown in Fig. 82. For $\varphi_0^* \gg 1$ the flux of attracted particles becomes constant – it no longer depends on the potential of the probe:

$$\left.\frac{J^+}{J_0}\right|_{\varphi_0^* \gg 1} = 1{,}47. \tag{10.47}$$

This is true, however, only at not too high values of the potential on the surface of the probe, values which satisfy condition (10.44). The case of very high potentials on the surface of the probe, when

$$|\,\varphi_0\,| \gtrsim \frac{\varkappa T}{e}\left(\frac{R_0}{D}\right)^{4/3}, \tag{10.48}$$

was considered in §39. Using in formula (10.32) the expression for the potential of the field obtained in this section, we find that

$$j^+(R_0) = j_0\left\{\frac{R_c^2}{R_0^2}\varphi_\infty^* + \frac{\varrho_0^2(R_0)}{R_0^2} + \frac{2}{R_0^2}\int\limits_{\rho_0(R_0)}^{\infty} r_0\left[1 - \exp\left(\varphi^* + \frac{r_0}{2}\frac{d\varphi^*}{dr_0}\right)\right]dr_0\right\}$$

$$= j_0\left\{\frac{R_c^2}{R_0^2}\varphi_\infty^* + \frac{R_c^2}{R_0^2} + \frac{2}{R_0^2}\int\limits_{R_c}^{\infty} r_0\left[1 - \exp\left(\varphi^* + \frac{r_0}{2}\frac{d\varphi^*}{dr_0}\right)\right]dr_0\right\} = 1.47 j_0\left(\frac{R_c}{R_0}\right)^2. \tag{10.49}$$

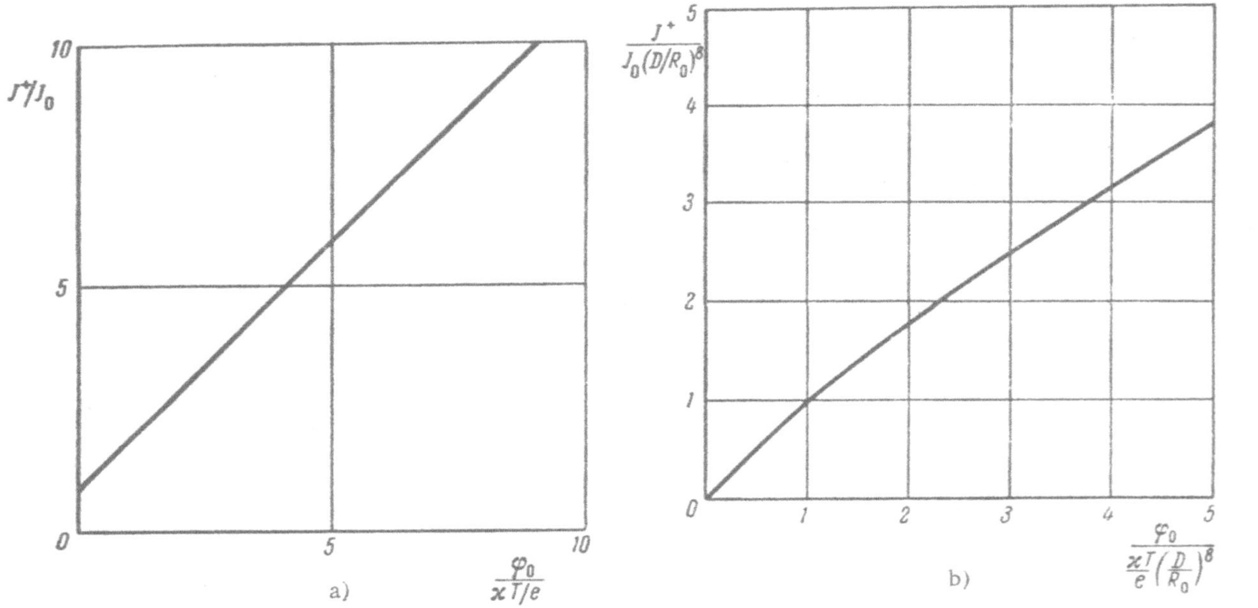

Fig. 84. Dependence of the total attracted particle flux on the potential φ_0 in the case of a small body ($R_0 \ll D$).

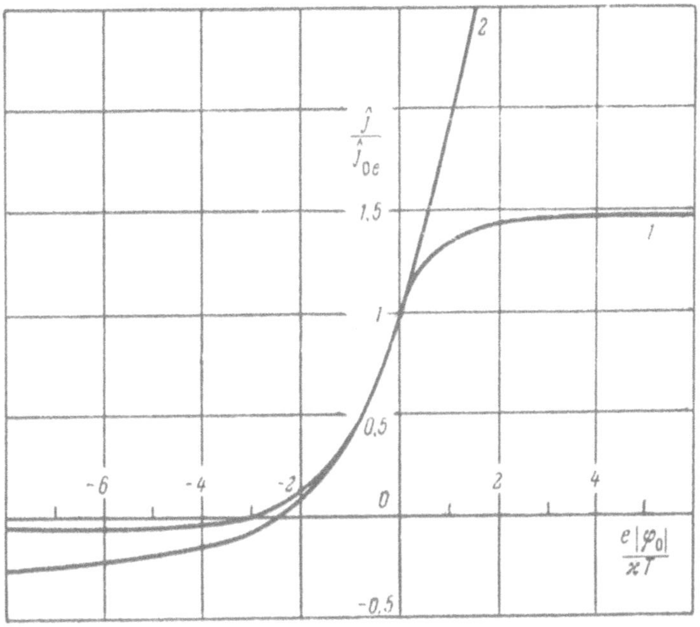

Fig. 85. Probe characteristics for the case of large ($R_0 \gg D$) (curve 1) and small ($R_0 \ll D$) (curve 2) probes. φ_θ is the potential on the surface of the probe, $\hat{J}_{0e} = -eN_0R_0^2(8\pi\varkappa T/3M)^{1/2}$ is the flux of electrons on the uncharged probe. The ratio $M/m \approx 1.84 \cdot 10^3$ (hydrogen).

Here the fact that $|\varphi^*(\rho_0)+(\rho_0/2)(d\varphi^*/dr)(\rho_0)| \gtrsim (R_0/D)^{4/3} \gg 1$ [see Eq. (9.50)] and that for $r \geq R_c$ the function $\varphi^*(\xi)$ is given by exactly the same expressions as in §37, only $\xi = r/R_c$, have been used. Further, R_c is the radius of the space charge region. The dependence of R_c/R_0 on $\varphi_0^*(D/R_0)^{4/3}$ is shown in Fig. 72.

For the total flux of attracted particles on the surface of the probe, we obtain

$$J^+ = 1.47J_0\left(\frac{R_c}{R_0}\right)^2 = 1.47\left(\frac{8\pi\varkappa T}{M}\right)^{1/2} N_0R_c^2. \tag{10.50}$$

The dependence of the flux J^+/J_0 on $\varphi^*(D/R_0)^{4/3}$ is shown in Fig. 83 and in Table XXVI. As $\varphi^*(D/R_0)^{4/3} \to 0$ the ratio $R_c/R_0 \to 1$ and the flux J^+ is equal to 1.47 J_0, as it should be. For very high values of the potential, $|\varphi_0| \gg (\varkappa T/e)(R_0/D)^{4/3}$

$$J^+ = 0.951J_0\left[\frac{e|\varphi_0|}{\varkappa T}\left(\frac{D}{R_0}\right)^{4/3}\right]^{6/7} = 1.12\frac{(\varkappa T)^{3/14}N_0^{3/7}\varphi_0^{6/7}R_0^{8/7}}{M^{1/2}e^{2/7}}. \tag{10.51}$$

as is clear from Eqs. (9.44) and (10.50). Thus, as $\varphi_0 \to \infty$ the particle flux increases like $\varphi_0^{6/7}$. As already mentioned (see p. 66) a series of measurements indicate that the potential of the body in a number of cases possibly reaches $-(5$ to $10)$ volts. Therefore, for $KT/e \sim 0.1$ eV and the often realized values $R_0/D \sim 10$, we have $(e\varphi_0/\varkappa T)(R_0/D)^{-4/3} \sim 2$ to 5, and from Fig. 83 it follows that $J^+/J_0 \sim 6$ to 10. Thus, in these cases the particle flux J^+ may differ considerably from the particle flux J_0, calculated without taking the influence of the field into consideration.

b. Flux of Attracted Particles on the Surface of a Small Probe. The dependence of the potential φ on r in the case $R_0 \ll D$ for not very high surface potentials,

$$|\varphi_0| < \frac{\varkappa T}{e}\frac{D}{R_0} \tag{10.52}$$

was obtained in §40a. Using the corresponding expression for $\varphi(r)$ in formula (10.32), we have the following expression for the flux:

$$J^+ = J_0(1+\varphi_0^*). \tag{10.53}$$

The simple formula (10.53) coincides with the formula of Langmuir and Mott-Smit (10.38). This is understandable, since upon fulfillment of condition (10.52) the field falls off according to Coulomb's law, that is, more slowly than $1/r^2$, up to distances of order D from the body's surface. Since $D \gg R_0$ in the case under consideration, then the flux is given by expression (10.38), which is valid in the case of a slowly decreasing field.

For values $\varphi^* > D/R_0$, as is clear from the results of §40c, the field as before falls off more slowly than $1/r^2$, and consequently the flux J^+ as before is determined by expression (10.53). This is valid, however, only for $\varphi_0^* \ll (D/R_0)^8$, that is, for

$$|\varphi_0| \ll \frac{\varkappa T}{e}\left(\frac{D}{R_0}\right)^8. \tag{10.54}$$

For very high values of the potential on the surface of the body, when the condition opposite to condition (10.54) is satisfied, that is, when

$$|\varphi_0| \gg \frac{\varkappa T}{e}\left(\frac{D}{R_0}\right)^8, \tag{10.55}$$

the attracted particle flux is given by the formula

$$J^+ = 0.951 J_0\left[\frac{e|\varphi_0|}{\varkappa T}\left(\frac{D}{R_0}\right)^{4/3}\right]^{6/7}. \tag{10.51}$$

The dependence of the flux J^+ on the potential of the field for a small body is shown in Fig. 84. For convenience, the dependence of J^+/J_0 on $e\varphi_0/\varkappa T$ for small values of $e\varphi_0/\varkappa T \sim 1$ is shown separately on this figure. Of course, in the case of very high potentials, $\varphi_0 \sim (\varkappa T/e)(D/R_0)^8$, only the asymptotic behavior of the function J^+/J_0 under conditions (10.54) and (10.55) is known exactly. However, expressions (10.53) and (10.51) join together well, which enables one to construct the complete curve.*

We note here that in [80] it is shown that in the case $R_0 \ll D$ considered here, if the body is moving slowly with velocity $V_0 \ll (2\varkappa T/M)^{1/2}$, we have the following expression for attracted particles, upon fulfillment of the criterion (10.52):

$$J^+ = J_0\left\{(1+\phi_0^*) + \frac{1}{3}\frac{MV_0^2}{2\varkappa T}(1+\phi_0^*)\right\},$$

and for the repelled particles we have

$$J^- = J_0 \exp(+\phi^*)\left\{1 + \frac{MV_0^2}{2\varkappa T}\left(\frac{1}{3} + \frac{2}{3}\phi_0^*\right)\right\}.$$

c. Probe Characteristics. As we saw above, for a sufficiently large mean free path of the particles, $l \gg R_0$, the flux of bounded particles on the surface of a body is always small in comparison with the flux of unbounded particles. Considering therefore the expressions obtained above for the flux of attracted particles and for the flux of repelled particles (10.42), one can construct the dependence of the total flux (10.43) on the probe as a function of the potential φ_0 on its surface, that is, the so-called probe characteristics. They are shown, for example, in the case of not very high surface potentials in Fig. 85. From the figure it is clear that the zone characteristics differ markedly in the limiting cases of large and small bodies.

*During the calculation, it is assumed everywhere that the particles are completely absorbed on the surface of the probe. Under actual conditions, the reflection of particles not infrequently plays an important role (especially for positive potentials φ_0) and secondary processes on the surface of the probe are also often important.

As already mentioned in §31 above, a considerable number of articles have been devoted to the calculation of the magnitude of the flux of attracted particles on a probe. In the majority of these calculations, however, the case of a heterotemperature plasma, $T_e \gg T_i$, was considered. Only the results of the approximate theory of Langmuir [39, 42, 60] and Bohm, Burhop, and Massey [40] exist for the case of a single-temperature plasma.

According to Langmuir, the particle flux on the surface of a large body ($R_0 \gg D$) for not very high surface potentials, satisfying condition (10.44), does not depend on the surface potential and is given by

$$J^+ = J_0 = (8\pi\varkappa T/M)^{1/2} N_0 R_0^2. \tag{10.56}$$

The influence of the field, penetrating into the quasi-neutral plasma on the attracted particle flux was not considered in the articles by Langmuir. In the article by Bohm, Burhop, and Massey the influence of this field is taken into account in a rough approximation (for more details, see §31). Here different approximations give different values for the flux. According to [40] the values are

$$J_1 = 1.77 J_0, \quad J_2 = 1.35 J_0. \tag{10.57}$$

The results of a rigorous calculation of $J^+(\varphi_0)$ are presented above in Fig. 84. According to Eq. (10.47) the saturation value of the flux is equal to $1.47\, J_0$.

LITERATURE CITED

1. Ya. L. Al'pert, Usp. Fiz. Nauk 71:369 (1960), Soviet Phys. – Usp. (English translation) 3:479 (1961); Iskusstvennye Sputniki Zemli (Collection: Artificial Earth Satellites), No. 7:125 (1961). This article is published in Am. Rocket Soc. J. 32:151 (1962) and Plan. Space Science 9:391 (1962).
2. A. V. Gurevich, Tr. IZMIR Akad. Nauk SSSR (Trans. Inst. Terr. Magnet., Ionosph., and Radio), No. 17(27):173 (1960). This article is published in Am. Rocket Soc. J. 32, (7): 1168 (1962) and Plan.Space Science 9:321 (1962).
3. A. V. Gurevich, Iskusstvennye Sputniki Zemli (Collection: Artificial Earth Satellites), No. 7:101 (1961).
4. L. P. Pitaevskii and V. Z. Kresin, Zh. Eksperim. i Teor. Fiz. 40:271 (1961); Soviet Phys. JETP (English translation) 13:185 (1961).
5. L. P. Pitaevskii, Geomagnetizm i Aeronomiya 1:194 (1961). This article is published in AIAA Journal 1(4): 994 (1963).
6. Ya. L. Al'pert and L. P. Pitaevskii, Geomagnetizm i Aeronomiya 1:709 (1961). This article is published in AIAA Journal 1(4): 1001 (1963).
7. L. P. Pitaevskii, Zh. Eksperim. i Teor. Fiz. 42:1326 (1962); Soviet Phys. JETP (English translation) 15:919 (1962).
8. A. V. Gurevich, Geomagnetizm i Aeronomiya 3:184 (1963).
9. L. P. Pitaevskii, Zh. Eksperim. i Teor. Fiz. 44:969 (1963); Soviet Phys. JETP (English translation) 17 : 658 (1963).
10. Ya. L. Al'pert, A. V. Gurevich, and L. P. Pitaevskii, Usp. Fiz. Nauk 79:233 (1963). This article is published in Space Science Rev. 2:680 (1963).
11. K. P. Chorpa, Rev. Mod. Phys. 33:153 (1961).
12. L. D. Jaffe and J. B. Rittenhouse, Rocket Technology [Russian translation], p. 3 (1962).
13. Ya. L. Al'pert, Propagation of Radiowaves and the Ionosphere, Izd. Akad. Nauk (1960).
14. S. K. Mitra, The Upper Atmosphere [Russian translation], IL (1959); second revised ed. Hafner Publishing Co., New York (1952).
15. Z. Triskova, Just. für Radiotechnik und Elektr., Tschechoslovakische Akad. d. Wissen., No. 16 (1961).
16. V. P. Dokuchaev, Zh. Eksperim. i Teor. Fiz. 43:595 (1962); Soviet Phys. JETP (English translation) 16:426 (1963).
17. V. P. Silin and A. A. Rukhadze, Electromagnetic Properties of Plasma and Plasma-Like Media, Atomizdat (1961).
18. S. P. Kapitsa, Zh. Eksperim. i Teor. Fiz. 39:1367 (1960); Soviet Phys. JETP (English translation) 12:954 (1961).
19. V. L. Granovskii, Electric Current in Vacuum and in Gases, Gostekhizdat (1953).
20. H. S. W. Massey and E. H. S. Burhop, Electrons and Ionic Impact Phenomena [Russian translation], IL (1958), p. 484, Oxford Univ. Press, Fair Lawn, New Jersey (1952).
21. L. D. Landau and E. M. Lifshitz, Mechanics of Continuous Media, Gostekhizdat (1953).
22. L. D. Landau, Zh. Eksperim. i Teor. Fiz. 7:203 (1937).
23. L. É. Gurevich, Foundations of Physical Kinetics, Gostekhizdat (1940), Chapt. 7.
24. B. I. Davydov, Zh. Eksperim. i Teor. Fiz. 7:1069 (1937).
25. V. L. Ginzburg and A. V. Gurevich, Usp. Fiz. Nauk 70:201 (1960); Soviet Phys. JETP (English translation) 3:115 (1960).
26. A. M. Moskalenko, Geomagnetizm i Aeronomiya 2:407 (1962).
27. L. D. Landau and E. M. Lifshitz, Mechanics, Fizmatgiz (1959).

28. M. A. Miller, Motion of Charged Particles in Weakly Inhomogeneous High-Frequency Electromagnetic Fields, Dissertation, Gor'kii (1960); Proc. Int. Conf. High-Energy Accel. and Instr. CERN (1959), p. 661.
29. G. Getmantsev and N. G. Denisov, Geomagnetizm i Aeronomiya 2:691 (1962).
30. L. P. Pitaevskii, Zh. Eksperim. i Teor. Fiz. 39:1450 (1960); Soviet Phys. 12:1008 (1961).
31. V. L. Ginzburg, Electromagnetic Waves in Plasma, Fizmatgiz (1960).
32. L. D. Landau, Zh. Eksperim. i Teor. Fiz. 16:574 (1946).
33. L. D. Landau and E. M. Lifshitz, Electrodynamics of Continuous Media, Gostekhizdat (1957).
34. A. V. Gurevich, Zh. Eksperim. i Teor. Fiz. 44:1302 (1963); Soviet Phys. JETP (English translation) 17:878 (1963).
35. S. Chapman and T. G. Cowling, Mathematical Theory of Nonuniform Gases [Russian translation], IL (1960); second ed. Cambridge Univ. Press, New York (1959).
36. G. Budker and V. Belyaev, Fizika Plazmy (Collection: Plasma Physics), Vol. II, Izd. Akad. Nauk SSSR (1958), p. 330.
37. V. Belyaev, Fizika Plazmy (Collection: Plasma Physics), Vol. III, Izd. Akad. Nauk SSSR (1958), p. 50.
38. L. D. Landau and E. M. Lifshitz, Mechanics of Continuous Media, Gostekhizdat (1953), p. 279.
39. J. Langmuir and K. Blodgett, Phys. Rev. 22:347 (1923); 24:49 (1924).
40. D. Bohm, E. Burhop, and H. Massey, The Characteristics of Electrical Discharges in Magnetic Fields, ed. by A. Guthrie and R. Wakerling, McGraw-Hill Book Co., Inc. New York (1949).
41. I. Bernstein and I. Rabinowitz, Phys. Fluids 2:112 (1959).
42. H. M. Mott-Smit and J. Langmuir, Phys. Rev. 28:727 (1926).
43. Yu. M. Kagan and V. I. Perel', Dokl. Akad. Nauk SSSR 95:765 (1954); Zh. Eksperim. i Teor. Fiz. 29:261 (1955); Soviet Phys. JETP (English translation) 2:326 (1956); Zh. Eksperim. i Teor. Fiz. 32:1479 (1957); Soviet Phys. JETP (English translation) 32:1479 (1957).
44. L. Kraus and K. Watson, Phys. Fluids 1:480 (1958).
45. A. I. Akhiezer and Ya. B. Fainberg, Zh. Eksperim. i Teor. Fiz. 21:971 (1951).
46. G. V. Gordeev, Zh. Eksperim. i Teor. Fiz. 24:524 (1954).
47. A. A. Galeev, V. N. Oraevskii, and R. Z. Sagdeev, Zh. Eksperim. i Teor. Fiz. 44:903 (1963); Soviet Phys. JETP (English translation) 17:615 (1963).
48. A. B. Mikhailovskii and L. I. Rudakov, Zh. Eksperim. i Teor. Fiz. 44:912 (1963); Soviet Phys. (English translation) 17:621 (1963).
49. V. P. Silin, Zh. Eksperim. i Teor. Fiz. 44:1271 (1963);Soviet Phys. JETP (English translation) 17:857 (1963).
50. S. S. Moiseev and R. Z. Sagdeev, Zh. Eksperim. i Teor. Fiz. 44:763 (1963); Soviet Phys. JETP (English translation) 17:515 (1963).
51. P. S. Greifinger, Phys. Fluids 4:104 (1961).
52. S. Rand, Phys. Fluids 2 : 649 (1959); 3 : 265, 588 (1960).
53. J. R. Reitz and L. L. Foldy, J. Fluid Mech 11: 133 (1961).
54. L. I. Dorman and Yu. M. Mikhailov, Zh. Eksperim. i Teor. Fiz. 43:752 (1962); Soviet Phys. JETP (English translation) 16:531 (1963).
55. R. Jastrow and C. A. Pearse, J. Geophys. Res. 62, 413 (1957).
56. M. E. Gintsburg, Tr. IZMIR Akad. Nauk SSSR (Trans. Inst. Terr. Magnet., Ionosph., and Radio), Vol. 17(27):187 (1960).
57. A. H. Davis and I. Harris, Rarefied Gas Dynamics, Acad. Press, New York—London (1961) p. 661.
58. K. L. Dura Sengupta, Can. J. of Phys. 41:132 (1963).
59. G. S. S. Ludford and J. D. Murray, J. Fluid Mech. 7:516 (1960).
60. J. Langmuir, Phys. Rev. 34:876 (1929).
61. F. Wensel, Z. Angew. Phys. 2(2):59 (1950).
62. J. Allen, R. Boyd, and P. Raynolds, Proc. Roy. Soc. A 70 : 297 (1957).
63. A. V. Gurevich and L. P. Pitaevskii, Geomagnetizm i Aeronomiya 3:823 (1963).
64. A. V. Gurevich and L. P. Pitaevskii, Zh. Eksperim. i Teor. Fiz. 45:1243 (1963); Soviet Phys. JETP (English translation) 18:855 (1964).
65. A. V. Gurevich, Geomagnetizm i Aeronomiya 3:1021 (1963).

66. L. P. Pitaevskii, Geomagnetizm i Aeronomiya 3:1036 (1963).
67. Radio Research 1962, Report of the Director of Radio Research, London, 1963.
68. Yu. V. Zonov, Iskusstvennye Sputniki Zemli (Collection: Artificial Earth Satellites), No. 3:1187 (1959).
69. Yu. S.Sayanov and L. A. Zhizhimov, Radiotekhn. i Elektron. 8:499 (1963).
70. Yu. M. Panchenko and L. P. Pitaevskii, Geomagnetizm i Aeronomiya 4:256 (1954).
71. A. V. Gurevich, Geomagnetizm i Aeronomiya 4:3 (1964).
72. A. V. Gurevich, Kosmicheskie Issledovaniya 2:68 (1964).
73. C. L. Brundin, AIAA Journal 1(11):2529 (1963).
74. V. P. Silin, Zh. Eksperim. i Teor. Fiz. 23, 649 (1952).
75. G. G. Chernyi, Gas Flow with Large Supersonic Velocity, Fizmatgiz (1954).
76. T. F. Volkov, Collection: Plasma Physics and Problems of Thermonuclear Reactions, Vol. 3, Moscow (1958), 336 pages.
77. C. L. Dolph and H. Weil, Plan. Space Science 6:123-132 (1961).
78. G. A. Bird, Rarefied Gas Dynamics, ed. by I. L. Talbot, Academic Press, New York (1961), pp. 245-260.
79. W. Sawchuk, Rarefied Gas Dynamics, ed. by J. A. Laurnani, Academic Press, New York (1963), pp. 33-44.
80. A. M. Moskalenko, Geomagnetizm i Aeronomiya 4(2, 3) (1964).
81. A. V. Gurevich, Geomagnetizm i Aeronomiya 4:247 (1964).
82. W. A. Gustafson and R. E. Kiel, I. A. Journal 1:2869 (1963).
83. R. L. Boyd, Proc. Roy. Soc., Ser. A 201:329 (1950).

GPSR Compliance

The European Union's (EU) General Product Safety Regulation (GPSR) is a set of rules that requires consumer products to be safe and our obligations to ensure this.

If you have any concerns about our products, you can contact us on ProductSafety@springernature.com

In case Publisher is established outside the EU, the EU authorized representative is:

Springer Nature Customer Service Center GmbH
Europaplatz 3
69115 Heidelberg, Germany

Batch number: 10838288

Printed by Printforce, the Netherlands